Groundwater Quality and Contamination at Regional Scales

Groundwater Quality and Contamination at Regional Scales

Guest Editors

Guanxing Huang
Liangping Li

MDPI

Basel • Beijing • Wuhan • Barcelona • Belgrade • Novi Sad • Cluj • Manchester

Guest Editors
Guanxing Huang
Institute of Hydrogeology
and Environmental Geology
Chinese Academy of
Geological Sciences
Shijiazhuang
China

Liangping Li
Department of Geology and
Geological Engineering
South Dakota School of Mines
and Technology
Rapid City
USA

Editorial Office
MDPI AG
Grosspeteranlage 5
4052 Basel, Switzerland

This is a reprint of the Special Issue, published open access by the journal *Water* (ISSN 2073-4441), freely accessible at: https://www.mdpi.com/journal/water/special_issues/5E10UA4N69.

For citation purposes, cite each article independently as indicated on the article page online and as indicated below:

Lastname, A.A.; Lastname, B.B. Article Title. *Journal Name* **Year**, *Volume Number*, Page Range.

ISBN 978-3-7258-6676-2 (Hbk)
ISBN 978-3-7258-6677-9 (PDF)
https://doi.org/10.3390/books978-3-7258-6677-9

Contents

About the Editors vii

Lili Hou, Qiuju Qi, Quanping Zhou, Jinsong Lv, Leli Zong, Zi Chen, et al.
Shallow Groundwater Quality Assessment and Pollution Source Apportionment: Case Study in Wujiang District, Suzhou City
Reprinted from: *Water* **2024**, *16*, 3139, https://doi.org/10.3390/w16213139 1

Jiayi Zhao, Lingxia Liu, Yaru Wang, Bingyan Li, Wenjing Lin and Haihui Yao
Hydrogeochemical Characteristics and Genesis of Hot Springs in Da Qaidam Area, Northern Qaidam Margin of the Qaidam Basin
Reprinted from: *Water* **2024**, *16*, 3509, https://doi.org/10.3390/w16233509 19

Lei Wang, Yi Li, Yanpeng Zhang, Wei Liu and Hongxin Zhang
Tracing Sulfate Sources of Surface Water and Groundwater in Liuyang River Basin Based on Hydrochemistry and Environmental Isotopes
Reprinted from: *Water* **2025**, *17*, 2105, https://doi.org/10.3390/w17142105 40

Wenfeng Gao, Yawei Kou, Hao Dong, Haoran Liu and Simin Jiang
A Robust Optimization Framework for Hydraulic Containment System Design Under Uncertain Hydraulic Conductivity Fields
Reprinted from: *Water* **2025**, *17*, 2617, https://doi.org/10.3390/w17172617 61

Dev Sen Gupta, Ashwani Raju, Abhinav Patel, Surendra Kumar Chandniha, Vaishnavi Sahu, Ankit Kumar, et al.
Integrated Assessment of the Hydrogeochemical and Human Risks of Fluoride and Nitrate in Groundwater Using the RS-GIS Tool: Case Study of the Marginal Ganga Alluvial Plain, India
Reprinted from: *Water* **2024**, *16*, 3683, https://doi.org/10.3390/w16243683 81

Emmanuel Nabala Hyeroba and Robert M. Kalin
Assessment of Groundwater Quality and Vulnerability in the Nakivale Sub-Catchment of the Transboundary Lake Victoria Basin, Uganda
Reprinted from: *Water* **2024**, *16*, 3386, https://doi.org/10.3390/w16233386 106

Wagari Mosisa Kitessa, Adisu Befekadu Kebede, Fayera Gudu Tufa, Beekan Gurmessa Gudeta, Alemu Yenehun, Bekan Chelkeba, et al.
Hydrogeochemical Characterization and Processes Controlling Groundwater Chemistry of Complex Volcanic Rock of Jimma Area, Ethiopia
Reprinted from: *Water* **2024**, *16*, 3470, https://doi.org/10.3390/w16233470 129

Giancarla Viviana Santa Cruz Mérida, Juliana Ikebe Otomo, Diego Ricardo Araoz Prado, Elaine Aparecida Rodrigues, Delvonei Alves de Andrade and Oscar Vega Bustillos
Advanced Analytical Approaches for Phenolic Compounds in Groundwater: A PRISMA Systematic Review
Reprinted from: *Water* **2025**, *17*, 1173, https://doi.org/10.3390/w17081173 154

About the Editors

Guanxing Huang

Guanxing Huang is a Professor within the Institute of Hydrogeology and Environmental Geology, Chinese Academy of Geological Sciences, China. He received his Ph.D. degree from the Chinese Academy of Geological Sciences, China, in 2014. He worked as a Visiting Scholar at the University of Massachusetts, Amherst, in 2015–2016. He has published over 70 refereed journal articles, with a total citation exceeding 2,000. Dr. Huang is a Guest Editor of the journal *Water*, and he is a Youth Editorial Board Member of the Carbon Research and China Geology journals. He served as an Editorial Board Member of *Frontiers in Water* (2021–2022), and as a Guest Editor of the *International Journal of Environmental Research and Public Health* (2021). He is a member of both the International Association of GeoChemistry (IAGC) and International Association of Hydrology (IAH). He has received several awards/honors, including the Leading Talent of Science and Technology of the Ministry of Natural Resources of China in 2021, Excellence in Review Award of Applied Geochemistry in 2020, Young Scientist Award of the Chinese Society of Environmental Sciences in 2020, and Outstanding Youth Science and Technology Talent Cultivation Plan, Ministry of Land and Resources of China in 2015. His current research focuses on regional groundwater contamination, the evolution of groundwater chemistry, groundwater quality assessment, and natural background levels in groundwater.

Liangping Li

Liangping Li is an Associate Professor (since 2022) within the Department of Geology and Geological Engineering, South Dakota School of Mines and Technology, Rapid City, USA, where he has been actively involved in teaching and research since 2015. Dr. Li received his Ph.D. degree from the Technical University of Valencia, Spain, in 2011, and he then worked at post-doc positions at the University of Texas, Austin, USA, during 2012–2015. Dr. Li's current research includes groundwater modeling, contaminant migration, groundwater sampling investigations, groundwater supplies, geochemistry, analytical methods in groundwater, and field methods. With his students and colleagues, he has published over 60 refereed articles—most in top journals such as *Water Resources Research*, *Advances in Water Resources*, and the *Journal of Hydrology*, many of which are frequently cited by his peers. His h-index is currently 22, and he has a total citation of over 1700. He has been invited to present his research results at several universities and institutions. Dr. Li served as an Associate Editor for the *Hydrogeology Journal*, the *Journal of Hydraulic Engineering*, and *Stochastic Environmental Research and Risk Assessment*, as a Guest Editor for *Geofluids*, *Mathematical Geosciences*, and *Water*, and as a Co-Chair of the Organizational Committee for the 20th Annual Meeting of the International Association of Mathematical Geosciences. Dr. Li obtained funds from NSF, USGS, NASA, and the South Dakota Board of Regents over the last five years. His research program is recognized internationally.

water

MDPI

Article

Shallow Groundwater Quality Assessment and Pollution Source Apportionment: Case Study in Wujiang District, Suzhou City

Lili Hou [1,2], Qiuju Qi [1,2], Quanping Zhou [1,2], Jinsong Lv [1,2,*], Leli Zong [1,*], Zi Chen [1,2], Yuehua Jiang [1,2,3], Hai Yang [1,2], Zhengyang Jia [1,2], Shijia Mei [1,2], Yang Jin [1,2,3], Hong Zhang [1,2], Jie Li [1] and Fangfei Xu [4]

1 Nanjing Center, China Geological Survey, Nanjing 210016, China; hlili@mail.cgs.gov.cn (L.H.)
2 Key Laboratory of Watershed Eco-Geological Processes, Ministry of Natural Resources, Nanjing 210016, China
3 Chinese Academy of Geological Sciences, Beijing 100037, China
4 Tonglu Environmental Monitoring Station, Hangzhou 311500, China
* Correspondence: ljinsong@mail.cgs.gov.cn (J.L.); zongleli@mail.cgs.gov.cn (L.Z.)

Abstract: Groundwater serves as a crucial resource, with its quality significantly impacted by both natural and human-induced factors. In the highly industrialized and urbanized Yangtze River Delta region, the sources of pollutants in shallow groundwater are more complex, making the identification of groundwater pollution sources a challenging task. In this study, 117 wells in Wujiang District of Suzhou City were sampled, and 16 groundwater quality parameters were analyzed. The fuzzy synthetic evaluation method was used to assess the current status of groundwater pollution in the study area; the principal component analysis (PCA) was employed to discern the anthropogenic and natural variables that influence the quality of shallow groundwater; and the absolute principal component scores–multiple linear regression (APCS-MLR) model was applied to quantify the contributions of various origins toward the selected groundwater quality parameters. The results indicate that the main exceeding indicators of groundwater in Wujiang District are *I* (28%), NH_4-*N* (18%), and Mn (14%); overall, the groundwater quality is relatively good in the region, with localized heavy pollution: class IV and class V water are mainly concentrated in the southwest of Lili Town, the north of Songling Town, and the south of Qidu Town. Through PCA, five factors contributing to the hydrochemical characteristics of groundwater in Wujiang District were identified: water–rock interaction, surface water–groundwater interaction, sewage discharge from the textile industry, urban domestic sewage discharge, and agricultural non-point source pollution. Additionally, the APCS-MLR model determined that the contributions of the three main pollution sources to groundwater contamination are in the following order: sewage discharge from the textile industry (10.63%) > urban domestic sewage discharge (8.69%) > agricultural non-point source pollution (6.26%).

Keywords: shallow groundwater; fuzzy synthetic evaluation; PCA/APCS-MLR; source apportionment

1. Introduction

Groundwater stands as one of the most invaluable natural resources found on our planet. It not only maintains ecological balance but also serves as an indispensable source of water for residential life, industrial production, and agricultural irrigation [1,2]. Hence, the preservation of water resources from quality deterioration is a fundamental and historical concern for human societies and is necessary for life, ecological diversity, and economic stability [3,4]. The quality of groundwater can be influenced by a multitude of factors and processes encompassing both natural and anthropogenic factors [5–8]. Certain natural factors, including climatic conditions and aquifer geological composition, may lead to the accumulation of specific ions in groundwater; for instance, arsenic, magnesium, and iodine [9–11]. Moreover, anthropogenic factors can have substantial detrimental effects on groundwater quality [12,13]. For instance, domestic sewage can contribute to elevated concentrations of ammonia in groundwater [14]; agricultural behaviors such as fertilization and livestock breeding serve as primary contributors to the excessive accumulation of nitrogen, phosphorus, and potassium in the natural environment [15–17]; and industrial activities have facilitated the ingress of substantial quantities of noxious chemicals, which include heavy metals and organic pollutants, into the groundwater system via wastewater effluents, inadvertent leakage, or accidental spills, thereby significantly compromising the purity and integrity of water sources [18–20]. For effective pollution control, accurately pinpointing the primary sources of groundwater pollution and quantitatively assessing their respective contributions have emerged as a pivotal concern in recent research endeavors [21,22].

To achieve this goal, numerous methods have been conceptualized and refined, including the utilization of stable isotope methodologies, numerical inversion procedures, and sophisticated multivariate statistical analyses [23–27]. The use of stable isotope approaches enables precise allocation of sources; however, their application is limited to instances involving isotope-related contaminants. Numerical inversion procedures demonstrate exceptional capabilities in precisely tracing the origins and pathways of groundwater pollutants at contaminated sites, thereby rendering them an ideal tool for conducting research at a localized scale [28–30]. Meanwhile, despite overlooking intricate details like meteorological and topographical factors that influence the emission process and the migration paths or transformations of pollutants, multivariate statistical methods provide a convenient and efficient way to interpret complex data matrices and tackle regional-scale challenges; they have been extensively employed for evaluating groundwater quality through pinpointing the origins of pollution and their controlling elements [31–36]. This group of methodologies encompasses various techniques such as factor analysis (FA), principal component analysis (PCA), cluster analysis (CA), and absolute principal component score-multiple linear regression (APCS-MLR), among others. FA and PCA are particularly useful in analyzing groundwater pollution sources as they allow for the efficient reduction in data dimensionality without requiring a deep understanding of the intricate evolution of hydrochemical constituents or the detailed composition of the pollutants [37,38]. Cluster analysis serves as a valuable tool for examining water quality parameters by identifying patterns and relationships related to their sources and contributing factors. However, relying solely on the aforementioned methods may present limitations in achieving an accurate and quantitative evaluation of the degree to which different pollution factors affect specific water quality indicators.

To tackle this problem, Thurston and Spengler [39] devised the APCS-MLR approach in 1985. This innovative method initially utilizes PCA to distill the essential components from the data, making it easier to pinpoint the source. Following this, the APCS method steps in to calculate the absolute scores for each of these components. Ultimately, the MLR model is leveraged to forge a quantitative link between pollution sources and pollutants, enabling the precise quantification of the individual contributions made by each pollution source. Through these procedures, the APCS-MLR method combines the strengths of PCA with the ability to quantify contributions from contaminant sources. Meanwhile, since its inception, this method has been effectively applied to a range of pollution source apportionment challenges across atmospheric, surface water, and sedimentary environments [40–44]. Its successful utilization in these contexts underscores its versatility and robustness in environmental research and assessment. More recently, the APCS-MLR approach has gained traction in groundwater quality research, demonstrating its effectiveness in pinpointing the influence of both natural and anthropogenic factors on groundwater quality [45–47]. The authors in Zhang et al. [48] utilized the PCA and APCS-MLR methods to pinpoint the sources of groundwater contamination and their respective contributions to the Hutuo River alluvial–pluvial fan. The authors in Meng et al. [49] employed the APCS-MLR receptor model to evaluate potential contamination origins of groundwater and quantify the relative contributions of various pollution sources within the Limin Groundwater Source Area in Harbin, which span the period from 2006 to 2016. The authors in Yu et al. [50] utilized the APCS-MLR technique for apportioning nitrate pollution sources and subsequently contrasted their findings with those obtained from a Bayesian isotope mixing model. The authors in Sheng et al. [51] adopted the APCS-MLR model as a sophisticated analytical tool for assessing the proportional contributions of diverse sources to heavy metal pollution within an arid oasis region situated in Northwest China. The authors in Haji Gholizadeh et al. [52] evaluated the water quality and quantified the potential pollution sources that affect the water quality of three major rivers in South Florida via the APCS-MLR receptor model. The authors in Cheng et al. [53] employed correlation analysis and the APCS-MLR model to explore the potential pollution sources impacting the Jinsha River Basin from 2016 to 2018, highlighting that the concentration of water pollutants in the Jinsha River mainstream basin is primarily influenced by environmental, agricultural, and demographic variables. The authors in Zhang et al. [54] assessed the suitability of both the PMF and APCS-MLR models for evaluating groundwater pollution sources by comparing their performance using groundwater sample data from typical mixed land use areas in Southwest China. The authors in Chen et al. [55] discovered that both the APCS-MLR and PMF models produced comparable estimates of the proportions of contamination source contributions when applied to evaluate groundwater pollution in Northeastern China. In conclusion, APCS-MLR presents a quantitative framework to assess the specific impacts of various pollution sources on water quality metrics. This versatile approach has the potential to be broadly implemented in quantitatively assessing regional groundwater pollution origins.

Wujiang District, Suzhou City, situated in the southeastern region of Jiangsu Province, stands as one of the pilot areas for the ecological green integrated development of the Yangtze River Delta. Endowed with a distinct geographical location and abundant natural resources, the district has outlined explicit plans and policies for green development and has an ambition to establish a world-class paradigm of livable waterfront communities. In recent years, the relentless advancements in modern industrialization and urbanization have inevitably impacted the groundwater environment within the region. The shallow groundwater, primarily replenished by precipitation and surface river networks, is highly susceptible to contamination from surface pollution sources such as industrial sewage canals and drains. Past research has shown that multiple contaminants, such as heavy-metal ions, nitrogen-based compounds, and phosphorus-based compounds, have been identified in the groundwater of this region [56–59]. Although extensive targeted research was conducted on several specific pollutants, the overarching issue of the holistic distribution of groundwater pollution sources remains unresolved.

This research endeavors to conduct a quantitative evaluation of groundwater pollution origins within a regional area characterized by diverse land-use types. The findings offer an impartial assessment of the pollution intensity associated with various groundwater pollution sources specific to each administrative region. Specifically, the present study, based on an in-depth analysis of the spatial distribution of inferior groundwater components in Wujiang District, employs Principal Component Analysis (PCA) to discern the natural and anthropogenic factors influencing groundwater quality. Furthermore, the Absolute Principal Component Scores–Multiple Linear Regression (APCS-MLR) model is adopted to quantitatively assess the contributions of potential water sources to various groundwater quality parameters. This approach offers an advanced methodology for regional groundwater pollution research and provides valuable technical support for administrative management efforts.

2. Materials and Methods

2.1. Study Area

Wujiang District, a jurisdictional entity of Suzhou City within Jiangsu Province, occupies a strategic position in the southeastern region of the province. It is bordered by Shanghai to the east, Taihu Lake to the west, Zhejiang Province to the south, and the primary urban hub of Suzhou to the north. Geographically, it is situated at latitudes ranging from 30°46′ to 31°14′ north and longitudes spanning from 120°21′ to 120°54′ east, encompassing a total area of 1176 square kilometers. The climatic profile of Wujiang District is characterized as a North subtropical monsoon marine climate, exhibiting distinct seasonal variations throughout the year. Summer is typically associated with southeasterly winds, whereas winter brings northwesterly winds. The climate is generally temperate, with ample precipitation. The average annual temperature hovers around 16 °C, and the mean annual rainfall stands at approximately 1121 mm. The district is devoid of mountainous terrain, instead exhibiting a predominantly low and flat landscape. The terrain gradually slopes from northeast to southwest, with a minimal height difference of approximately 2 m between the northern and southern extremities.

Wujiang District is situated within the expanse of the Taihu Lake Plain, a geological region characterized by a pronounced Quaternary stratigraphy. The Quaternary loose layer within this district exhibits a remarkable thickness, which attains a peak depth of over 300 m in specific regions and consistently exceeds 200 m in other regions. The phreatic aquifer, featuring a thickness ranging from 4 to 10 m, is composed of various layers, including alluvial and lacustrine silty clay, silt, and silty soil. Meanwhile, the groundwater within this aquifer is typically found buried at a depth of 1 to 1.5 m. Based on comprehensive remote sensing surveys, the land use types in this region are cultivated farmlands, residential settlements, and surface water bodies (Figure 1). Notably, 267 square kilometers of this territory is composed of water bodies, constituting approximately 22.70% of the overall district area. There exist over 300 lakes, each boasting an area exceeding 3000 square meters. As a result, the interplay between surface water and groundwater is exceedingly intricate. What's more, multiple types of human activities, including factory pollution, municipal sewage, and agricultural contamination, etc., jeopardize the safety of groundwater.

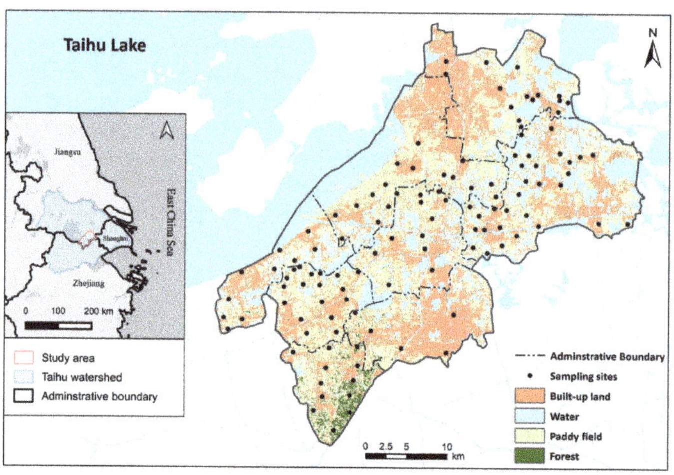

Figure 1. Location of Wujiang District and sampling sites for groundwater with land use.

2.2. Sampling Collection and Analysis

Shallow groundwater samples were systematically collected from 117 wells at depths ranging from 0 to 30 cm in May 2022 and May 2023, with sampling sites carefully selected considering diverse land-use types, including cultivated land and residential areas, along with their spatial distribution. In each well, the total dissolved solids (TDS), pH, and dissolved oxygen (DO) were meticulously determined on-site using high-precision instruments, namely the Hanna DiST, SX-620 pH Testor and SX-630 ORP Testor, respectively. For the preservation and transportation of the groundwater samples, polyethylene containers with a capacity of 1.5 L were utilized. Subsequently, these samples were transported to the laboratory for a comprehensive analysis that encompassed the determination of calcium (Ca^{2+}), magnesium (Mg^{2+}), sodium (Na^{+}), potassium (K^{+}), ammonium nitrogen (NH_4-N), nitrate nitrogen (NO_3-N), nitrite nitrogen (NO_2-N), chloride (Cl^{-}), sulfate

(SO_4^{2-}), manganese (Mn), iodine (I), antimony (Sb), and total phosphorus (TP). The analysis was performed utilizing a diverse array of specialized instruments, which guaranteed the precision and dependability of the obtained results (Table 1). These results are subsequently employed to assess the quality of the groundwater.

Table 1. Overview of quality metrics, units, and analytical equipment used for groundwater samples.

Parameter	Abbreviation	Unit	Analytical Equipment	Manufacturers
Calcium	Ca^{2+}	mg/L	ICAP 6300Duo	Thermo Fisher Scientific, USA
Magnesium	Mg^{2+}	mg/L	ICAP 6300Duo	/
Sodium	Na^{+}	mg/L	ICAP 6300Duo	/
Potassium	K^{+}	mg/L	ICAP 6300Duo	/
Ammonical nitrogen	NH_4-N	mg/L	AutoAnalyzer3	Seal Analytical, USA
Nitrate nitrogen	NO_3-N	mg/L	Dionex-2500	Thermo Fisher Scientific, USA
Nitrite nitrogen	NO_2-N	mg/L	TU-1950	PERSEE, China
Chloride	Cl^{-}	mg/L	Dionex-2500	/
Sulfate	SO_4^{2-}	mg/L	Dionex-2500	/
Manganese	Mn	μg/L	ICAP Q	Thermo Fisher Scientific, USA
Iodine	I	μg/L	ICAP Q	/
Antimony	Sb	μg/L	ICAP Q	/
Total phosphorus	TP	μg/L	ICAP 6300Duo	/
Dissolved oxygen	DO	mg/L	SX-630 ORP Testor	Apera Instruments, German
Pondus Hydrogenii	pH	pH unit	SX-620 pH Testor	Apera Instruments, German
Total dissolved solids	TDS	mg/L	Hanna DiST	Hanna Instruments, USA

2.3. Fuzzy Synthetic Evaluation

As the single-factor water quality assessment mainly depends on the worst single pollution indicator in water quality to determine the overall water quality level, it sometimes imposes overly strict requirements on water quality, resulting in a lower evaluation result. Moreover, it is difficult to consider the interaction and comprehensive impact of various factors, which makes it impossible to comprehensively and scientifically reflect the water quality. Fuzzy sets and fuzzy optimization methods offer a valuable framework for managing the inherent uncertainty in groundwater quality assessment, enabling more precise and nuanced evaluations [60,61]. For instance, the fuzzy synthetic evaluation method effectively incorporates diverse parameters into the assessment process by quantifying and integrating the vague information from different evaluation indicators [62], which enables a more accurate portrayal of the primary sources of uncertainty that arise in large-scale, intricate decision-making scenarios [63]. In this research, the fuzzy membership function is adopted to evaluate the quality of groundwater. Based on the Chinese Groundwater Quality Standard [64], the fuzzy membership function is articulated across 5 distinct levels of groundwater quality for water quality I (j = 1):

$$r_{ij} = \begin{cases} 1, & C_i \leq S_{i,j} \\ \frac{C_i - S_{i,j+1}}{S_{i,j} - S_{i,j+1}}, & S_{i,j} < C_i < S_{i,j+1} \\ 0, & C_i \geq S_{i,j+1} \end{cases} \tag{1}$$

for water quality II to IV (j = 2–4):

$$r_{ij} = \begin{cases} 0, & C_i \leq S_{i,j-1} \\ \frac{C_i - S_{i,j-1}}{S_{i,j} - S_{i,j-1}}, & S_{i,j-1} < C_i < S_{i,j} \\ \frac{C_i - S_{i,j+1}}{S_{i,j} - S_{i,j+1}}, & S_{i,j} < C_i < S_{i,j+1} \end{cases} \tag{2}$$

for water quality V (j = 5):

$$r_{ij} = \begin{cases} 0, & C_i \leq S_{i,j-1} \\ \frac{C_i - S_{i,j-1}}{S_{i,j} - S_{i,j-1}}, & S_{i,j-1} < C_i < S_{i,j} \\ 1, & C_i \geq S_{i,j} \end{cases} \tag{3}$$

where r_{ij} represents the degree of fuzzy membership of the water quality indicator i belonging to the class j, C_i indicates the analytically determined value of the water quality parameter i, $S_{i,j}$ stands for the threshold value of the water quality parameter i in class j. The water quality indicators and classes r_{ij} are then employed to generate the fuzzy membership matrix **R**.

The fuzzy synthetic evaluation results for each sample can be determined by matrix **B** through the following equation:

$$\mathbf{B} = \mathbf{A} \cdot \mathbf{R} \tag{4}$$

where **A** represents the weight matrix, which is composed of the normalized weight of each indication derived using Equation (5),

$$a_i = \frac{C_i}{S_i} / \sum_{i=1}^{n} \frac{C_i}{S_i} \tag{5}$$

where a_i is the normalized weight of indicator i, S_i stands for the average value of indicator i across 5 thresholds.

2.4. Principal Component Analysis

Initially, the Kaiser–Meyer–Olkin (KMO) criterion and Bartlett's test of sphericity were executed to evaluate the appropriateness of the dataset for PCA [65]. For PCA to be considered valid, it is necessary that the KMO value surpasses 0.5 and the significance level of Bartlett's test of sphericity falls below 0.05. After implementing the PCA process, the key principal components corresponding to relevant groundwater quality indicators were successfully extracted.

$$(A_z)_{ij} = a_{i1}C_{1j} + a_{i2}C_{2j} + \ldots + a_{im}C_{mj}, \tag{6}$$

where A_z denotes the component score, i represents the index of components number, j stands the index of samples, a represents the component loading, C is the measured concentration of each groundwater quality parameter and m indicates the total number of groundwater quality parameters.

The principal components that exhibit eigenvalues exceeding 1.0 in the variance computation are deemed capable of providing qualitative insights about the probable contamination [66]. To facilitate a more straightforward understanding, it is typically necessary to rotate the original loadings principal component loadings are reallocated and polarized. Typically, this process is referred to as varimax rotation, resulting in the creation of new variables known as varifactors (VFs). For each of these VFs, the loadings of their components reflect the proportional contribution of groundwater quality metrics. Specifically, absolute loading values surpassing 0.75 are classed as high, those between 0.75 and 0.5 are termed medium, and those ranging from 0.5 to 0.3 are defined as weak [67,68].

2.5. APCS-MLR Model

Utilizing the principal components derived from PCA, the Absolute Principal Component Score-Multiple Linear Regression (APCS-MLR) methodology was employed to quantify the individual contributions of each component. Specifically, the methodology involves transforming the standardized principal component scores into absolute principal component scores, followed by the integration of a multivariate linear regression analysis to comprehensively evaluate and model the concentrations of the measured contaminants [69,70]. The quantity of each groundwater quality parameter at each sampling location is assessed using a multiple linear regression analysis, as outlined in Equation (7):

$$C_{kj} = r_{k0} + \sum_{i=1}^{p} r_{ki} \times APCS_{ij}, \tag{7}$$

where r_{k0} and r_{ki} denotes the intercept term and the coefficient in the multiple linear regression model for pollutant k, respectively; $APCS_{ij}$ represents the absolute principal component scores, which can be derived by following the computational steps outlined in Equations (8)–(10).

$$(Z_0)_k = \frac{-\bar{C}_k}{\sigma_k}, \tag{8}$$

$$(A_0)_i = \sum_{k=1}^{m} S_{ki} \times (Z_0)_k, \tag{9}$$

$$APCS_{ij} = (A_z)_{ij} - (A_0)_i, \tag{10}$$

where $(Z_0)_k$ denotes the normalized concentration of contaminant k found in an area that is not subject to pollution, $\bar{C}_k$ and σ_k stands for the mean and standard deviation concentration of contaminant k respectively, $(A_0)_i$ represents the principal component score in the non-pollution site, S_{ki} denotes the score coefficient of component i for pollutant k and $(A_z)_{ij}$ is the principal component score of sample j in principal component i. It is noteworthy that during the calculation process, negative values may arise, potentially resulting in a cumulative contribution of all pollutants exceeding 100%. Consequently, to address this issue, Haji Gholizadeh et al. [52] introduced a method utilizing absolute values to compute the contribution of contaminant sources to groundwater quality parameters. This approach is outlined in Equations (11) and (12), ensuring a more accurate and meaningful assessment of the contaminant sources' influence on groundwater quality.

$$PC_{ki} = \frac{\left|r_{ki} \times \overline{APCS_i}\right|}{\left|r_{k0}\right| + \sum_{i=1}^{p}\left|r_{ki} \times \overline{APCS_i}\right|} \times 100\%, \tag{11}$$

$$PC_k = \frac{\left|r_{k0}\right|}{\left|r_{k0}\right| + \sum_{i=1}^{p}\left|r_{ki} \times \overline{APCS_i}\right|} \times 100\%, \tag{12}$$

where PC_{ki} denotes the relative contribution rate of contaminant source i to the pollutant k, PC_k stands for the relative contribution rate of unidentified source to the pollutant k, $\overline{APCS_i}$ indicates the mean value of absolute principal component scores across all samples.

3. Results

3.1. Hydrochemical Characteristics

A comprehensive summary of the descriptive statistics pertaining to the physicochemical parameters of all groundwater samples is presented in Table 2. According to the Level III of Chinese Groundwater Quality Standard, the main exceeding indicators of groundwater in Wujiang District are I (28%), NH_4-N (18%), and Mn (14%). For the hydrochemical test data of these three types of excessive ions, the inverse distance method is used for regional interpolation to generate a fishing net map with a grid size of 1 square kilometer (Figure 2). The groundwater exhibits a noteworthy maximum I content of 605 μg/L, with a mean value of 82.4 μg/L. The overall exceedance rate is conspicuously high, leading to the general classification of this region's water quality as Class IV. Additionally, the groundwater with elevated levels of I is predominantly found in the middle part of the investigated region, specifically Pingwang Town, as well as in the eastern region encompassing Lili Town. The groundwater in Wujiang District exhibits a maximum NH_4-N content of 8.51 mg/L, accompanied by an average concentration of 0.64 mg/L. Furthermore, the coefficient of variation stands at a remarkable 228%, indicating significant variability in NH_4-N levels. Notably, limited regions within the district have been classified as Class V water, a category denoting substandard water quality. Analysis of the concentration distribution map reveals that these Class V water areas are primarily concentrated in the northern section of Tongli Town, the southern portion of Qidu Town, and the eastern segment of Lili Town. The maximum concentration of Mn in groundwater is recorded at 2050 μg/L, whereas the mean concentration averages out to 117.2 μg/L. The coefficient of variation stands at a considerable value of 314%, indicating significant variability in Mn content. In comparison to other elements present in the region, the Mn content exceeds the established standards at a relatively lower frequency, with approximately 14% of groundwater samples surpassing the permissible limit. Nevertheless, the occurrence of numerous high-value instances is noteworthy, and the region encompassing Class V water exhibits a comparatively vast extent, primarily concentrated in the southwestern section of the study area.

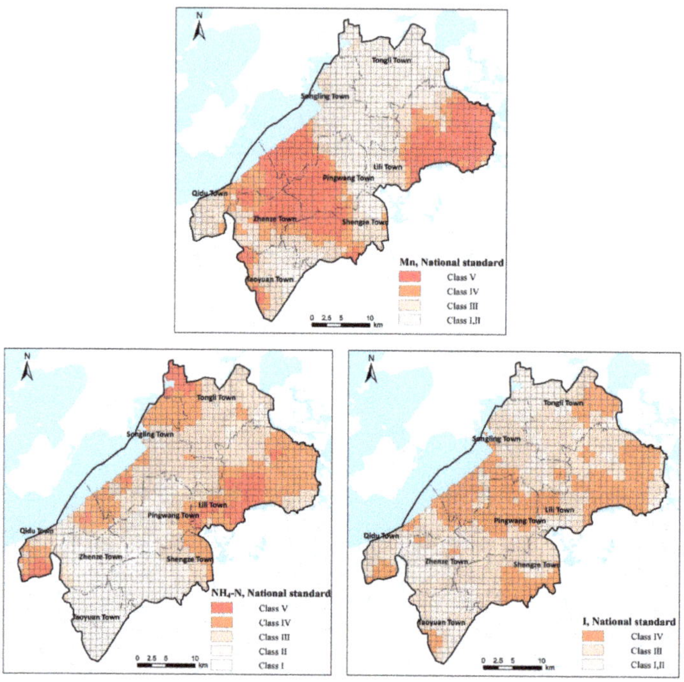

Figure 2. Spatial distributions of the main exceeding indicators of groundwater in Wujiang District.

Table 2. Overview of groundwater physicochemical characteristics in Wujiang District.

Parameter	Min	Max	Mean	Standard Deviation	Coefficients of Variation (%)	National Standard, Class III	Exceeding Standard Rate (%)
Ca^{2+}	13.60	158.50	70.08	27.56	39	/	/
Mg^{2+}	0.83	75.20	17.75	12.41	70	/	/
Na^+	2.57	139.40	49.55	24.31	49	200	0
K^+	0.42	90.30	17.71	18.05	102	/	/
NH_4-N	/	6.62	0.50	1.14	228	0.50	18
NO_3-N	/	23.26	4.89	5.44	111	20	3
NO_2-N	/	2.98	0.25	0.52	207	1	8
Cl^-	2.14	138.60	46.29	29.44	64	250	0
SO_4^{2-}	3.00	171.10	46.28	32.31	70	250	0
Mn	0.10	2050.00	117.21	368.12	314	100	14
I	1.81	605.00	82.42	110.55	134	80	28
Sb	/	5.89	2.31	1.40	61	5	4
P	6.37	3740.00	530.38	625.94	118	/	/
DO	0.2	21.0	3.1	2.3	75	/	/
pH	6.57	8.67	7.19	0.33	5	6.5–8.5	1
TDS	40	801	406	153	38	1000	0

To perform a more logical and comprehensive quality assessment of groundwater, a Fuzzy Synthetic Evaluation method is employed, and the outcome is presented in Figure 3. The groundwater quality in Wujiang District appears to be generally favorable. The Class I-III water accounts for 21.7%, 30.1%, 35.1%, respectively, of the total area, and is widely distributed throughout the region. Among them, the water quality in Tongli Town is the best as a whole, with almost all the water in the town belonging to Class I. Water of Class IV and V accounts for 4.7% and 8.4%, respectively, of the total area and is mainly concentrated in the southwest of Lili Town, the north of Songling Town, and the south of Qidu Town in the east.

The hydrochemical evolution of groundwater is influenced by numerous factors, and the analysis of the correlation among dissolved constituents within groundwater serves as a powerful tool for elucidating the origins of dissolved substances as well as the hydrochemical processes occurring during transportation. To achieve a thorough comprehension of the origins and contributions of diverse hydrochemical constituents found in the groundwater of Wujiang District, a correlation coefficient graph was constructed, focusing on the primary hydrochemical indicators specific to this region (Figure 4). A notable positive correlation exists between TDS and the ionic concentrations of Ca^{2+}, Mg^{2+}, Na^+ and Cl^- with correlation coefficients ranging from 0.55 to 0.77. This correlation suggests a common material source or formation mechanism underlying the observed variations in these parameters. Meanwhile, a positive correlation with a coefficient of 0.54 is observed between NH_4-N and NO_2-N in groundwater samples collected from Wujiang District. The strength of this correlation strongly suggests that the primary source of NH_4-N and NO_2-N in these groundwater samples is the discharge of domestic sewage. There exists a substantial positive correlation (0.57) between P and NO_3-N concentrations in groundwater. Given that NO_3-N typically originates from anthropogenic activities, including the discharge of domestic sewage and the contamination from agricultural fertilizers, the strong correlation observed between P and NO_3-N suggests a common source. Based on these findings, it is hypothesized that the primary source of P and NO_3-N in groundwater is agricultural fertilizers.

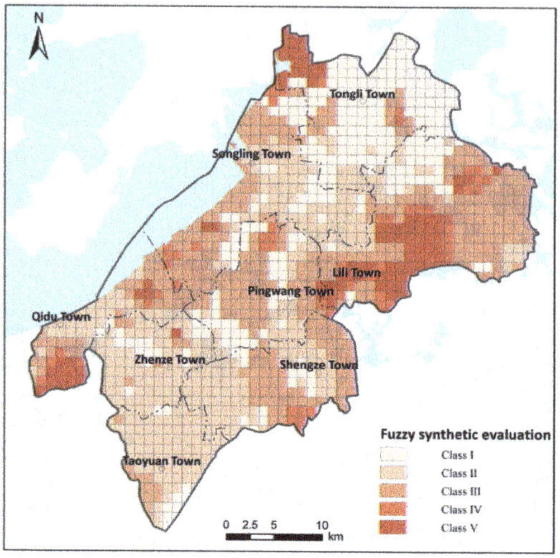

Figure 3. The fuzzy synthetic evaluation result of shallow groundwater in Wujiang District.

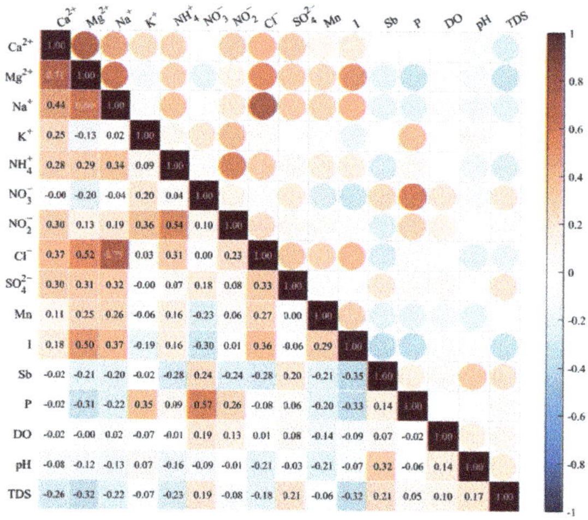

Figure 4. Correlation analysis of groundwater quality parameters in Wujiang District. The colored circles represents the degree of positive and negative correlation.

3.2. Source Identification

3.2.1. PCA

To evaluate the appropriateness of the dataset for conducting principal component analysis, the Kaiser–Meyer–Olkin (KMO) measure of sampling adequacy and Bartlett's test of sphericity were utilized in the present academic investigation. The resulting KMO value stands at 0.75, exceeding the commonly accepted threshold of 0.5, thereby indicating a sufficient level of common variance among the variables. Furthermore, Bartlett's test of sphericity yielded a probability value of 0.0001, which is significantly below the critical threshold of 0.001. This finding satisfies the prerequisites for the application of principal component analysis, thus validating the subsequent results derived from this analytical technique. Five principal components with eigenvalues greater than 1 were selected as the focus of this study, with a cumulative variance contribution rate of 67.88% (Table 3 and Figure 5).

Table 3. Varimax rotated varifactor loadings of 16 groundwater quality parameters. Values that are highlighted in bold represent parameters with strong loadings.

Parameters	VF1	VF2	VF3	VF4	VF5
Ca^{2+}	**0.79**	−0.03	0.24	0.15	−0.24
Mg^{2+}	**0.80**	−0.38	0.07	−0.10	0.01
Na^{+}	**0.76**	−0.19	0.15	−0.25	0.12
K^{+}	0.08	0.34	0.51	0.28	−0.47
NH_4-N	0.25	−0.02	0.67	−0.24	0.08
NO_3-N	0.07	**0.82**	0.03	−0.09	0.19
NO_2-N	0.18	0.14	**0.84**	0.05	0.06
Cl^{-}	0.67	−0.09	0.19	−0.40	0.14
SO_4^{2-}	0.63	0.32	−0.23	0.01	0.14
Mn	0.18	−0.33	0.06	−0.44	−0.19
I	0.29	−0.61	0.10	−0.27	0.05
Sb	0.09	0.39	−0.49	0.53	−0.04
P	−0.13	**0.78**	0.26	−0.05	−0.14
DO	0.04	0.10	0.14	0.24	**0.84**
pH	−0.08	−0.17	0.01	**0.81**	0.12
TDS	**0.89**	−0.02	0.20	0.01	−0.09
Eigenvalues	4.40	2.43	1.69	1.21	1.12
% of Variance	27.47	15.23	10.57	7.59	7.02
Cumulative %	27.47	42.70	53.27	60.86	67.88

The variance contribution rate of VF1, which stands at 27.47%, constitutes the primary determinant of the groundwater chemical characteristics observed in Wujiang District. In VF1, the elements Ca^{2+}, Mg^{2+}, Na^{+}, Cl^{-}, SO_4^{2-}, and TDS exhibit significant loadings ranging from 0.63 to 0.89. This combination of ion characteristics, when analyzed collectively, suggests that VF1 reflects hydrochemical processes of natural origin, particularly those attributed to water–rock interactions.

The variance contribution rate of VF2 stands at 15.23%, with primary loading ions comprising P at a loading of 0.78 and NO_3-N at 0.82. It is widely acknowledged that NO_3-N in groundwater often originates from anthropogenic activities, including the disposal of domestic sewage and the contamination stemming from agricultural fertilizer application. Given the abundance of farmland and the dispersed rural population in Wujiang District, coupled with the intensive use of pesticides, fertilizers, and livestock feed, it is plausible to hypothesize that VF2 serves as an indicator of agricultural non-point source pollution.

The variance contribution rate of VF3 is 10.57%, and the main loading ions are NH_4-N(0.67) and NO_2-N(0.84). The most common sources of NH_4-N and NO_2-N in groundwater are domestic sewage and livestock manure. In addition, studies have shown that the artificial paving of asphalt pavements and the coverage of poorly porous fine clay in urban areas can cause a lack of oxygen in groundwater, promote the reduction conditions of aquifers, and thus prevent the transformation of NH_4-N(such as nitrification), leading to the accumulation of NH_4-N and NO_2-N in groundwater. In this study area, the main coverage area of VF3 is mostly urban areas, so VF3 should refer to urban domestic sewage discharge.

The variance contribution rate of VF4 has been determined to be 7.59%, with Sb (0.53) and pH (0.81) identified as the primary loading ions. The enrichment of Sb in groundwater is predominantly attributed to primary minerals or the discharge of wastewater from textile factories. Similarly, the observed abnormal increase in pH, indicating an alkaline environment, is associated with wastewater from printing and dyeing factories. Given the presence of significant textile factories in Zhenze Town and Songling Town within the research area, it is reasonable to deduce that VF4 represents the pollution discharge stemming from the textile industry in Wujiang District.

The variance contribution rate of VF5 is 7.02%, and the main loading ion is DO (0.84). High DO concentrations in groundwater are typically associated with industrial discharge or interactions with surface water. Given the observation that the loading of other hydrochemical indicators in VF5 is relatively low, and considering the well-developed surface water system within the study area, a logical inference is that VF5 serves as an indicator of interactions with surface water.

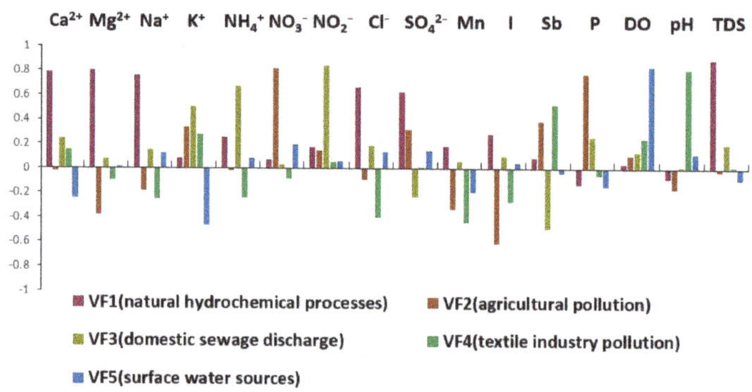

Figure 5. Component loadings for 16 groundwater quality parameters after varimax rotation in Wujiang District.

3.2.2. APCS-MLR

On the basis of principal component analysis, this study also used the APCS-MLR receptor model to conduct a quantitative assessment of the respective contributions of diverse pollution sources to the groundwater quality indicators in Wujiang District. The derived principal component coefficients and regression equation R^2 values, presented in Table 4, exhibit a commendable correlation between the predicted APCS-MLR concentrations and their corresponding measured values. Specifically, for all water quality indicators, except for Mn (0.37), I (0.54), SO_4^{2-} (0.57), and NH_4-N (0.58), the R^2 values exceed 0.6, averaging at 0.68. This observation signifies a substantial agreement between the modeled and observed data, thus validating the appropriateness and acceptability of the simulation results.

Table 4. Mean contribution of pollution sources to each groundwater quality metrics based on APCS-MLR model.

Parameters	Measured Mean	Predicted Mean	R^2	Source Contribution					
				VF1	VF2	VF3	VF4	VF5	UIS
Ca^{2+}	70.08	70.07	0.76	0.33	0.00	0.06	0.06	0.30	0.24
Mg^{2+}	17.75	17.75	0.79	0.64	0.11	0.03	0.07	0.03	0.12
Na^{+}	49.55	49.55	0.71	0.45	0.04	0.05	0.14	0.22	0.11
K^{+}	17.71	17.71	0.67	0.03	0.04	0.10	0.08	0.44	0.31
NH_4-N	0.50	0.50	0.58	0.16	0.00	0.24	0.14	0.15	0.31
NO_3-N	4.89	4.89	0.72	0.04	0.18	0.01	0.05	0.37	0.35
NO_2-N	0.25	0.25	0.76	0.10	0.03	0.29	0.03	0.10	0.45
Cl^{-}	46.29	46.29	0.66	0.37	0.02	0.06	0.20	0.22	0.12
SO_4^{2-}	46.28	46.28	0.57	0.34	0.06	0.07	0.00	0.23	0.29
Mn	117.21	117.21	0.37	0.08	0.05	0.01	0.18	0.25	0.43
I	82.42	82.42	0.54	0.24	0.18	0.05	0.21	0.11	0.21
Sb	2.31	2.25	0.67	0.06	0.10	0.21	0.36	0.08	0.20
P	530.38	530.37	0.71	0.09	0.18	0.10	0.03	0.28	0.32
DO	3.1	3.10	0.79	0.01	0.01	0.02	0.04	0.47	0.45
pH	7.19	7.19	0.71	0.01	0.01	0.00	0.09	0.04	0.84
TDS	406	406.00	0.84	0.59	0.00	0.08	0.01	0.18	0.14

Regarding the analysis, Figure 6 provides a detailed visualization of the spatial distribution of pollution sources within three distinct categories: VF2, representing agricultural non-point source pollution; VF3, pertaining to domestic sewage; and VF4, associated with pollution from the textile industry. Upon scrutiny of the figure, it becomes apparent that Qidu Town, Zhenze Town, Taoyuan Town, and Shengze Town are the regions most severely impacted by agricultural non-point source pollution. Conversely, Lili Town, Songling Town, Tongli Town, and Qidu Town exhibit significant pollution stemming from domestic sources. Furthermore, Zhenze Town, Taoyuan Town, Shengze Town, Songling Town, and Tongli Town are identified as the areas heavily contaminated by the textile industry. These insights offer a comprehensive understanding of the spatial patterns and impact of pollution sources within the study area.

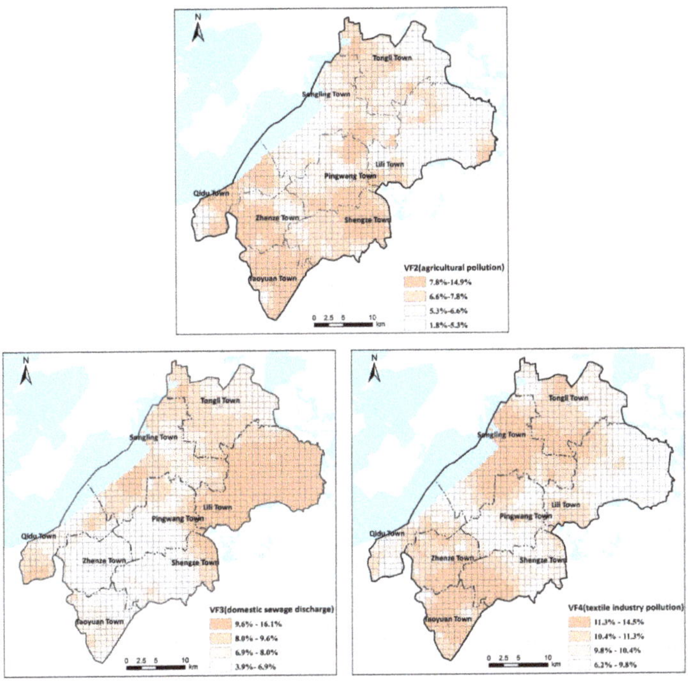

Figure 6. Spatial distribution of contributions of VF2, VF3, and VF4 in groundwater in Wujiang District.

Furthermore, Figure 7 illustrates the results of source assignment obtained by the APCS-MLR model, the relative contributions of each pollution source to various pollutants were calculated, and the average relative contribution of each pollution source was determined for a comprehensive evaluation. The hydrochemical characteristics of groundwater in Wujiang District are most significantly influenced by the natural hydrochemical processes (VF1), accounting for 22.11% of the total sources, with higher contributions to Mg^{2+} (64%) and TDS (59%). The second most significant influence comes from surface water sources (VF5), comprising 21.73% of the total sources, with the highest contribution to DO (47%). The sewage discharge from the textile industry (VF4) accounts for 10.63% of the total sources, with relatively high contributions to Sb (36%), I (21%), and Cl^- (20%) among the 16 water quality parameters. Additionally, urban domestic sewage discharge (VF3) constitutes 8.69% of all pollution sources, with higher contributions to $NO_2\text{-}N$ (29%) and $NH_4\text{-}N$ (24%). Agricultural non-point source pollution (VF2) accounts for 6.26% of all pollution sources, with a relatively high contribution to P (18%), $NO_3\text{-}N$ (18%), and I (18%). In addition, unknown pollution sources account for 30.59% of the total sources, primarily due to the complexity of pollutant evolution processes. Consequently, the order of contributions to groundwater pollution sources in Wujiang District, Suzhou City, is as follows: natural hydrochemical processes > surface water sources > sewage discharge from the textile industry > urban domestic sewage discharge > agricultural non-point source pollution.

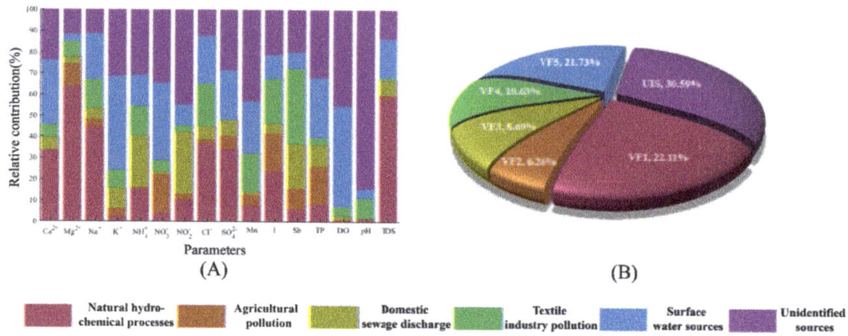

Figure 7. The contributions to groundwater quality parameters (**A**) and average contributions (**B**) of pollution sources in Wujiang District.

4. Discussion

Through the application of fuzzy synthetic evaluation and source apportionment of groundwater pollution conducted in this study, we successfully identify the primary pollutants and their respective origins of the shallow groundwater in Wujiang District. A comprehensive analysis of the distribution of pollution sources for the main exceeding indicators can be found in Table 4 and Figure 6. The main origins of I, NH_4-N, and Mn are natural hydrochemical processes, domestic sewage, and surface water sources, respectively. Furthermore, a comparative analysis of land use patterns alongside the spatial distribution of VF2, VF3, and VF4 reveals a notable consistency. Agricultural pollution is predominantly concentrated in paddy field areas. Pollution associated with the textile industry is primarily concentrated in the southern and central areas of Wujiang District, demonstrating a strong correlation with the geographical distribution of textile factories in the area. As for the domestic sewage discharge, the primary distribution does not concentrate within built-up areas. This phenomenon can be attributed to the highly systematic nature of the drainage infrastructure in such regions. In contrast, urban-rural fringe areas are more vulnerable to contamination from domestic sewage discharge.

In addition to the aforementioned advantages, the results further indicate that for certain indicators (pH and DO), the proportion of unidentified pollution sources remains excessively high. This situation underscores the need to incorporate additional hydrochemical indicators into the APCS-MLR model, or alternatively, employing a combined approach with other methods to enhance the effectiveness of apportioning the contaminant sources.

5. Conclusions

In this research, a fuzzy comprehensive evaluation technique was implemented to evaluate the groundwater quality in Wujiang District, Suzhou City. Subsequently, Principal Component Analysis (PCA) was leveraged to discern the underlying natural and human-induced factors that may be influencing the groundwater quality. Based on these insights, the study adopted the Absolute Principal Component Scores–Multiple Linear Regression (APCS-MLR) model to quantitatively ascertain the contributions of potential pollution sources to 16 distinct groundwater quality parameters.

The findings suggest a generally favorable groundwater quality within the studied region, with Classes I-III comprising 21.7%, 30.1%, and 35.1% of the total area, respectively, and evenly dispersed across the area. Meanwhile, Class IV and Class V, accounting for 4.7% and 8.4%, respectively, are predominantly located in specific areas such as the southwest of Lili Town, the north of Songling Town, and the south of Qidu Town. Moreover, utilizing the correlation of hydrochemical components and principal component analysis, we have pinpointed five potential origins that impact groundwater quality in Wujiang District. These include natural hydrochemical processes, agricultural non-point source pollution, urban domestic sewage discharge, sewage discharge from the textile industry, and groundwater–surface interaction. Utilizing Principal Component Analysis (PCA) as a foundation, the APCS-MLR receptor model was employed to quantify the individual contributions of potential pollution sources within the study area. The major sources of pollution were identified as follows: sewage discharge from the textile industry (10.63%) > urban domestic sewage discharge (8.69%) > agricultural non-point source pollution (6.26%). The areas severely affected by sewage discharge from the textile industry are mainly distributed in Zhenze Town, Taoyuan Town, Shengze Town, Songling Town, and Tongli Town. The areas severely affected by domestic sewage discharge are primarily located in the northern part of the study area, including Lili Town, Songling Town, and Tongli Town. The areas severely affected by agricultural non-point source pollution are mainly distributed in the southern part of the study area, primarily in Qidu Town, Zhenze Town, Taoyuan Town, and Shengze Town. All in all, this study provides scientific evidence and reasonable recommendations for the pollution management of shallow groundwater resources by evaluating groundwater quality and quantitatively analyzing pollution source contributions in Wujiang District, Suzhou City.

Author Contributions: Conceptualization, L.H., J.L. (Jinsong Lv) and L.Z.; methodology, L.Z.; validation, Q.Q., H.Y., Z.J. and S.M.; formal analysis, L.Z.; investigation, L.H., Q.Q., J.L. (Jinsong Lv), Z.C., H.Y., Z.J. and Y.J. (Yang Jin); resources, J.L. (Jinsong Lv); data curation, J.L. (Jinsong Lv); writing—original draft preparation, L.H.; writing— review and editing, H.Z., J.L. (Jie Li) and F.X.; supervision, Q.Z. and Y.J. (Yuehua Jiang); project administration, Q.Z. and Y.J. (Yuehua Jiang). All authors have read and agreed to the published version of the manuscript.

Funding: This research was funded by China Geological Survey, grant number DD20190260 and DD20221728. Zi Chen acknowledges the Program of MOE Key Laboratory of Groundwater Circulation and Environmental Evolution, China University of Geosciences (Beijing) (2023-002).

Data Availability Statement: The data that support the findings of this study are available from the corresponding author upon reasonable request.

Conflicts of Interest: The authors declare no conflict of interest.

References

1. Li, J.; Yang, Y.; Huan, H.; Li, M.; Xi, B.; Lv, N.; Wu, Y.; Xie, Y.; Li, X.; Yang, J. Method for screening prevention and control measures and technologies based on groundwater pollution intensity assessment. *Sci. Total Environ.* **2016**, *551–552*, 143–154. [CrossRef]
2. Wegahita, N.K.; Ma, L.; Liu, J.; Huang, T.; Luo, Q.; Qian, J. Spatial Assessment of Groundwater Quality and Health Risk of Nitrogen Pollution for Shallow Groundwater Aquifer around Fuyang City, China. *Water* **2020**, *12*, 3341. [CrossRef]
3. Xu, Z.; Hu, B.X.; Davis, H.; Cao, J. Simulating long term nitrate-N contamination processes in the Woodville Karst Plain using CFPv2 with UMT3D. *J. Hydrol.* **2015**, *524*, 72–88. [CrossRef]
4. Xin, J.; Wang, Y.; Shen, Z.; Liu, Y.; Wang, H.; Zheng, X. Critical review of measures and decision support tools for groundwater nitrate management: A surface-to-groundwater profile perspective. *J. Hydrol.* **2021**, *598*, 126386. [CrossRef]
5. Huang, G.; Sun, J.; Zhang, Y.; Chen, Z.; Liu, F. Impact of anthropogenic and natural processes on the evolution of groundwater chemistry in a rapidly urbanized coastal area, South China. *Sci. Total Environ.* **2013**, *463–464*, 209–221. [CrossRef]
6. Boateng, T.K.; Opoku, F.; Acquaah, S.O.; Akoto, O. Groundwater quality assessment using statistical approach and water quality index in Ejisu-Juaben Municipality, Ghana. *Environ. Earth Sci.* **2016**, *75*, 1–14. [CrossRef]
7. Bouderbala, A.; Gharbi, B.Y. Hydrogeochemical characterization and groundwater quality assessment in the intensive agricultural zone of the Upper Cheliff plain, Algeria. *Environ. Earth Sci.* **2017**, *76*, 1–17. [CrossRef]
8. Dugga, P.; Pervez, S.; Tripathi, M.; Siddiqui, M.N. Spatiotemporal variability and source apportionment of the ionic components of groundwater of a mineral-rich tribal belt in Bastar, India. *Groundw. Sustain. Dev.* **2020**, *10*, 100356. [CrossRef]
9. Wen, D.; Zhang, F.; Zhang, E.; Wang, C.; Han, S.; Zheng, Y. Arsenic, fluoride and iodine in groundwater of China. *J. Geochem. Explor.* **2013**, *135*, 1–21. [CrossRef]
10. He, X.; Li, P.; Ji, Y.; Wang, Y.; Su, Z.; Elumalai, V. Groundwater Arsenic and Fluoride and Associated Arsenicosis and Fluorosis in China: Occurrence, Distribution and Management. *Expo. Health* **2020**, *12*, 355–368. [CrossRef]

11. Podgorski, J.; Berg, M. Global threat of arsenic in groundwater. *Science* **2020**, *368*, 845–850. [CrossRef] [PubMed]
12. Purushotham, D.; Prakash, M.R.; Rao, A.N. Groundwater depletion and quality deterioration due to environmental impacts in Maheshwaram watershed of R.R. district, AP (India). *Environ. Earth Sci.* **2011**, *62*, 1707–1721. [CrossRef]
13. Kurunc, A.; Ersahin, S.; Sonmez, N.K.; Kaman, H.; Uz, I.; Uz, B.Y.; Aslan, G.E. Seasonal changes of spatial variation of some groundwater quality variables in a large irrigated coastal Mediterranean region of Turkey. *Sci. Total Environ.* **2016**, *554–555*, 53–63. [CrossRef]
14. McCance, W.; Jones, O.A.; Edwards, M.; Surapaneni, A.; Chadalavada, S.; Currell, M. Contaminants of Emerging Concern as novel groundwater tracers for delineating wastewater impacts in urban and peri-urban areas. *Water Res.* **2018**, *146*, 118–133. [CrossRef] [PubMed]
15. Datta, P.S.; Deb, D.L.; Tyagi, S.K. Assessment of groundwater contamination from fertilizers in the Delhi area based on 18O, NO3/- and K+ composition. *J. Contam. Hydrol.* **1997**, *27*, 249–262. [CrossRef]
16. Gu, B.; Ge, Y.; Chang, S.X.; Luo, W.; Chang, J. Nitrate in groundwater of China: Sources and driving forces. *Glob. Environ. Chang.* **2013**, *23*, 1112–1121. [CrossRef]
17. Tao, Y.; Deng, Y.; Du, Y.; Xu, Y.; Leng, Z.; Ma, T.; Wang, Y. Sources and enrichment of phosphorus in groundwater of the Central Yangtze River Basin. *Sci. Total Environ.* **2020**, *737*, 139837. [CrossRef] [PubMed]
18. Bhagure, G.R.; Mirgane, S.R. Heavy metal concentrations in groundwaters and soils of Thane Region of Maharashtra, India. *Environ. Monit. Assess.* **2011**, *173*, 643–652. [CrossRef]
19. Bhutiani, R.; Kulkarni, D.B.; Khanna, D.R.; Gautam, A. Water Quality, Pollution Source Apportionment and Health Risk Assessment of Heavy Metals in Groundwater of an Industrial Area in North India. *Expo. Health* **2016**, *8*, 3–18. [CrossRef]
20. Sharma, M.K.; Kumar, M. Sulphate contamination in groundwater and its remediation: An overview. *Environ. Monit. Assess.* **2020**, *192*, 1–10. [CrossRef]
21. Han, D.; Currell, M.J.; Cao, G. Deep challenges for China's war on water pollution. *Environ. Pollut.* **2016**, *218*, 1222–1233. [CrossRef] [PubMed]
22. Gómez-Hernández, J.J.; Xu, T. Contaminant Source Identification in Aquifers: A Critical View. *Math. Geosci.* **2021**, *54*, 437–458. [CrossRef]
23. Lapworth, D.J.; Baran, N.; Stuart, M.E.; Ward, R.S. Emerging organic contaminants in groundwater: A review of sources, fate and occurrence. *Environ. Pollut.* **2012**, *163*, 287–303. [CrossRef] [PubMed]
24. Zamani, A.A.; Yaftian, M.R.; Parizanganeh, A. Multivariate statistical assessment of heavy metal pollution sources of groundwater around a lead and zinc plant. *Iran. J. Environ. Health Sci. Eng.* **2012**, *9*, 29. [CrossRef]
25. Nijenhuis, I.; Schmidt, M.; Pellegatti, E.; Paramatti, E.; Richnow, H.H.; Gargini, A. A stable isotope approach for source apportionment of chlorinated ethene plumes at a complex multi-contamination events urban site. *J. Contam. Hydrol.* **2013**, *153*, 92–105. [CrossRef]
26. Zhou, H.; Gómez-Hernández, J.J.; Li, L. Inverse methods in hydrogeology: Evolution and recent trends. *Adv. Water Resour.* **2014**, *63*, 22–37. [CrossRef]
27. Cui, J.; Zhou, F.; Gao, M.; Zhang, L.; Zhang, L.; Du, K.; Leng, Q.; Zhang, Y.; He, D.; Yang, F.; et al. A comparison of various approaches used in source apportionments for precipitation nitrogen in a mountain region of southwest China. *Environ. Pollut.* **2018**, *241*, 810–820. [CrossRef]
28. Xu, T.; Zhang, W.; Gómez-Hernández, J.J.; Xie, Y.; Yang, J.; Chen, Z.; Lu, C. Non-point contaminant source identification in an aquifer using the ensemble smoother with multiple data assimilation. *J. Hydrol.* **2022**, *606*, 127405. [CrossRef]
29. Chen, Z.; Xu, T.; Gómez-hernández, J.J.; Zanini, A.; Zhou, Q. Reconstructing the release history of a contaminant source with different precision via the ensemble smoother with multiple data assimilation. *J. Contam. Hydrol.* **2022**, *252*, 104115. [CrossRef]
30. Chen, Z.; Zong, L.; Gómez-Hernández, J.J.; Xu, T.; Jiang, Y.; Zhou, Q.; Yang, H.; Jia, Z.; Mei, S. Contaminant source and aquifer characterization: An application of ES-MDA demonstrating the assimilation of geophysical data. *Adv. Water Resour.* **2023**, *181*, 104555. [CrossRef]
31. Cloutier, V.; Lefebvre, R.; Therrien, R.; Savard, M.M. Multivariate statistical analysis of geochemical data as indicative of the hydrogeochemical evolution of groundwater in a sedimentary rock aquifer system. *J. Hydrol.* **2008**, *353*, 294–313. [CrossRef]
32. Li, P.; Tian, R.; Liu, R. Solute Geochemistry and Multivariate Analysis of Water Quality in the Guohua Phosphorite Mine, Guizhou Province, China. *Expo. Health* **2019**, *11*, 81–94. [CrossRef]
33. Ravikumar, P.; Somashekar, R.K. Principal component analysis and hydrochemical facies characterization to evaluate groundwater quality in Varahi river basin, Karnataka state, India. *Appl. Water Sci.* **2017**, *7*, 745–755. [CrossRef]
34. Gulgundi, M.S.; Shetty, A. Groundwater quality assessment of urban Bengaluru using multivariate statistical techniques. *Appl. Water Sci.* **2018**, *8*, 1–15. [CrossRef]
35. Wu, J.; Li, P.; Qian, H.; Duan, Z.; Zhang, X. Using correlation and multivariate statistical analysis to identify hydrogeochemical processes affecting the major ion chemistry of waters: A case study in Laoheba phosphorite mine in Sichuan, China. *Arab. J. Geosci.* **2014**, *7*, 3973–3982. [CrossRef]
36. Wu, J.; Li, P.; Wang, D.; Ren, X.; Wei, M. Statistical and multivariate statistical techniques to trace the sources and affecting factors of groundwater pollution in a rapidly growing city on the Chinese Loess Plateau. *Hum. Ecol. Risk Assess.* **2020**, *26*, 1603–1621. [CrossRef]

37. Ielpo, P.; Cassano, D.; Lopez, A.; Pappagallo, G.; Uricchio, V.F.; De Napoli, P.A. Source apportionment of groundwater pollutants in Apulian agricultural sites using multivariate statistical analyses: Case study of Foggia province. *Chem. Cent. J.* **2012**, *6*, 1–12. [CrossRef]
38. Huang, G.; Liu, C.; Sun, J.; Zhang, M.; Jing, J.; Li, L. A regional scale investigation on factors controlling the groundwater chemistry of various aquifers in a rapidly urbanized area: A case study of the Pearl River Delta. *Sci. Total Environ.* **2018**, *625*, 510–518. [CrossRef]
39. Thurston, G.D.; Spengler, J.D. A quantitative assessment of source contributions to inhalable particulate matter pollution in metropolitan Boston. *Atmos. Environ.* **1985**, *19*, 9–25. [CrossRef]
40. Wang, H.; An, J.; Cheng, M.; Shen, L.; Zhu, B.; Li, Y.; Wang, Y.; Duan, Q.; Sullivan, A.; Xia, L. One year online measurements of water-soluble ions at the industrially polluted town of Nanjing, China: Sources, seasonal and diurnal variations. *Chemosphere* **2016**, *148*, 526–536. [CrossRef]
41. Rahman, M.S.; Bhuiyan, S.S.; Ahmed, Z.; Saha, N.; Begum, B.A. Characterization and source apportionment of elemental species in PM2.5 with especial emphasis on seasonal variation in the capital city "Dhaka", Bangladesh. *Urban Clim.* **2021**, *36*, 100804. [CrossRef]
42. Choi, H.; Cho, Y.C.; Kim, S.H.; Yu, S.J.; Kim, Y.S.; Im, J.K. Water Quality Assessment and Potential Source Contribution Using Multivariate Statistical Techniques in Jinwi River Watershed, South Korea. *Water* **2021**, *13*, 2976. [CrossRef]
43. Cho, Y.c.; Choi, H.; Lee, M.G.; Kim, S.-H.; Im, J.-K. Identification and Apportionment of Potential Pollution Sources Using Multivariate Statistical Techniques and APCS-MLR Model to Assess Surface Water Quality in Imjin River Watershed, South Korea. *Water* **2022**, *14*, 793. [CrossRef]
44. Proshad, R.; Kormoker, T.; Al, M.A.; Islam, M.S.; Khadka, S.; Idris, A.M. Receptor model-based source apportionment and ecological risk of metals in sediments of an urban river in Bangladesh. *J. Hazard. Mater.* **2022**, *423*, 127030. [CrossRef]
45. Shen, D.; Huang, S.; Zhang, Y.; Zhou, Y. The source apportionment of N and P pollution in the surface waters of lowland urban area based on EEM-PARAFAC and PCA-APCS-MLR. *Environ. Res.* **2021**, *197*, 111022. [CrossRef]
46. Zhang, H.; Xu, Y.; Cheng, S.; Li, Q.; Yu, H. Application of the dual-isotope approach and Bayesian isotope mixing model to identify nitrate in groundwater of a multiple land-use area in Chengdu Plain, China. *Sci. Total Environ.* **2020**, *717*, 137134. [CrossRef]
47. Chen, Z.; Zhou, Q.; Lv, J.; Jiang, Y.; Yang, H.; Yang, H.; Mei, S.; Jia, Z.; Zhang, H.; Jin, Y.; et al. Assessment of Groundwater Quality Using APCS-MLR Model: A Case Study in the Pilot Promoter Region of Yangtze River Delta Integration Demonstration Zone, China. *Water* **2023**, *15*, 225. [CrossRef]
48. Zhang, Q.; Wang, H.; Wang, Y.; Yang, M.; Zhu, L. Groundwater quality assessment and pollution source apportionment in an intensely exploited region of northern China. *Environ. Sci. Pollut. Res.* **2017**, *24*, 16639–16650. [CrossRef]
49. Meng, L.; Zuo, R.; Wang, J.-s.; Yang, J.; Teng, Y.-g.; Shi, R.-t.; Zhai, Y.-z. Apportionment and evolution of pollution sources in a typical riverside groundwater resource area using PCA-APCS-MLR model. *J. Contam. Hydrol.* **2018**, *218*, 70–83. [CrossRef]
50. Yu, L.; Zheng, T.; Yuan, R.; Zheng, X. APCS-MLR model: A convenient and fast method for quantitative identification of nitrate pollution sources in groundwater. *J. Environ. Manag.* **2022**, *314*, 115101. [CrossRef]
51. Sheng, D.; Meng, X.; Wen, X.; Wu, J.; Yu, H.; Wu, M. Contamination characteristics, source identification, and source-specific health risks of heavy metal(loid)s in groundwater of an arid oasis region in Northwest China. *Sci. Total Environ.* **2022**, *841*, 156733. [CrossRef] [PubMed]
52. Haji Gholizadeh, M.; Melesse, A.M.; Reddi, L. Water quality assessment and apportionment of pollution sources using APCS-MLR and PMF receptor modeling techniques in three major rivers of South Florida. *Sci. Total Environ.* **2016**, *566-567*, 1552–1567. [CrossRef]
53. Cheng, G.; Wang, M.; Chen, Y.; Gao, W. Source apportionment of water pollutants in the upstream of Yangtze River using APCS-MLR. *Environ. Geochem. Health* **2020**, *42*, 3795–3810. [CrossRef] [PubMed]
54. Zhang, H.; Cheng, S.; Li, H.; Fu, K.; Xu, Y. Groundwater pollution source identification and apportionment using PMF and PCA-APCA-MLR receptor models in a typical mixed land-use area in Southwestern China. *Sci. Total Environ.* **2020**, *741*, 140383. [CrossRef]
55. Chen, R.; Teng, Y.; Chen, H.; Hu, B.; Yue, W. Groundwater pollution and risk assessment based on source apportionment in a typical cold agricultural region in Northeastern China. *Sci. Total Environ.* **2019**, *696*, 133972. [CrossRef]
56. Xing, G.; Cao, Y.; Shi, S.; Sun, G.; Du, L.; Zhu, J. N pollution sources and denitrification in waterbodies in Taihu Lake region. *Sci. China Ser. B Chem.* **2001**, *44*, 304–314. [CrossRef]
57. Li, D.; Jiang, X.; Wang, J.; Wang, K.; Zheng, B. Effect of sewage and industrial effluents on bacterial and archaeal communities of creek sediments in the Taihu Basin. *Water* **2017**, *9*, 373. [CrossRef]
58. Yin, X.; Jiang, B.; Feng, Z.; Yao, B.; Shi, X.; Sun, Y.; Wu, J. Comprehensive evaluation of shallow groundwater quality in Central and Southern Jiangsu Province, China. *Environ. Earth Sci.* **2017**, *76*, 1–16. [CrossRef]
59. Jiang, Y.; Ni, H.; Zhou, Q.; Cheng, Z.; Duan, X.; Zhu, Z.; Wu, J.; Ren, H.; Fan, C.; Yang, J.; et al. Key technology of ecological restoration demonstration in the Yangtze River Economic Zone and its application. *Geol. China* **2021**, *48*, 1305–1333.
60. Mujumdar, P.P.; Sasikumar, K. A fuzzy risk approach for seasonal water quality management of a river system. *Water Resour. Res.* **2002**, *38*, 3653–3664. [CrossRef]

61. Zhang, B.; Song, X.; Zhang, Y.; Han, D.; Tang, C.; Yu, Y.; Ma, Y. Hydrochemical characteristics and water quality assessment of surface water and groundwater in Songnen plain, Northeast China. *Water Res.* **2012**, *46*, 2737–2748. [CrossRef] [PubMed]
62. Dahiya, S.; Singh, B.; Gaur, S.; Garg, V.K.; Kushwaha, H.S. Analysis of groundwater quality using fuzzy synthetic evaluation. *J. Hazard. Mater.* **2007**, *147*, 938–946. [CrossRef] [PubMed]
63. Chang, N.B.; Chen, H.W.; Ning, S.K. Identification of river water quality using the fuzzy synthetic evaluation approach. *J. Environ. Manag.* **2001**, *63*, 293–305. [CrossRef] [PubMed]
64. *GB/T 14848-2017*; Standard for Groundwater Quality. National Resources and Territory Spatial Planning Standardization Technical Committee (SAC/TC 93). General Administration of Quality Supervision, Inspection and Quarantine of the People's Republic of China, Standardization Administration of China: Beijing, China, 2017.
65. Shrestha, S.; Kazama, F. Assessment of surface water quality using multivariate statistical techniques: A case study of the Fuji river basin, Japan. *Environ. Model. Softw.* **2007**, *22*, 464–475. [CrossRef]
66. Helena, B.; Pardo, R.; Vega, M.; Barrado, E.; Fernandez, J.M.; Fernandez, L. Temporal evolution of groundwater composition in an alluvial aquifer (Pisuerga River, Spain) by principal component analysis. *Water Res.* **2000**, *34*, 807–816. [CrossRef]
67. Liu, L.; Tang, Z.; Kong, M.; Chen, X.; Zhou, C.; Huang, K.; Wang, Z. Tracing the potential pollution sources of the coastal water in Hong Kong with statistical models combining APCS-MLR. *J. Environ. Manag.* **2019**, *245*, 143–150. [CrossRef]
68. Li, W.; Wu, J.; Zhou, C.; Nsabimana, A. Groundwater Pollution Source Identification and Apportionment Using PMF and PCA-APCS-MLR Receptor Models in Tongchuan City, China. *Arch. Environ. Contam. Toxicol.* **2021**, *81*, 397–413. [CrossRef]
69. Li, Q.; Zhang, H.; Guo, S.; Fu, K.; Liao, L.; Xu, Y.; Cheng, S. Groundwater pollution source apportionment using principal component analysis in a multiple land-use area in southwestern China. *Environ. Sci. Pollut. Res.* **2020**, *27*, 9000–9011. [CrossRef]
70. Zhang, H.; Li, H.; Yu, H.; Cheng, S. Water quality assessment and pollution source apportionment using multi-statistic and APCS-MLR modeling techniques in Min River Basin, China. *Environ. Sci. Pollut. Res.* **2020**, *27*, 41987–42000. [CrossRef]

water

MDPI

Article

Hydrogeochemical Characteristics and Genesis of Hot Springs in Da Qaidam Area, Northern Qaidam Margin of the Qaidam Basin

Jiayi Zhao [1,2], Lingxia Liu [1,2,*], Yaru Wang [1,2], Bingyan Li [1], Wenjing Lin [1,2] and Haihui Yao [3]

1 Institute of Hydrogeology and Environmental Geology, Chinese Academy of Geological Sciences, Shijiazhuang 050061, China; zhaojiayi@mail.cgs.gov.cn (J.Z.); yaru_wang@foxmail.com (Y.W.); libingyan@mail.cgs.gov.cn (B.L.); linwenjing@mail.cgs.gov.cn (W.L.)
2 Technical Innovation Center for Geothermal and HDR Exploration and Development, Ministry of Natural Resources, Shijiazhuang 050061, China
3 School of CIvil Engineering, Xuchang University, Xuchang 461000, China; yaohaihui05@163.com
* Correspondence: liulingxia@mail.cgs.gov.cn

Abstract: Hydrogeochemical research on fluids is an effective method to understand the formation mechanism, occurrence environment, and circulation process of groundwater. The groundwater sampling sites are located in the town of Dachaidan on the northeastern edge of the Tibetan Plateau, which was selected as the study object. Samples were collected from hot and cold springs and surface water in the area. This study is based on the analysis of water chemistry and isotopes, and aims (1) to discuss the chemical characteristics of groundwater in Da Qaidam, (2) to estimate the deep reservoir temperatures, recharge elevation and circulation depth of geothermal waters, and (3) to figure out the heat source beneath the geothermal area and its genetic mechanism. The result showed the following: The hydrochemical type of the hot spring is $Cl{\cdot}SO_4$-Na and Cl-Na, and the hydrochemical type of cold spring is $SO_4{\cdot}HCO_3$-Na·Ca and $Cl{\cdot}HCO_3{\cdot}SO_4$-Ca·Na. The main source of groundwater recharge is snow and ice melt water. The recharge elevation ranges from 4666.8 m to 5755.9 m. The geothermal reservoir temperature is about 119.15–126.6 °C. Ice and snow melt water infiltrate into the high mountainous areas on the north side of Da Qaidam and circulate underground through the developed deep and large fractures. Part of the groundwater migrates upwards under the water conduction of the Da Qaidam fault fracture zone to form cold springs, while another part is heated by deep circulation and exposed to the surface in the form of medium to low temperature tectonic hot springs. The research results can provide a scientific basis for geothermal resource exploitation and utilization in Qinghai Province.

Keywords: Da Qaidam groundwater; hydrogeochemistry; isotopes; geothermal reservoir temperature; geothermal temperature scale

1. Introduction

Economic and social advancement is driving significant global environmental shifts, with climate change being a central concern, profoundly affecting human societies across multiple facets [1]. Consequently, the scarcity of energy and environmental degradation are now critical barriers to sustainable global progress. The production and utilization of geothermal resources are paramount for achieving carbon peak and neutrality objectives, bolstered by research into groundwater genesis which underpins the conservation of the ecological milieu.

The Qaidam Basin, situated in arid to semi-arid regions, has limited surface water, making groundwater the chief water supply. In 1957, China's Salt Lake Scientific Survey Team conducted analyses of the Qaidam Basin's thermal springs [2]. Subsequent studies by Zhang et al. [3] on the basin's salt lakes indicated the meteoric origin of thermal spring water, confirmed via hydrogen and oxygen isotope assessments. Additionally, Liu et al. [4] conducted an early geochemical simulation of the link between the evolution of the basin's salt lakes and their water sources, implying a strong association with both surface and deep-source water.

Located in the Qaidam Basin's northern sub-basin and adjacent to the southern edge of the Qilianshan fold belt, Da Qaidam Lake epitomizes the typical salt lakes of Northwest China, characterized by an arid climate and scant rainfall. The over-exploitation of groundwater in the under-examined Da Qaidam area has precipitated issues such as water mineralization and soil salinization [5]. Therefore, hydrogeochemical studies of Da Qaidam's groundwater are essential for local aquatic environmental conservation. Xiao [6–8] examined isotopic compositions including lithium, boron, and chlorine in Da Qaidam Lake. Li et al. [2] probed the geochemical origins of the study area's thermal springs, deducing their sources from isotopic data. Zhang et al. [5] analyzed the hydrochemical properties of groundwater and the water–rock interactions during subterranean flow. Kong et al. [9] studied radium isotopes in the salt lakes to trace radium migration and delineate water body interactions near Da Qaidam based on the hydrogeochemical characteristics of surface water and groundwater (hot and cold springs); this study explored the hydrogeochemical characteristics and origin of the study area's water, with implications for the production, utilization, and protection of water resources in this area.

Energy shortage and environmental pollution are the two major challenges that currently constrain global sustainable development. With the increasing depletion of traditional energy and the growing awareness of human environmental protection, new and renewable energy are receiving more and more attention from the world. Geothermal resources, as clean and renewable energy sources, have attracted widespread attention [10].

Qinghai Province is located in the northeast of the Qinghai Tibet Plateau, with complex geological structures and strong neotectonic movements, and abundant geothermal resources in the region. Over the years, scholars both domestically and internationally have conducted extensive research on geothermal resources in Qinghai Province, achieving fruitful results related to the hydrogeochemical characteristics of geothermal fields [11–14], the occurrence patterns and heat source mechanisms of geothermal resources [15,16], thermal structures [17,18], and as the types and heat accumulation patterns of dry hot rock resources [19,20]. However, these research results are mainly concentrated in the areas of Gonghe and Guide in the eastern part of Qinghai Province, with relatively less research on other regions within the province. Da Qaidam Hot Spring is located about 10 km north of Da Qaidam Town in northern Qinghai Province, with nearly a hundred hot spring discharge points developed within a range of less than 1 km^2. However, due to its remote location and sparse population, there is a relative lack of research on geothermal resources in the region. Only Zhang Pengxi et al. [3], Li Jiansen [2], and Wang Bing [21] used isotope technology to preliminarily explore the genesis of hot springs in the area, revealing that atmospheric precipitation is the main source of supply for hot springs, and fault zones are the main factors controlling hot springs. Based on previous research, this article uses hydrochemical and isotopic techniques to analyze the geochemical properties and evolutionary characteristics of underground hot water. The equilibrium diagram method is applied to determine the water rock equilibrium state of hot water. Combined with the geochemical temperature scale, multi mineral equilibrium, and the silicon enthalpy model, reservoir temperature and thermal storage cycle depth are analyzed. Based on the analysis of the chemical characteristics of groundwater, combined with the recharge, runoff, and discharge characteristics of atmospheric precipitation, surface water, groundwater, and geothermal water, a preliminary genetic model of the geothermal system in the Da Qaidam area has been constructed. This has important scientific significance for the systematic evaluation of geothermal resource potential in the area and the promotion of geothermal research in the region.

2. Study Area

The Da Qaidam thermal springs are situated in a thermal spring gully approximately 8.5 km north of Da Qaidam Town, Haixi Prefecture, Qinghai Province. Accessible transportation is available, with national highways Nos. 315 and 215 intersecting the area, along with the Dunhuang-Golmud and Qinghai-Xinjiang highways. The study area exhibits a typical inland basin climate, dominated by arid conditions and scant rainfall. It records an average annual temperature of 1.53 °C, annual precipitation of 81.8 mm, and annual evaporation rates reaching 2154.64 mm—more than 26 times the precipitation volume, leading to the formation of numerous salt marshes and alkali flats [22]. Geographically, the northeast contains mountains while the central zone is characterized by an alluvial-proluvial-lacustrine plain. The terrain slopes from the higher northeast to the lower southwest, culminating at the center of the lacustrine basin. Climatic conditions are heavily influenced by the topography, with temperature and evaporation decreasing, and precipitation increasing with elevation; meteorological analysis shows a 12–15 mm increase in precipitation for each 100 m rise in altitude, making the high mountain areas vital for groundwater recharge. The area features diverse geomorphic types, including high mountains, mid-altitude proluvial-lacustrine plains, chemical sedimentary plains, outwash accumulation platforms, and hills.

In the study area, the Da Qaidam, Zongwulongshan, and Lvliangshan faults primarily extend in the NWW direction, marking the division between the Qaidam Meso-Cenozoic depression basin and the Qilianshan fold uplift area. Based on their geometric distribution, activity patterns, and dynamics, the Da Qaidam fault is categorized into northern, central, and southern segments, delineated by the thermal spring gully and Tatalin Gol gully [23]. The thermal springs appear at the nexus of the northern and central fault segments in the thermal spring gully (Figure 1). Exposed strata consist of Indosinian porphyritic granites, grayish-white to light-gray, with plagioclase phenocrysts and a matrix composed of plagioclase, potassium feldspar, quartz, and muscovite. Neotectonic activity along the Da Qaidam fault is characterized by compressive thrusting and dextral strike-slip movement [24].

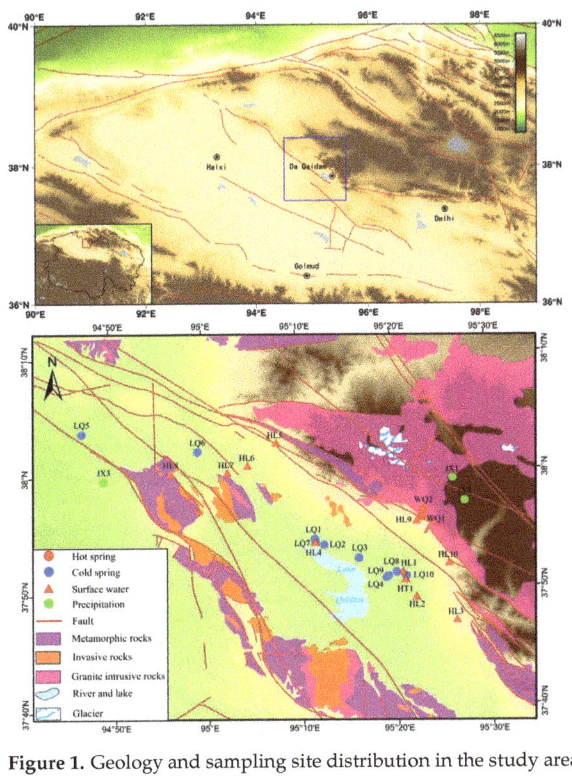

Figure 1. Geology and sampling site distribution in the study area.

3. Sampling and Testing

In the hydrogeological assessment of the study area and the surrounding geothermal geological context, 26 sample sets were collected, comprising 11 sets of surface water, two sets of thermal spring water, 10 sets of cold spring water, and 3 sets of rain and snow. The locations of the sampling sites are depicted in Figure 1. Prior to sampling, on-site measurements of physicochemical parameters including pH, electrical conductivity (EC), and temperature were conducted, and temperatures until stable readings were obtained. Appropriate preservatives were added to each sample following stabilization as necessitated by the test protocols. To maintain data integrity, all samples were stored in chilled Teflon bottles at 4 °C and promptly transported to the laboratory for analysis.

The comprehensive analysis, including trace element quantification and hydrogen–oxygen isotope measurement, was carried out at the Key Laboratory of Groundwater Sciences and Engineering under the Ministry of Natural Resources, part of the Institute of Hydrogeology and Environmental Geology, Chinese Academy of Geological Sciences. Major cations (K^+, Na^+, Ca^{2+}, Mg^{2+}) and trace elements were measured using an inductively coupled plasma mass spectrometer Optima 8000 ICP-OES from PerkinElmer (Waltham, MA, USA), with a testing error margin of ±1%. Chloride (Cl^-) and sulfate ($SO_4{}^{2-}$) concentrations were determined by ion chromatography on an ICS-2500 from Dionex (Sunnyvale, CA, USA), with a testing accuracy of ±1%. Bicarbonate ($HCO_3{}^-$) concentrations were quantified via titration, with an error margin of ±2%. Silicate concentrations were gauged using a UV-2550 ultraviolet spectrophotometer from SHIMADZU (Kyoto, Japan), ensuring a testing error below 0.0003%. Deuterium (δD) and oxygen-18 ($\delta^{18}O$) isotopic ratios were assessed with a L2130-i ultra-precision isotope analyzer from Picarro (Santa Clara, CA, USA), achieving an accuracy of 0.025%. The results of these tests are tabulated in Table 1.

Table 1. Main ion concentration in the study area.

Type	Sample No.	pH	Temperature (°C)	Hydrochemical Component/(mg/L)									δD/‰	$\delta^{18}O$/‰	Hydrochemical Type
				K^+	Na^+	Ca^{2+}	Mg^{2+}	Cl^-	$SO_4{}^{2-}$	$HCO_3{}^-$	SiO_2	TDS			
Surface water	HL1	7.56	-	3.13	55.52	48.52	13	87.95	76.99	119	5.20	363.9	-	-	Cl·HCO₃·SO₄-Ca·Na
	HL2	7.65	-	3.31	122.1	80.49	28.19	147.2	231.5	146.4	6.14	697.3	-	-	HCO₃-Na·Ca
	HL3	7.79	-	3.54	275.6	85.75	28.65	135.5	553.7	197	10.19	1195	−69.0	−10.1	SO₄-Na
	HL4	7.7	-	3.99	223.6	49.4	16.91	114.5	370.3	145.3	10.76	870.1	-	-	SO₄-Na
	HL5	7.63	-	1.56	82.11	47.5	10.8	53.57	143.9	136.9	5.20	421.5	−70.0	−10.4	SO₄·HCO₃-Na·Ca
	HL6	7.68	-	1.86	116.6	52.34	12.03	64.77	217.9	154.9	6.14	559.6	−68.0	−10.1	SO₄·HCO₃-Na·Ca
	HL7	7.51	-	1.9	90.49	54.97	13	67.58	165.4	156.1	6.45	488.1	−66.0	−9.7	SO₄·HCO₃-Na·Ca
	HL8	7.61	-	1.41	102.2	41.78	7.27	36.76	191.4	130.3	5.77	456.6	−61.0	−9.8	SO₄·HCO₃-Na·Ca
	HL9	7.8	-	0.91	93.05	31.15	4.4	16.46	179.4	100.9	4.94	385.7	−52.0	−8.7	SO₄·HCO₄-Na·Ca
	HL10	7.48	-	1.96	143.3	49.82	17.8	103.3	252.4	97.29	5.72	629.5	−54.0	−8.9	SO₄·Cl-Na
	HT1	7.35	-	3.15	104.8	36.75	17.35	116.7	156.9	73.22	10.76	488.8	−5.0	−3.3	Cl·SO₄-Na
Thermal spring	WQ1	7.22	73.5	14.29	431.4	29.76	0.45	327.4	347.8	213.2	77.92	1543	−77.0	−10.6	Cl·SO₄-Na
	WQ2	7.85	72	14.02	353.0	26.85	5.92	330.17	220.35	224.03	81.18	1266	-	-	Cl-Na
	WQ3 *	8.45	52	12.17	389.2	35.82	0.22	229.15	131.25	211.8	79.7	1188.4	−83.1	−12.1	Cl-Na
	WQ4 *	8.04	74	11.27	352.3	34.82	0.27	209.28	121.22	168.5	72.66	959.8	−84.0	−12.6	Cl-Na
	WQ5 *	8.15	61	11.60	283.3	36.53	0.26	217.76	147.55	205.9	82.49	1074.0	−86.0	−12.7	Cl-Na
	WQ6 *	8.37	74	11.55	283.2	33.68	0.28	217.16	151.89	255.4	82.87	1061.3	−85.0	−12.7	Cl-Na
Cold spring	LQ1	7.5		3.89	157.9	57.69	18.24	92.79	280.3	204.2	10.86	737.5	-	-	SO₄·HCO₃-Na·Ca
	LQ2	7.61	17.9	2.61	194.5	45.02	9.16	88.58	256.4	228.2	9.01	738.4	-	-	SO₄·HCO₃-Na
	LQ3	7.33	13.9	2.62	122.9	52.79	16.6	71.78	141.9	261.2	9.77	565.3	-	-	HCO₃·SO₄-Na·Ca
	LQ4	7.64	9.7	3.9	167.2	63.95	15.48	106.8	239.4	222.2	28.7	748.9	-	-	SO₄·HCO₃-Na·Ca
	LQ5	7.73	-	104.4	3248	356.8	684	4202	4158	433.6	37.12	13034	−55.0	−5.1	Cl·SO₄-Na·Mg
	LQ6	7.62	-	2.1	157.3	45.71	13.68	59.87	275.8	162.7	9.74	655.8	-	-	SO₄·HCO₃-Na
	LQ7	7.7	-	3.99	223.6	49.4	16.91	114.5	370.3	145.3	7.3	870.1	-	-	SO₄-Na
	LQ8	7.7	-	2.66	128	36.01	7.82	71.78	159.9	162.1	33.42	535.8	-	-	Cl·HCO₃·SO₄-Na
	LQ9	7.5	-	4.13	184.9	69.07	17.52	119.1	272.3	240.2	33.08	836	-	-	Cl·HCO₃·SO₄-Na
	LQ10	8.13	-	3.37	64.84	57.47	11.03	57.07	93.44	174.2	9.58	397.6	-	-	Cl·HCO₃·SO₄-Ca·Na
Meteoric water	JX1	-	-	2.2	7	15.8	0.7	10.8	4.6	55.1	-	69.2	−59.1	−9.7	HCO₃-Ca·Na
	JX2	-	-	0.9	6.8	15.3	0.9	6.5	7.8	49	-	63.6	−46.0	−7.0	HCO₃-Ca·Na
	JY3	-	-	1	9.6	15.3	3.8	8.2	9	46.4	-	83.5	−78.0	−11.0	HCO₃-Ca·Na

Note: * after Wang et al [21]; "-"no data.

4. Results

4.1. Hydrogeochemical Characteristics

4.1.1. Total Dissolved Solids (TDS) Content

TDS content, reflecting the concentration of major ions in groundwater, serves as a key metric for evaluating hydrochemical characteristics. In the study area, TDS content in surface water predominantly ranged from 363.9 to 697.3 mg/L, indicative of freshwater. However, samples from HL4 near Da Qaidam Lake's north and HL3 at the western Datouyang Gully mountain pass had higher TDS contents of 870.1 mg/L and 1195 mg/L, respectively, due to elevated Na+ and Cl- from evaporation, compared to other surface water sites. For cold springs, aside from the LQ5 site upstream of Kushui River with TDS content of 13,034 mg/L, five sites had TDS contents from 565.3 to 870.1 mg/L, suggesting freshwater. LQ5's water, uniquely high in Na^+ at 3248 mg/L and Cl^- and SO_4^{2-} at 4200 mg/L, is influenced by evaporation and rock mineral dissolution over its roughly 80 km journey from Tuergendaban Mountain. Thermal spring samples had TDS contents of 959.8–1543 mg/L, likely from extensive water–rock interaction such as the dissolution of potassium feldspar, sodium feldspar, and quartz and ion exchange within the deep thermal cycle, enriching the thermal springs' ion content.

4.1.2. Distributions of Hydrochemical Types

The ionic composition data of surface, thermal spring, and cold spring water samples from the Da Qaidam area were analyzed, and a Piper diagram was constructed (Figure 2) to delineate the hydrochemical classes of the area. Excluding meteoric water, Na+ is the predominant cation across the sample set, with Ca^{2+} and Mg^{2+} also present; as for anions, Cl^- leads the anion group, followed by HCO_3^- and SO_4^{2-}. The surface water at HT1 is classified as the $Cl{\cdot}SO_4$-Na type; the thermal spring water is classified as the Cl-Na and $Cl{\cdot}SO_4$-Na types; HL3, HL4, and LQ07 are classified as the SO_4-Na type; the majority of other surface and cold spring water are classified as $SO_4{\cdot}HCO_3$-Na·Ca or $Cl{\cdot}HCO_3{\cdot}SO_4$-Ca·Na; and meteoric water is classified as HCO_3-Ca·Na. High HCO_3^- concentrations in HL1 and LQ3 can be attributed to their proximity to Tuergendaban Mountain, which leads to a greater direct recharge from meteoric water and less ion exchange from water–rock interaction, resulting in elevated bicarbonate concentrations compared to other samples. In contrast, LQ5's cold spring exhibits a significantly higher Cl^- concentration, a characteristic typically associated with the high solubility and migration ability of salts formed with sodium, calcium, and magnesium ions in groundwater, with concentrations generally rising alongside TDS content. Thermal spring and LQ5's long circulation path and slow flow rate intensify evaporation and water–rock reactions, thereby increasing their ion concentrations.

SO_4^{2-} ions in geothermal water are commonly derived from gypsum or other sulfate mineral dissolution, and sulfur and sulfide oxidation. The high SO_4^{2-} concentrations of WQ1 can be linked to substantial dissolution influenced by the adjacent Yuka coal mine area and the surface water's oxidizing conditions, which favor the oxidation of sulfates.

4.1.3. Trace Element Characteristics

Groundwater typically contains trace elements in low concentrations, which are primarily derived from rock leaching during water–rock interactions and from deep-seated geological processes. Overall, thermal springs exhibit higher concentrations of trace elements compared to surface water and cold springs, with elements such as K, B, and Li being particularly elevated. This suggests that these trace elements originate from distinct sources. It is widely accepted that boron and lithium in groundwater are primarily sourced from deep magmatic processes.

The geothermal water samples from adjacent zones and their similar chemical components correlate strongly with hydraulic conditions and, thus, serve as indirect indicators of groundwater runoff conditions. Therefore, a hydrochemical correlation analysis of these samples can offer vital insights into groundwater circulation. Chlorine (Cl) is known to maintain a stable presence in groundwater, as it tends not to react with minerals or be readily adsorbed, making it an ideal tracer for ions that are highly correlated with it. This property allows for the examination of groundwater runoff conditions in the study area. However, due to the anomalously high ion concentration in sample LQ5, it was omitted from the Cl correlation plot (Figure 3) to prevent skewing the analysis. Findings indicate a low correlation of Cl with other ions within the study area, suggesting that the trace elements predominantly originate from mineral dissolution rather than from deep geological sources.

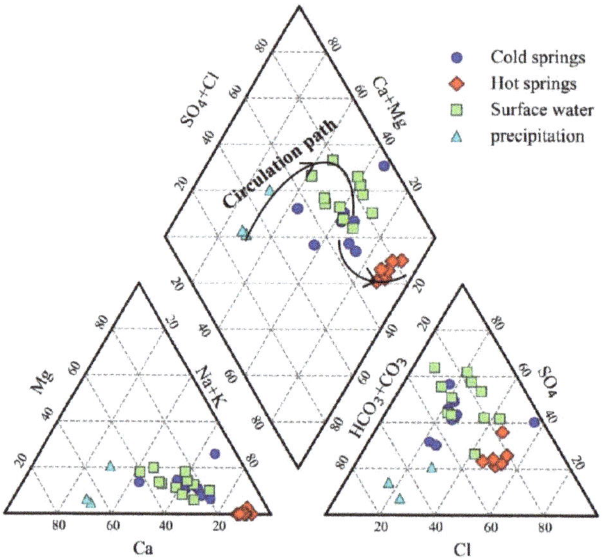

Figure 2. Piper trilinear diagram for the study area.

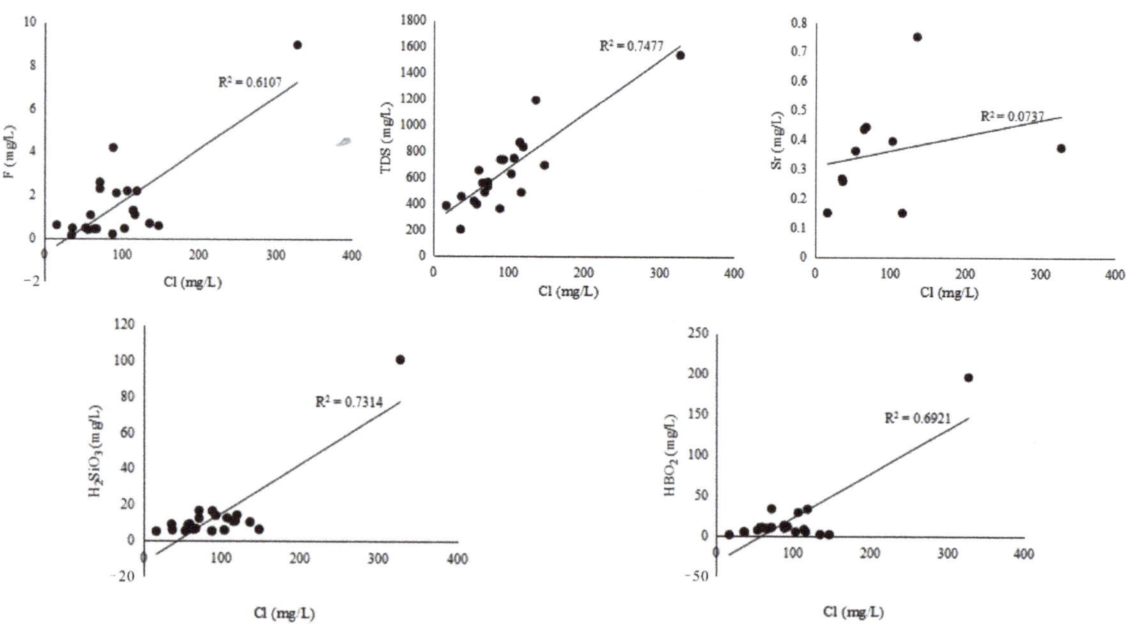

Figure 3. Relationships between major ions and Cl in the study area.

4.2. Hydrogeochemical Processes

Elucidating the hydrogeochemical processes that groundwater and surface water undergo during their circulation is essential for understanding the formation mechanisms of chemical components in these water bodies.

4.2.1. Evaporative Concentration

Gibbs diagrams are instrumental in illustrating the relationship between ions (such as Na^+, Ca^{2+}, Cl^-, and HCO_3^-) and TDS content, aiding in the investigation of the main factors that shape the formation of major ions in water. These factors include concentrations from evaporation, rock weathering, and precipitation. The Gibbs diagrams for various samples from the study area (Figure 4) reveal that cold springs and rivers predominantly fall within the rock weathering zones, suggesting extensive weathering and water–rock interaction. Contrastingly, LQ5 and six thermal springs are distinctly within the evaporative concentration zones, underscoring that evaporative concentration primarily influences them, with LQ5 exhibiting particularly high ion concentrations.

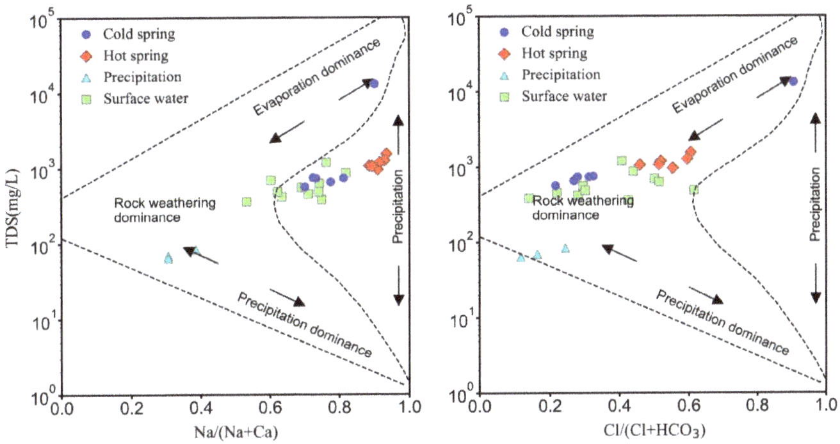

Figure 4. Gibbs diagrams of the study area.

The data points in the Na/(Na + Ca) vs. TDS content diagram are positioned to the right of those in the Cl/(Cl + HCO_3) vs. TDS content diagram. This may be attributed to a decrease in Ca^{2+} due to precipitation out of solution, or an increase in Na^+ and a decrease in Ca^{2+} resulting from cation exchange reactions between Na^+ and Ca^{2+}.

4.2.2. Hydrogeochemical Processes of Ion Formation

Water–rock interactions govern the transformation of ion components in groundwater. By examining the interrelationships between these ions and mineral element ratios, the ion origins can be deduced. Specifically, the influence of minerals like dolomite, calcite, and gypsum on Ca^{2+} and Mg^{2+} in geothermal water can be analyzed through the interplay of Ca^{2+}, Mg^{2+}, SO_4^{2-}, and HCO_3^- ions.

In the $\gamma(Ca^{2+} + Mg^{2+})$ vs. $\gamma(SO_4^{2-} + HCO_3^-)$ diagram (Figure 5), all samples from the study area sit below the equilibrium line of $\gamma(Ca^{2+} + Mg^{2+})/\gamma(HCO_3^- + SO_4^{2-}) = 1$. This suggests an overabundance of HCO_3^- and SO_4^{2-} relative to Ca^{2+} and Mg2$^+$. The presence of Ca^{2+} and Mg^{2+} is primarily due to the dissolution of dolomite, calcite, and gypsum, while the high SO_4^{2-} is largely influenced by the nearby Yuka mining area. Apart from thermal springs, all samples lie above the isoline in the $\gamma(Ca^{2+} + Mg^{2+})$ vs. γHCO_3^- diagram (Figure 5), signifying that HCO_3^- predominantly originates from carbonate mineral dissolution. Thermal spring samples, however, fall below this line towards the HCO_3^- end, indicating an excess of HCO_3^-. This excess hints that bicarbonate ions stem not only from carbonate dissolution but also from the biodegradation of organic carbon and the dissolution of geologically sourced CO_2.

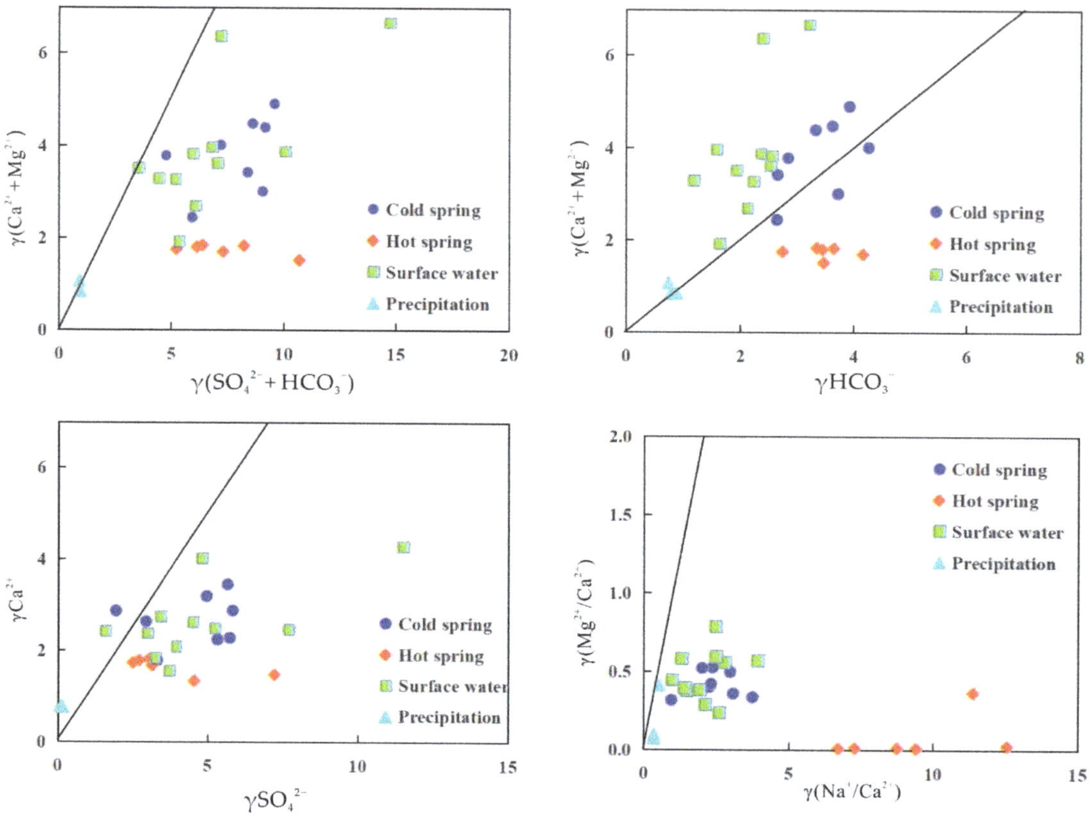

Figure 5. Relationships among major ions in the study area.

The $\gamma(Ca^{2+})$ vs. $\gamma(SO_4{}^{2-})$ diagram (Figure 5) indicates that samples are generally positioned below the line where $\gamma(Ca^{2+})/\gamma(SO_4{}^{2-}) = 1$, pointing to a disproportionally high $SO_4{}^{2-}$ concentration compared to Ca^{2+}, which implies that $SO_4{}^{2-}$ in geothermal water arises not just from gypsum dissolution but also from additional sources such as the Yuka mining area's influence. The $\gamma(Mg^{2+}/Ca^{2+})$ vs. $\gamma(Na^+/Ca^{2+})$ diagram (Figure 5) demonstrates that the $\gamma(Mg^{2+}/Ca^{2+})$ values for all samples from the study area are less than one, signifying a more significant source of Ca^{2+} than Mg^{2+}. Ca^{2+} is derived not just from dolomite but also from calcite, gypsum, and fluorite. Meanwhile, the $\gamma(Na^+/Ca^{2+})$ ratio exceeds one for all samples, and the near-constant $\gamma(Mg^{2+}/Ca^{2+})$ ratio suggests a comparatively low Mg^{2+} concentration.

In summary, the Ca^{2+} and Mg^{2+} ions in the water samples from the study area predominantly arise from the dissolution of dolomite, calcite, and gypsum. The $SO_4{}^{2-}$ ions are derived not just from gypsum dissolution but are also significantly influenced by the surrounding Yuka mining area. The $HCO_3{}^-$ ions in cold springs primarily originate from carbonate mineral dissolution. Conversely, the $HCO_3{}^-$ ions in thermal springs stem from both carbonate dissolution and the biodegradation of organic carbon, as well as from the dissolution of geologically sourced CO_2.

4.3. Characteristics of Hydrogen and Oxygen Isotopes

4.3.1. Recharge Sources

Understanding the hydrogen and oxygen isotopic characteristics of groundwater is pivotal for pinpointing the sources and recharge conditions of groundwater and determining the relationships of meteoric water with groundwater and surface water, thereby illuminating groundwater flow paths [25]. Deuterium (D) and oxygen-18 (^{18}O) isotopes, which have lower vapor pressures, tend to be enriched in the liquid phase and depleted in the gas phase. This differential distribution results in distinct hydrogen and oxygen isotopic concentrations across varying groundwater circulation systems [26]. Variations in δD are influenced by mixing but are primarily contingent upon recharge temperatures and elevations, whereas $\delta^{18}O$ changes are mostly determined by the extent of water–rock exchange and the water/rock mass ratio [27].

The data of stable hydrogen and oxygen isotopes in water samples can be plotted alongside the global meteoric water line (GMWL) and the local meteoric water line (LMWL) on a diagram (Figure 6). By analyzing the sample positions relative to the GMWL, defined by the equation $\delta D = 8\delta^{18}O + 10$ [28], insights into the source of groundwater recharge can be ascertained.

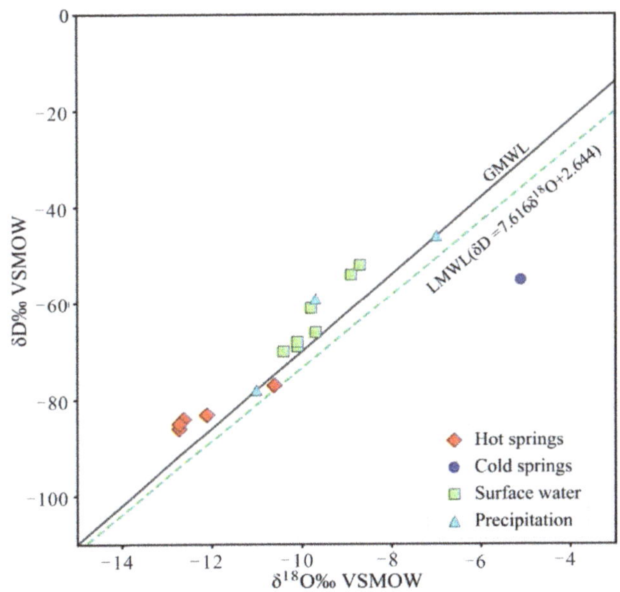

Figure 6. δD-$\delta^{18}O$ diagram of water samples from the study area.

Figure 6 indicates that the meteoric water line of the study area, represented by certain snowfalls, lies to the right of the GMWL. This positioning reflects the intricate relationship between stable isotope compositions in meteoric water and both the original state of water vapor and subsequent changes during vapor transport. Factors such as the continental effect, temperature, precipitation amounts, and latitude cause hydrogen and oxygen isotope values in meteoric water to vary by region and over time. In the study area, the surface water and thermal springs align closely with the local meteoric water line, approximating the isotope values of snow. Conversely, the cold spring sample LQ5 deviates to the right of the national meteoric water line with an evident oxygen isotope shift, suggesting that while the area's surface water and thermal springs are primarily fed by meteoric water, LQ5 has undergone extensive water–rock interaction. Normally, oxygen isotope shifts are more common in thermal springs. However, the occurrence of this shift in the study area's cold springs could be due to the thermal springs' proximity to Tuergendaban Mountain, which would mean a shorter post-recharge circulation path, while cold springs farther from the recharge source have longer groundwater paths. This extended contact promotes more thorough water–rock exchange and isotopic exchange between groundwater and oxygen-bearing rocks (such as limestones or silicates), consequently increasing the ^{18}O concentration in the groundwater.

By creating a Cl-D model diagram (Figure 7), it can be observed that the δD and $\delta^{18}O$ values of the sample water in the study area increase continuously with the increasing Cl content. Due to the enrichment of $\delta^{18}O$ isotopes in rocks, the water–rock interaction process leads to a significant drift amplitude of $\delta^{18}O$ in underground hot water. Moreover, most of the underground hot water in the area falls between the cold water end element and the deep fluid, exhibiting different characteristics with different proportions of cold water mixing, indicating the contribution of deep fluid to the composition of hot water and roughly estimating the proportion of cold water mixing.

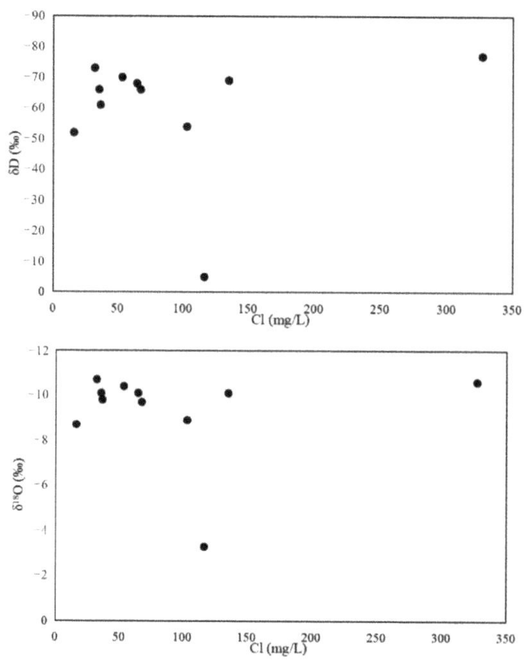

Figure 7. Relations of δD, δ ^{18}O, and Cl concentrations of water in study area.

4.3.2. Recharge Elevations

Generally, the δD and $\delta^{18}O$ values in meteoric water exhibit a linear relationship with temperature. The elevation effect implies that higher altitudes are associated with lower temperatures, and thus, with lower δD and $\delta^{18}O$ values [29]. Specifically, increased elevation results in temperature declines, which in turn cause water vapor to condense and isotopes to fractionate, along with a reduction in raindrop evaporation. This leads to enhanced rainfall in the presence of upslope winds, creating a preferential rainfall zone along windward mountain slopes. As a result, isotopic concentrations diminish with the ascent of cloud masses, contributing to the elevation effect [30]. This effect allows for the estimation of groundwater recharge elevations based on hydrogen and oxygen isotope data. The recharge elevation can be calculated using the available data and the following equation:

$$H = \frac{\delta_G - \delta_P}{k} + h \quad (1)$$

where δG represents the δD or $\delta^{18}O$ value of the water sample, δP is the δD or $\delta^{18}O$ value of meteoric water, k denotes the elevation gradient of the δD or $\delta^{18}O$ value in meteoric water, and h is the elevation of the sampling point, m. According to the data in [22], the δD elevation gradient in atmospheric precipitation in the study area is 2.02‰/100 m, and the $\delta^{18}O$ elevation gradient is 0.35‰/100 m.

Due to the occurrence of oxygen isotope shifts in some water samples from the study area, utilizing δD data yields a more precise calculation of recharge elevation. Applying the equation, the determined recharge elevation for cold springs is 4666.8 m, while for thermal springs the elevations are between 4666.8 m and 5755.9 m. These values align with the altitude of Tuergendaban Mountain. Considering the regional topography, it is inferred that the primary sources of groundwater recharge in the study area are meteoric water and meltwater. After the recharge, the water travels deep into the study area, then runs off and discharges, eventually surfacing to form springs.

4.3.3. Characteristics of Tritium Isotopes

Radioisotopes are useful for dating young groundwater to ascertain its origin and age of formation. Groundwater with a longer recharge circulation will display reduced tritium concentrations, whereas shorter runoff paths and faster circulation are associated with increased tritium concentrations. In the study area, tritium concentrations in groundwater and rivers vary between 9.1 TU and 37.8 TU, suggesting brief recharge paths and rapid circulation times. Specifically, cold springs have a tritium concentration of 9.1 TU, while thermal springs' tritium concentration measures 23.6 TU, indicating that cold springs have longer circulation paths and times, in contrast to the shorter paths and times of thermal springs. This result is consistent with the observed ^{18}O shift. Figure 8 illustrates an inverse relationship between Cl and tritium concentrations in water samples; lower Cl concentrations correlate with higher tritium concentrations. Since Cl predominantly originates from deep-seated geological sources, a lower Cl concentration coupled with a higher tritium concentration signifies a reduced contribution from deep fluids, and thus a shorter groundwater circulation period.

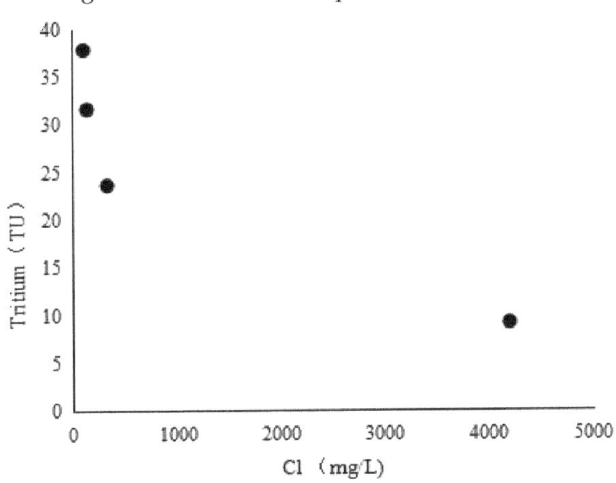

Figure 8. Tritium vs. Cl concentration diagram for water samples from the study area.

Tritium is commonly utilized for dating groundwater over relatively short timescales. Yet, real-world groundwater runoffs are intricate, with varied recharge sources and mixing of waters from different periods. Additionally, since the cessation of atmospheric nuclear testing, atmospheric tritium concentrations have not risen and are now near natural background levels. These factors complicate the quantitative estimation of groundwater age using tritium. However, tritium can still provide a qualitative assessment of the age and rough circulation time of groundwater for determining the groundwater recharge age.

Typically, groundwater with tritium concentrations of less than 1 TU is thought to have been recharged by water predating 1952, termed sub-modern water. Groundwater with tritium concentrations ranging from 1 to 4 TU is likely geothermal water resulting from a mix of sub-modern and modern water [31]. The tritium concentrations in this study's groundwater samples, ranging from 9.1 TU to 23.6 TU, suggest a predominance of modern water as the recharge source.

4.4. Deep Circulation Characteristics

4.4.1. Analysis of the Water–Rock Equilibrium State

The Na-K-Mg triangle diagram, introduced by Giggenbach in 1988 [32], is an effective tool to evaluate the suitability of the cationic geothermometer (Na/K/Mg) and the state of fluid–mineral equilibrium in geothermal fluids. This diagram classifies geothermal fluids into three categories: fully equilibrated water, partially equilibrated water, and immature water. It is extensively used to assess the water–rock equilibrium state in geothermal water and to differentiate water sample types. The diagram's strength is its ability to determine the equilibrium state of numerous water samples simultaneously and to clearly identify mixed water from equilibrated water. Based on experimental and analytical data, the Na-K-Mg triangle diagram was constructed, as shown in Figure 9. "One thermal spring" does not fall within the partially equilibrated water zone but rather within the immature zone as indicated in Figure 8. This indicates that it has not yet achieved water–rock ion equilibrium, implying ongoing dissolution processes or potential mixing with shallower, colder water during the ascension of geothermal water. Since the water–rock interaction in these thermal spring samples has not attained equilibrium and the ions are not fully saturated, utilizing the cationic geothermometer to estimate the geothermal reservoir temperature may result in substantial errors. Consequently, the temperature estimates thus obtained should be treated as indicative.

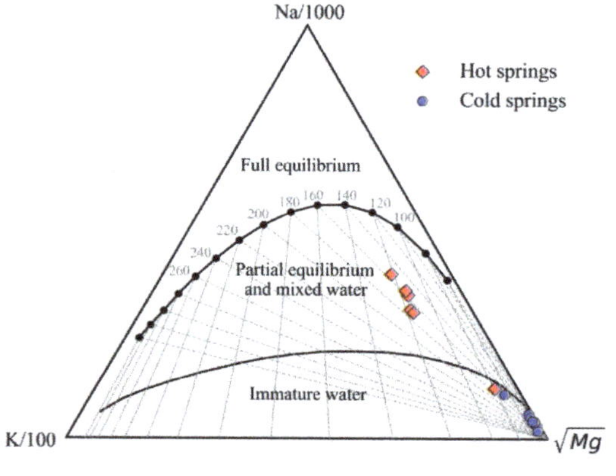

Figure 9. Na-K-Mg triangle diagram for springs in the Da Qaidam area.

4.4.2. Saturation Indices (SIs) of Multiple Minerals

During migration, groundwater will lead to the dissolution and precipitation of minerals within surrounding rocks, followed by water–rock interactions. Consequently, the mineral saturation state in groundwater can reveal trends in these processes. The SIs for different minerals in groundwater indicate mineral saturation states and can be determined at actual temperatures using the PHREEQC 3.3.7 software.

Figure 10 demonstrates that the SIs of anhydrite, aragonite, calcite, quartz, and gypsum in the study area's surface water are generally below zero, except for dolomite, which also has an SI below zero, except in samples HL1–HL7. This suggests that anhydrite, aragonite, calcite, quartz, and gypsum are dissolving, while dolomites are precipitating. For groundwater, the SIs for anhydrite, aragonite, and gypsum are mostly below zero, indicating an ongoing dissolution of sulfate minerals in the geothermal water. Calcite and dolomite show supersaturation, leading to the precipitation of carbonate minerals and the consequent reduction in Ca^{2+}, Mg^{2+}, and $CO3^{2-}$ concentrations in the geothermal water. Excluding samples LQ4 and LQ5, the SIs for quartz are approximately zero, signifying that quartz, a mineral that governs the silicate composition in groundwater, is in a state of equilibrium. This equilibrium indicates that silicate minerals in geothermal water are relatively stable. These minerals are significant in the hydrogeochemical processes of groundwater. The Ca^{2+} to Mg^{2+} ratio analysis further aids in understanding the dissolution and precipitation dynamics of calcite and dolomite [33]. The Ca^{2+} vs. Mg^{2+} diagram (Figure 11) shows that the water samples from the study area mainly lie below the 1:1 line, implying that calcite is more prevalent than dolomite in the study area.

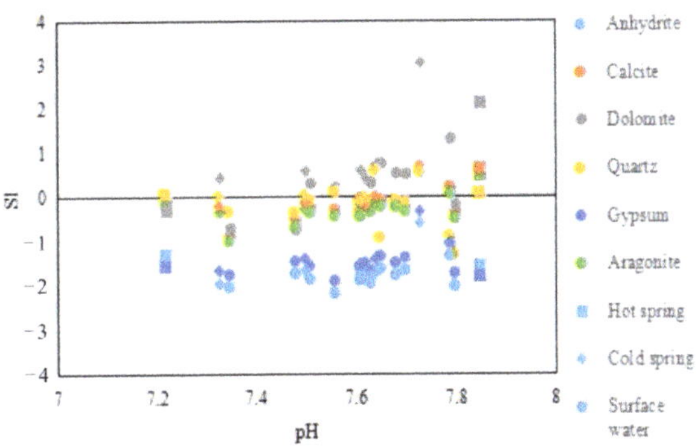

Figure 10. Saturation index diagram of water samples from the study area.

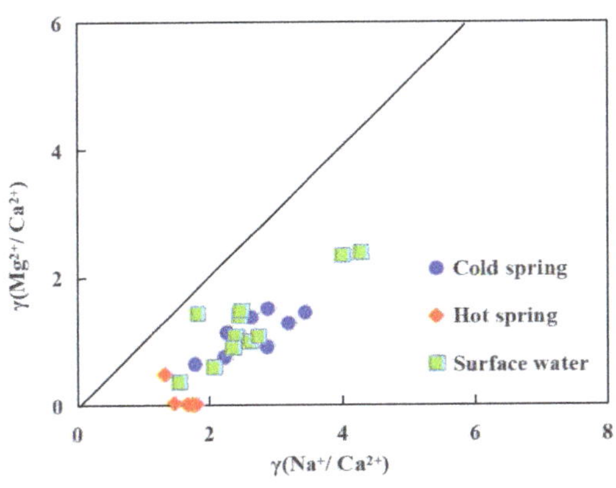

Figure 11. Relationship between Ca^{2+} and Mg^{2+} of the study area.

The study area's rock mineral survey data facilitated the construction of saturation index–temperature (SI-T) diagrams (Figure 12) for two springs, incorporating eight prevalent minerals: quartz, chalcedony, K-feldspar, illite, fluorite, diopside, magnesite, and gypsum. The diagrams reveal that diopside in the two thermal springs is supersaturated, signifying that silicate minerals have dissolved to a point of saturation, with precipitation ensuing when the thermal springs emerge at the surface. The gypsum mineral curves, in conjunction with hydrochemical characteristics, show that $SO_4{}^{2-}$ dissolves in the reservoir during circulation but does not reach saturation. The WQ1 curves exhibit relatively high convergence, suggesting a geothermal reservoir temperature of 122 °C, whereas the WQ2 curves are less well-defined, yielding an approximate geothermal reservoir temperature of 130 °C.

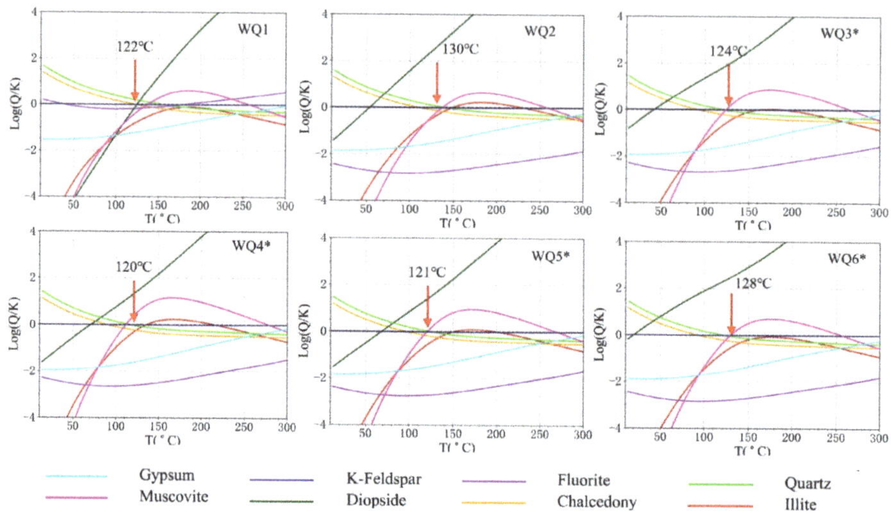

Figure 12. SI–T diagrams of minerals in springs in the Da Qaidam area. * after Wang et al. [21].

4.4.3. Geothermal Reservoir Temperature Estimation Using Geothermometers

The objective of the chemical geothermometer method is to estimate geothermal reservoir temperatures using the ratios of mineral components after geothermal water and aquifer minerals have reached chemical equilibrium under temperature influence. Moreover, these chemical components must be abundant and not diluted by cold water during ascent; if dilution occurs, the impact of cold water must be negated using the silicon-enthalpy model prior to geothermal reservoir temperature calculation. Currently, widely-used chemical geothermometer methods include the SiO_2 geothermometer (quartz and chalcedony) method and the cationic geothermometer (Na-K and Na-K-Ca) method [34]. These methods allow for the calculation of the deep geothermal reservoir temperatures in the study area, as shown in Table 2.

Table 2. Temperature and circulation depth of geothermal reservoirs estimated using different methods.

No.	T_{quartz}	$T_{chalcedony}$	$T_{Na\text{-}K}$	$T_{Na\text{-}K\text{-}Ca}$	$T_{K\text{-}Mg}$	Multiple Mineral Equilibrium	Silicon-Enthalpy Model	Cold Water Mixing Ratios (%)
WQ1	121.4	95.6	137.6	103.6	117.4	122	170	60
WQ2	123.2	98.0	148.8	110.4	81.7	130	175	58
WQ3 *	122.4	96.9	134.2	91.4	123.4	124	200	78
WQ4 *	118.3	91.7	135.5	91.7	117.9	120	160	55
WQ5 *	123.9	98.9	150.8	100.6	119.3	121	190	70
WQ6 *	124.1	99.2	150.5	100.8	118.1	128	185	60

Note: * after Wang et al [21].

The Na-K-Ca geothermometer is typically most applicable for calcium-rich, moderate to low-temperature hot water or shallow geothermal water [35], while the Na/K geothermometer is better suited for high-temperature (above 150 °C) groundwater in calcium-deficient areas, particularly for borehole hot water [36]. The K/Mg geothermometer is temperature-sensitive and tends to approach mineral fluid equilibrium at lower temperatures, supporting the theory that geothermal water mixes with shallow cold water during ascent. Temperatures deduced using the K/Mg geothermometer are often higher than the actual discharge temperatures from geothermal wells. In the case of the study area, where geothermal water is non-equilibrated or partially equilibrated, there are low calcium ion concentrations, and there is a relatively low temperature. The SI of SiO_2 in the research area approaches zero, and therefore, the SiO_2 temperature scale method and multi mineral equilibrium method are relatively more accurate.

4.4.4. Silicon-Enthalpy Model

The silicon-enthalpy model and its associated equation [37] were employed to negate the influence of cold water, allowing for the estimation of cold water mixing ratios and pre-mixing geothermal reservoir temperatures.

$$\begin{cases} S_c x + S_h(1-x) = S_s \\ SiO_{2c} x + SiO_{2h}(1-x) = SiO_{2s} \end{cases} \quad (2)$$

where S and SiO_2 represent the enthalpy (J/g) and SiO_2 (mg/l) concentrations of water, respectively. The variable c denotes the near-surface cold water, h signifies the initial concentration in hot water, s is the final concentration in spring water, and x is the mixing ratio of cold groundwater. The relationships between temperature and the enthalpy and SiO_2 concentrations of saturated water are detailed in Table 3. It is noted that when the temperature is below 100 °C, the enthalpy value for saturated water corresponds to its temperature value.

Table 3. Relationships between temperature, enthalpy, and SiO_2 concentrations.

Temperature (°C)	Enthalpy (J/g)	SiO_2 (mg/L)	Temperature (°C)	Enthalpy (J/g)	SiO_2 (mg/L)	Temperature (°C)	Enthalpy (J/g)	SiO_2 (mg/L)
50	50.0	13.5	150	151.0	125.0	250	259.2	486.0
75	75.0	26.6	175	177.0	185.0	275	289.0	614.0
100	100.1	48.0	200	203.6	265.0	300	321.0	692.0
125	125.1	80.0	225	230.9	365.0	—	—	—

The enthalpy and SiO_2 concentrations were modeled as functions of temperature in accordance with the silicon-enthalpy equation, with the intersection point indicating the cold water mixing ratio, as depicted in Figure 13. For the study area, it is determined that the cold water mixing ratios during geothermal water ascent were approximately ranging from 58 to 78%, and the deep geothermal reservoir temperatures range from 160 °C to 200 °C before mixing cold water.

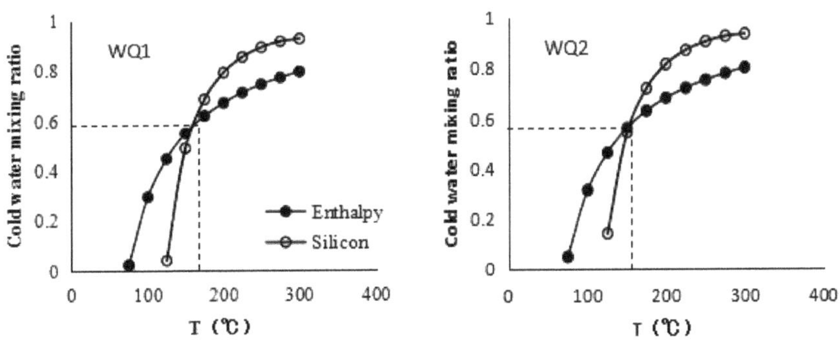

Figure 13. *Cont.*

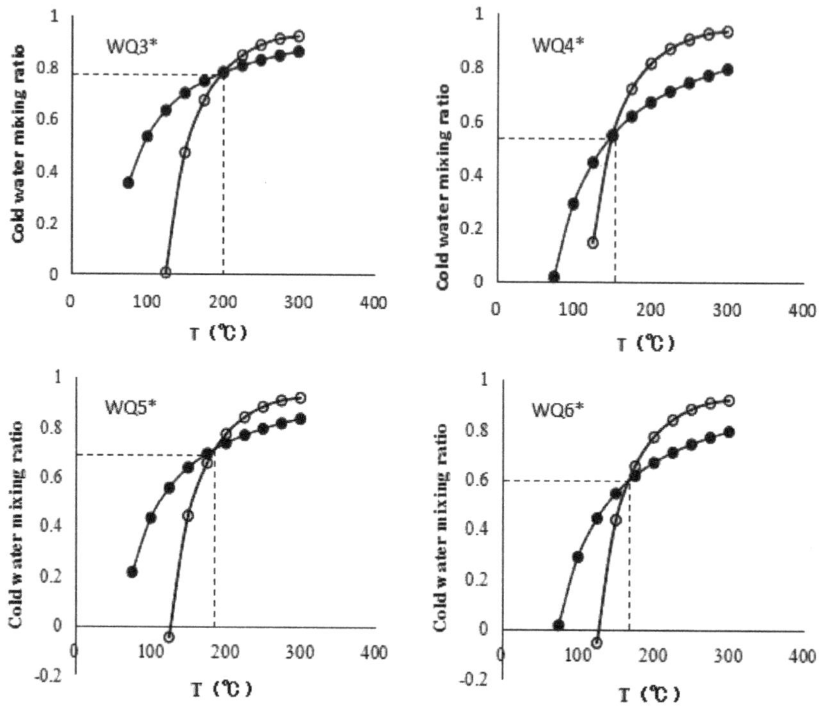

Figure 13. Silicon-enthalpy models of geothermal water in the study area. * after Wang et al. [21].

4.4.5. Circulation Depths

When comparing various methods for estimating the geothermal reservoir temperatures of thermal springs in the study area—namely water–rock equilibrium assessments, the SiO_2 geothermometer method, multiple mineral equilibrium diagrams, and the silicon-enthalpy model—it was determined that the temperatures given by the chalcedony geothermometer tended to be lower than the actual temperatures. Conversely, temperatures estimated using the quartz geothermometer and multiple mineral equilibrium diagrams were more in line with actual temperatures. Consequently, the average temperatures derived from these latter two methods were considered to reflect the true geothermal reservoir temperatures of the study area's thermal springs. These average values were then used to calculate geothermal reservoirs' circulation depths.

The study area, situated on the northern margin of the Qaidam Basin, is characterized by sedimentary basin geothermal resources. Here, geothermal water is typically heated through deep circulation, tapping into the Earth's geothermal energy. The circulation depth of thermal springs can be approximated using the equation provided.

$$H = \frac{t_1 - t_2}{I} + h \tag{3}$$

where H is the hot water circulation depth, m; t_1 is the geothermal reservoir temperature, °C; t_2 is the local annual average temperature, °C; I is the geothermal gradient, °C/100 m; and h is the thickness of the constant temperature zone, m. As per the findings of Liu et al. [38], the study area has an average annual temperature of 1.53 °C, a 2 m thick constant temperature zone, a permafrost thickness of 41.33 m, and a geothermal gradient of 3.243 °C/100 m. Using these parameters, the calculated depths for hot water circulation in the study area are shown in Table 4. The circulation depth before mixing with cold water is between 4888 and 6122 m, and after mixing with cold water, the circulation depth is between 3628 and 3859 m.

Table 4. Thermal cycle depths calculated for the geothermal reservoirs of study area.

No.	Temperature (°C)	Circulation Depth (m)	Final Temperature (°C)	Final Circulation Depth (m)
WQ1	121.7	3707.52	170	5196.88
WQ2	126.6	3858.61	175	5351.06
WQ3 *	123.2	3753.77	200	6121.95
WQ4 *	119.15	3628.89	160	4888.52
WQ5 *	122.45	3730.65	190	5813.59
WQ6 *	126.05	3841.65	185	5659.42

Note: * after Wang et al. [21].

4.4.6. Reverse Hydrogeochemical Pathway Simulation

PHREEQC software was used for hydrogeochemical simulation. Based on the topography, geomorphology, and geological structure characteristics of the Dachaidan area, the high-altitude sampling points were used as the starting point for reverse hydrogeochemical simulation, and the lower altitude sampling points were used as the endpoint for reverse hydrogeochemical simulation. Therefore, the path of hot springs is from WQ4* to WQ1, and the path of cold springs is from LQ3 to LQ4 (Table 5). Due to the fact that the lithology of the Dachaidan Formation is mainly granite, quartz, potassium feldspar, biotite, sodium feldspar, illite, fluorite, and gypsum were selected as possible reaction terms for path simulation. In addition, there may be cation alternating adsorption during the formation of spring water, thus adding two alternating terms, NaX and CaX_2.

Table 5. Reverse hydrogeochemical simulation of sample water chemical composition content.

ID		Constituent Concentration (mg/L)									T (°C)	pH
		K	Na	Ca	Mg	HCO_3	SO_4	Cl	SiO_2	F		
Thermal spring	WQ4 *	11.27	352.3	34.82	0.27	168.5	121.22	209.28	72.66	0.56	74	8.04
	WQ1	14.29	431.4	29.76	0.45	213.2	347.8	327.4	77.92	0.48	72	7.22
Cold spring	LQ3	2.62	122.9	52.79	16.6	261.2	141.9	71.78	9.77	0.88	13.9	7.33
	LQ4	3.9	167.2	63.95	15.48	222.2	239.4	106.8	28.7	0.52	9.7	7.64

Note: * after Wang et al. [21].

Based on the characteristics of geothermal conditions and the mineral composition of rock samples in the research area, the optimal solution was selected from the simulation results of PHREEQC software. The results are shown in Table 6.

Table 6. Reverse hydrogeochemical simulation results.

Path	Potassium Feldspar	Sodium Feldspar	Gypsum	Quartz	Biotite
Thermal spring	3.95×10^{-4}	1.97×10^{-4}	/	6.44×10^{-4}	/
Cold spring	/	-2.25×10^{-4}	1.74×10^{-5}	/	-1.52×10^{-4}
Path	**Illite**	**Calcium feldspar**	**Fluorite**	**NaX**	**CaX_2**
Thermal spring	/	-1.68×10^{-4}	-1.16×10^{-5}	/	/
Cold spring	1.95×10^{-4}	1.57×10^{-3}	/	2.7×10^{-3}	-1.35×10^{-3}

Note: The "/" in the table indicates that minerals do not participate in the reaction.

According to the simulation results, the hot spring geothermal fluid undergoes potassium feldspar dissolution during the migration process, resulting in an increase in K^+ content. The dissolution of sodium feldspar increases the Na^+ content. The precipitation of plagioclase and calcite leads to a decrease in the content of Ca^{2+} and HCO_3^-. The precipitation of fluorite leads to a decrease in F^- content. The dissolution of quartz leads to an increase in SiO_2 content. During the migration of the cold spring, illite dissolved into groundwater, leading to an increase in K^+ content, cation alternation, and Na^+ change. The dissolution of gypsum and plagioclase, as well as the alternating action of cations, leads to an increase in Ca^{2+} content. The content of dissolved Mg^{2+} in illite decreases. Gypsum dissolution leads to an increase in $SO_4{}^{2-}$ content. Due to the influence of silicates, the SiO_2 content increases.

5. Discussion

The study area, situated in the Qaidam inland arid basin, has its groundwater recharge, runoff, and discharge conditions predominantly influenced by factors such as aquifer media, geological structures, topography, and the development of fractures. Climate acts as a secondary influence. These conditions vary across different groundwater types.

As previously discussed, meteoric water and meltwater are the primary recharge sources for groundwater in the study area. The cold spring LQ5, located northwest of Da Qaidam in the upper Kushui River, receives its recharge from meteoric water in the northern Dakendaban Mountain. Here, meteoric water infiltrates and travels over 50 km downstream, undergoing extensive water–rock interaction. This process, combined with evaporative concentration effects, results in spring water with high TDS content emerging at the surface. Other cold springs surface in the silty areas at the front edge of the alluvial–proluvial fan. Due to relatively fine aquifer lithology, rising water levels, and gentler terrain slopes, pore phreatic water primarily receives recharge from the lateral runoff of upstream groundwater and vertical infiltration from meteoric water, ultimately discharging to form spring-fed rivers.

For geothermal water formation, both water and heat sources are crucial. The lower δH and $\delta^{18}O$ values in thermal springs relative to other water types in the study area suggest that the thermal springs are recharged by meteoric water from higher altitudes or from times when the climate was cooler. The estimated recharge elevation for the thermal springs, based on hydrogen and oxygen isotopic concentrations, is approximately 5755.9 m, indicating that the thermal spring water originates from meltwater in the Tuergendaban Mountain in the north, which rises above 5700 m in elevation.

The primary heat source for thermal springs is predominantly influenced by crustal tectonic heating [39]. In zones where thermal springs emerge, the basal geothermal aquifer is often composed of magmatic rocks, primarily Indosinian gray-white and light-gray porphyritic granites [24]. These rocks contain high contents of radioactive elements like U, K, and Th, whose decay contributes radiogenic heat to the local geothermal gradient. Consequently, alongside the Earth's normal geothermal gradient from the crust and mantle, the radiogenic decay in granitic rocks is a significant local heat source, resulting in a high heat flow background and elevated geothermal reservoir temperatures. Geothermal resources are formed as groundwater circulates through fault fracture zones in areas with such heat sources. The Da Qaidam thermal springs are located at the juncture of deep faults and minor feather fractures on the southern margin of Dakendaban Mountain. The interaction of these faults creates a series of NE-trending transtensional and transpressive faults. The resultant fault fracture zones, characterized by highly fractured strata, connect with the deep-seated faults at the southern base of Dakendaban Mountain through feather fractures. The deep circulating hot water, which obtains heat from the deep-seated faults in Dakendaban Mountain, ascends along these fault fracture zones, discharging at the surface to form thermal springs. Thus, the fractures provide a conduit linking the deep heat sources and facilitating groundwater circulation.

Overall, meltwater from the high mountains on the northern margin of Da Qaidam infiltrates downward, flows to the subsurface along deep-seated faults, contributes to surface rivers, and recharges groundwater through subterranean channels and bedrock fissures. After extensive evaporation and runoff, some of the groundwater engages in water–rock interactions and ion exchange, emerging as springs at the surface along the

Da Qaidam fault fracture zone. Meanwhile, other portions of the groundwater penetrate deeper, are warmed by the geothermal heat, and rise, driven by the hydraulic and thermal dynamics of the Da Qaidam fault fracture zone. This water, mixing with cooler water en route, surfaces as medium- to low-temperature thermal springs, as illustrated in Figure 14.

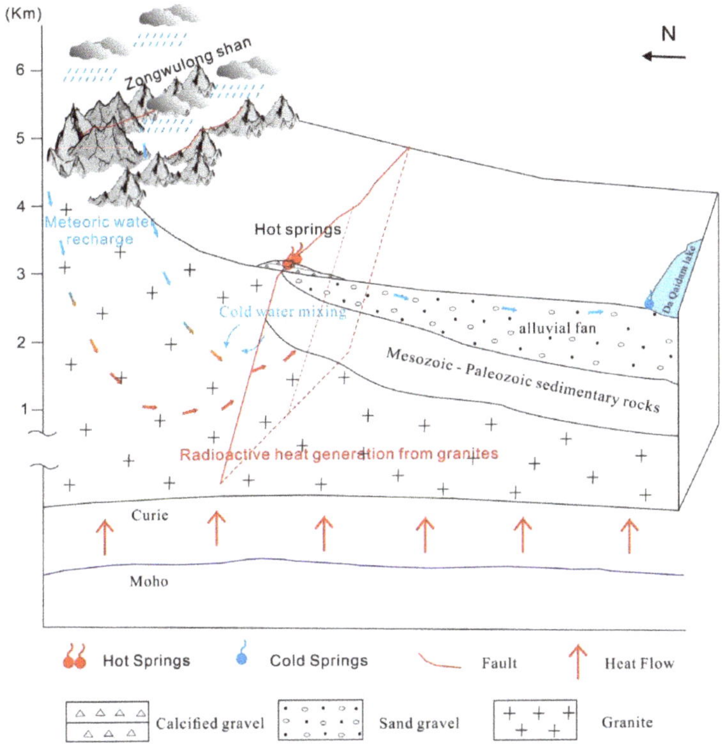

Figure 14. Geological section of thermal springs in northern Da Qaidam Town.

6. Conclusions

(1) As revealed by the test results of 30 sets of samples from the study area, most surface water and cold springs exhibit predominant hydrochemical types of $SO_4 \cdot HCO_3$-Na·Ca and Cl·HCO_3·SO_4-Ca·Na; meteoric water is characterized as the HCO_3-Ca·Na type, while the thermal spring is the Cl-Na type. The hydrochemical analysis indicates a low correlation between Cl and other ions, suggesting that the trace elements primarily originate from mineral dissolution rather than substances in deep Earth.

(2) The cold spring and river samples from the study area mostly fall within the rock weathering zones in the Gibbs diagrams, suggesting intense rock weathering and water–rock interactions. In contrast, sample LQ5 falls within the evaporation zone, hinting at significant evaporation. Ca^{2+} and Mg^{2+} ions in water samples are derived from the dissolution of dolomite, calcite, and gypsum, while SO_4^{2-} originates from both gypsum dissolution and the influence of the nearby Yuka mining area. The HCO_3^- is largely sourced from the dissolution of carbonate minerals, with thermal springs' HCO_3^- also resulting from organic carbon biodegradation and the dissolution of geologically sourced CO_2.

(3) The isotopic analysis shows that the D and ^{18}O values align closely with the meteoric water line, with a slight oxygen isotope shift in cold springs. Surface water and thermal springs are mainly recharged by meteoric water. In contrast, cold springs have a lengthy recharge path at elevations ranging from 4666.81 to 5755.92 m, indi-

cating modern water as the recharge source. The deep geothermal reservoirs have temperatures of around 120 °C, as estimated using an SiO_2 geothermometer. The cold water mixing ratios range from 58 to 78%; in addition, the circulation depth before mixing with cold water is between 4888 and 6122 m, and after mixing with cold water, the circulation depth is between 3628 and 3859 m.

(4) Meltwater infiltrates downward in the high mountain area in the north of Da Qaidam Town, circulates to the subsurface along deep-seated faults, and then recharges groundwater via subterranean channels and bedrock fissures, forming groundwater. Some groundwater ascends along the Da Qaidam fault fracture zone, forming cold springs. In contrast, some groundwater reaches deep parts, is heated by deep heat sources, and then rises, driven by the Da Qaidam fault fracture zone and heat, finally emerging as medium–low-temperature thermal springs.

Author Contributions: Conceptualization, L.L. and W.L.; methodology, L.L. and J.Z.; software, L.L. and Y.W.; formal analysis, J.Z.; investigation, B.L.; resources, L.L.; writing—original draft preparation, J.Z.; writing—review and editing, L.L.; supervision, W.L.; funding acquisition, L.L., B.L. and H.Y. All authors have read and agreed to the published version of the manuscript.

Funding: This research was funded by the National Key Research and Development Program of China (Grant No. 2021YFB1507401), Qinghai Province Clean Energy Minerals Special Project (No. 2022013004qj004), Key scientific and technological project in Henan province (No. 212102310383), and Key scientific and technological project in Henan province in Anyang (No. 2022C01SF086).

Data Availability Statement: The original contributions presented in the study are included in the article, further inquiries can be directed to the corresponding author.

Conflicts of Interest: The authors declare no conflicts of interest.

References

1. Wang, G.L.; Zhang, W.; Liang, J.Y.; Lin, W.J.; Liu, Z.M.; Wang, W.L. Evaluation of Geothermal Resources Potential in China. *Acta Geosci. Sin.* **2017**, *38*, 449–459.
2. Li, J.S.; Dong, H.Q.; Jiang, Y.X.; Nie, Y.F.; Zhu, G.Q.; Li, L.M. Geochemical genesis of the springs in Wenquan Ditch of Da Qaidam area. *J. Salt Lake Res.* **2017**, *25*, 55–59.
3. Zhang, P.X. *Salt Lake in the Ghidam Basin*; Science Press: Beijing, China, 1987.
4. Liu, X.Q.; Cai, K.Q.; Yu, S.S. Geochemical simulation of formation and evolution of salt lakes and their water sources in Qardam Basin, Application of Pitzer's model. *Geochimica* **2002**, *5*, 501–507.
5. Zhang, J.T.; Shi, Z.M.; Wang, G.C.; Jiang, J.; Yang, B.C. Hydrochemical characteristics and evolution of groundwater in the Dachaidan area, Qaidam Basin. *Earth Sci. Front.* **2021**, *28*, 194–205.
6. Xiao, Y.K.; Sun, D.P. Lithium isotope composition of Da Qaidam Lake. *J. Salt Lake Res.* **1991**, *4*, 13–18.
7. Xiao, Y.K.; Sun, D.P. Boron isotope composition of Da Qaidam Lake. *J. Salt Lake Res.* **1991**, *1*, 1–7.
8. Xiao, Y.K.; Jin, L.; Liu, W.G.; Qi, P.H.; Wang, Y.H.; Sun, D.P. Chlorine isotope composition of Da Qaidam Lake. *Chin. Sci. Bull.* **1994**, *39*, 1319–1322.
9. Kong, F.C.; Yang, Y.K.; Ma, Y.J.; Sha, Z.J.; Ling, Z.Y.; Wang, J.P. The distribution and sources of radium isotopes in Da Qaidam salt lake. *Lake Sci.* **2021**, *33*, 632–646.
10. Wang, G.L.; Zhang, F.W.; Liu, Z.M. An Analysis of Present Situation and Prospects of Geothermal Energy Development and Utilization in the World. *Acta Geoscien. Sin.* **2000**, *21*, 134–139.
11. Wang, R.J. *Study on the Geoehemical Characteristies of Geothermal Water and Resources Assessment in the Qia Bu-Qia Geothermal Area of Gong He Basin in Qinghai Province*; Chang'an University: Xi'an, China, 2009.
12. Li, Y.G. *Hydrogeochemical Characteristics and Its Origin Analysis Geothermal Water in the Qia Bu-Qia Area, Gong He Basin, Qinghai Province*; East China University of Thechnology: Nanchang, China, 2016.
13. Lang, X.J.; Lin, W.J.; Liu, Z.M.; Xing, L.X.; Wang, G.L. Hydrochemical characteristics of geothermal water in Guide basin. *Earth Sci.* **2016**, *41*, 1723–1734.
14. Dai, W. *The Hydrogeochemical Characteristics and the Evolution of Geothermal Water in Guide Area, Qinghai*; China University of Geosciences: Beijing, China, 2020.
15. Li, L.L. *Research on the Preservative Law and Genetic Model of Geothermal Resources in Guide Basin, Qinghai Province*; East China University of Thechnology: Nanchang, China, 2016.
16. Li, X.L.; Wu, G.L.; Lei, Y.D.; Li, C.Y.; Zhao, J.C.; Bai, Y.G.; Zeng, Z.F.; Zhao, Z.; Zhang, S.S.; Zhao, A.J. Suggestions for geothermal genetic mechanism and exploiation of Zhacang Temple geothermal energy in Guide county Qinghai province. *J. Jilin Univ. Earth Sci. Ed.* **2016**, *46*, 220–229.

17. Lin, W.J.; Gan, H.N.; Zhao, Z.; Zhang, S.S. Lithospheric Thermal-rheological Structure and Geothermal Significance in Gonghe Basin, Qinghai Province. *Acta Geosci. Sin.* **2023**, *44*, 45–56.
18. He, B.Z.; Zheng, M.L.; Yun, X.R.; Cai, Z.H.; Jiao, C.L.; Chen, X.J.; Zheng, Y.; Ma, X.X.; Liu, R.H.; Chen, H.M.; et al. Structural architecture and energy resource potential of Gonghe Basin, NE Qinghai-Tibet Plateau. *Earth Sci. Front.* **2023**, *30*, 81–105.
19. Lin, W.J.; Wang, G.L.; Zhang, S.S.; Zhao, Z.; Xing, L.X.; Gan, H.N.; Tan, X.F. Heat Aggregation Mechanisms of Hot Dry Rocks Resources in the Gonghe Basin, Northeastern Tibetan Plateau. *Acta Geol. Sin.* **2021**, *95*, 1793–1804. [CrossRef]
20. Tang, X.C.; Wang, G.L.; Ma, Y.; Zhang, D.L.; Liu, Z.; Zhao, X.; Cheng, T.J. Geological model of heat source and accumulation for geothermal anomalies in the Gonghe basin, northeastern Tibetan Plateau. *Acta Geol. Sin.* **2020**, *94*, 2052–2065.
21. Wang, B.; Qin, X.W.; Ren, E.F.; Jiang, Z.W.; Feng, N.; Li, G.R.; Yang, S.; Li, W. Gensis of geothermal water in the Da Qaidam area, northern margin of the Qaidam Basin, Constraints from hydrochemistry and isotopes of hydrogen-oxygen-strontium. *Bull. Mineral. Petrol. Geochem.* **2024**, *43*, 167–177. [CrossRef]
22. Zhu, P.C. *Groundwater Circulation Patterns of Yuqia-Mahai Basin in the Middle and Lower Reach of Yuqia River*; Jilin University: Changchun, China, 2015.
23. Pang, W.; He, W.G.; Yuan, D.Y.; Zhang, B.; Wu, Z. Paleoseismic characteristics of Dachaidan fult in Qinghai. *J. Earth Sci. Environ.* **2015**, *37*, 87–103.
24. Li, X.Q.; Wang, Z.X.; Hou, X.W.; Chen, J.; Liu, L.X.; Gao, M.; Ma, J.F.; Zhang, C.C.; Zheng, Z.X.; Han, S.P.; et al. *Hydrogeology and Environmental Geology Survey Report of Important Energy Base in Qinghai Province*; Institute of Hydrogeology and Environmental Geology, Chinese Academy of Geological Sciences: Shijiazhuang, China, 2016.
25. Song, C.; Han, G.L.; Wang, P.; Shi, Y.C.; He, Z. Hydrochemical and isotope characteristics of spring water discharging from Qiushe Loess Section in Lingtai, northwestern China and their implication to groundwater recharge. *J. Groundw. Sci. Eng.* **2017**, *5*, 364–373. [CrossRef]
26. Zhang, X.G. Application of isotope geochemistry to geothermal exploration. *Hydrogeol. Eng. Geol.* **1988**, *1*, 28–31.
27. Craig, H. Isotopic variations in meteoric waters. *Science* **1961**, *133*, 1702–1703. [CrossRef]
28. Ni, G.Q.; Zhang, H.; Wei, Y.T.; Hu, Y.Z. Hydrogeochemical and isotope characteristics of geothermal fluid in Sichuan. *Adv. New Renew. Energy* **2016**, *4*, 184–194.
29. Sun, Z.X.; Li, X.L.; Shi, W.J. Isotopic hydrogeochemistry of mid-low temperature geothermal water in Jiangxi Province. *J. East China Geol. Inst.* **1992**, *15*, 243–248.
30. Chen, L.M. Isotopic characteristic analysis of hydrogen-oxygen environment in geothermal water in Fujian Province. *Geol. Fujian* **2019**, *1*, 61–68.
31. Liu, Z. *The Forming Mechanism of Typical High-Temperature Geothermal Systems in Nimu-Naqu Geothermal Belt, Tibet*; Chinese Academy of Geological Science: Beijing, China, 2014.
32. Giggenbach, W.F. Geothermal solute equilibria. Derivation of Na-K-Mg-Ca geoindicators. *Geochim. Cosmochim. Acta* **1988**, *52*, 2749–2765. [CrossRef]
33. Gao, Y.Y.; Qian, H.; Ren, W.H.; Wang, H.K.; Liu, F.X.; Yang, F.X. Hydrogeochemical characterization and quality assessment of groundwater based on integrated-weight water quality index in a concentrated urban area. *J. Clean. Prod.* **2020**, *260*, 121006. [CrossRef]
34. Wang, H.; Chai, R. Application of geothermometers in the geothermal system. *J. Hebei Univ. Eng.* **2009**, *26*, 54–58.
35. Zhao, Q.S. The application of geothermometers in the exploration of hydrothermal resources in western Sichuan. *J. Chengdu Univ. Sci. Technol.* **1988**, *37*, 44–50.
36. Wang, J.Y.; Xiong, L.P.; Pang, Z.H. *Low-Medium Temperature Geothermal System of Convective Type*; Science Press: Beijing, China, 1993.
37. Tong, W.; Zhang, T.M. *Geothermal Tibet*; Science Press: Beijing, China, 1981.
38. Liu, L.X.; Qi, S.H.; Zhang, Y.B.; Gao, M.; Hu, J.; Wang, W.Z. Isotopic and hydrogeochemical characteristics and genesis of warm springs in the Jiangcang Basin, Qinghai, China. *Front. Earth Sci.* **2022**, *10*, 905852. [CrossRef]
39. Zhang, Z.X.; Li, W.; Zhang, H.Z.; Wang, B. *Exploration Report on Underground Hot Water Resources in Dachaidan Hot Water Ditch Area, Qinghai Province*; Hydrogeological and Engineering Geological Survey Institute of Qinghai Provincial: Xining, China, 2014.

water

MDPI

Article

Tracing Sulfate Sources of Surface Water and Groundwater in Liuyang River Basin Based on Hydrochemistry and Environmental Isotopes

Lei Wang [1,2,3], Yi Li [4], Yanpeng Zhang [1,2,*], Wei Liu [3] and Hongxin Zhang [1,2]

1 Department of Hydrogeology and Water Resources, Wuhan Center of China Geological Survey, Wuhan 430205, China; 1202221884@cug.edu.cn (L.W.); whzhanghongxin@126.com (H.Z.)
2 Key Laboratory of Eco-Hydrogeological Process of River and Lake Wetlands in Changjiang Reaches, China Geological Survey, Wuhan 430205, China
3 Institute of Geological Survey, China University of Geosciences, Wuhan 430074, China; wliu@cug.edu.cn
4 Hydrogeology Engineering Geology Brigade of Hubei Geological Bureau, Jingzhou 434020, China; ly8920240119@163.com
* Correspondence: yanpeng1028@126.com; Tel.: +86-13114347608

Abstract

Sulfate as a potential pollution source in the water environment of the basin, identifying sulfate sources and migration mechanisms is essential for protecting the water environment and ensuring sustainable water management. Liuyang River is a primary tributary of the Xiangjiang River. It has experienced progressively intensifying anthropogenic influences in recent decades, manifested by sustained sulfate concentration increases. However, the sulfate sources and their contributions were not clear. This study used hydrochemistry and multi-isotopes methods combined with Simmr model to study the hydrochemical characteristics, sulfate sources, and migration–transformation processes of surface water and groundwater. The results showed that the hydrochemical types of surface water were HCO_3-Ca and $HCO_3 \cdot SO_4$-Ca·Mg, and groundwater were HCO_3-Ca, HCO_3-Ca·Mg, and $HCO_3 \cdot SO_4$-Ca. Ions in the water primarily originated from carbonate and silicate rocks dissolution and sulfide oxidation, augmented by mining operations, sewage discharge, and chemical production. The analyses of hydrochemistry, isotopes, and Simmr model revealed that surface water sulfate originated from soil sulfate (35.70%), sulfide oxidation (26.56%), sewage (16.58%), and atmospheric precipitation (12.45%). Groundwater sulfate was derived predominantly from sewage (34.96%), followed by soil sulfate (28.09%), atmospheric precipitation (17.35%), and sulfide oxidation (12.25%). Sulfate migration and transformation were controlled by the natural environment and anthropogenic impacts. When unaffected by human activities, sulfate mainly originated from soil and atmospheric precipitation, relating to topography, geological conditions, agricultural activities, and precipitation intensity. However, in regions with intense human activities, contributions from sewage and sulfide oxidation significantly increased due to the influences of mining and industrial activities.

Keywords: sulfate sources; hydrochemistry; sulfur and oxygen isotopes; Simmr model; Liuyang River Basin

1. Introduction

Surface water and groundwater in river basins serve as crucial water supply sources for human society and industrial-agricultural development, with their environmental quality being of paramount importance [1–3]. In recent decades, China's rapid economic growth has substantially increased water demands for industrial manufacturing and agricultural production. This has exacerbated irrational exploitation of water resources, resulting in extensive contamination of surface water and groundwater while severely threatening water security and sustainable environmental development [4,5]. Dissolved sulfate (SO_4^{2-}) is a key chemical component in water bodies that reflects water quality. Sulfate originates from natural sources (evaporite dissolution, sulfide oxidation, soil, seawater intrusion, and atmospheric precipitation) and anthropogenic pollutants (industrial and domestic sewage, agricultural fertilizers, and mining drainage) [6–8]. Variation in sulfate concentrations significantly impacts sulfur biogeochemical cycling, induces water acidification and acid rain formation [9–11], and generates toxic compounds (e.g., hydrogen sulfide) that endanger public health and water environment [12,13].

Due to the existence of multiple potential sources of sulfate, it is difficult to accurately distinguish the sources only by using conventional hydrochemical methods. For example, ion ratio relationships (e.g., SO_4^{2-}/Cl^-, SO_4^{2-}/Ca^{2+}) cannot accurately identify the influences of natural and anthropogenic factors on sulfate concentrations [14]. Sulfur isotope of sulfate ($\delta^{34}S_{SO4}$) serves as a natural tracer for sulfate source identification, undergoing minimal isotopic fractionation during biogeochemical processes [15,16]. The values of $\delta^{34}S_{SO4}$ in water preserve the inherent isotopic characteristics of their sources [17]. Therefore, $\delta^{34}S_{SO4}$ is widely used to identify sulfate sources [18]. For instance, Han et al. applied $\delta^{34}S_{SO4}$ to identify sulfate sources in the Daweijia area of Dalian, China, indicating fertilizers as primary sources in Quaternary groundwater [19]. However, overlapping $\delta^{34}S_{SO4}$ ranges among different sources limit the accuracy of sulfate isotope tracing [20]. Zhang et al. utilized $\delta^{34}S_{SO4}$ to trace sulfate sources in the Yiluo River, revealing overlapping isotopic ranges between gypsum and domestic sewage, which made these sources difficult to distinguish [21]. The value of $\delta^{18}O_{SO4}$ is related to the ratio of oxygen derived from water and oxygen [22,23]. Combining $\delta^{18}O_{SO4}$ with $\delta^{34}S_{SO4}$ can resolve these limitations and significantly improves sulfate source identification accuracy [24]. For example, Otero et al. applied $\delta^{34}S_{SO4}$ and $\delta^{18}O_{SO4}$ to groundwater in the Llobregat potash mining basin, identifying potassium fertilizers production and mine tailing drainage as predominant sulfate sources [25]. This dual-isotope approach has been extensively employed to investigate sulfate sources and migration processes in water environment [26,27].

The Liuyang River, a primary tributary of the Xiangjiang River, holds exceptional significance for regional water supply. The basin features fireworks manufacturing, and mining and agricultural activities are also active. Surface water and groundwater receive sulfate inputs from both natural processes and anthropogenic activities, with sulfate concentrations demonstrating significant influence from mining, industrial production, and agricultural activities [28]. However, few studies have addressed sulfate source tracing in the Liuyang River Basin, and pollution sources and contributions of sulfate remain unclear. To address these challenges, surface water and groundwater samples were collected from the Liuyang River Basin. Hydrochemistry combined with multi-isotopes ($\delta^{34}S_{SO4}$, $\delta^{18}O_{SO4}$, δD_{H2O}, $\delta^{18}O_{H2O}$) was employed to identify various sources of sulfate. The Stable Isotope Mixing Models in R (Simmr) was utilized to quantify contributions from different sources of sulfate. Sulfate migration–transformation was elucidated based on topography, land use types, and geological conditions. This study comprehensively revealed the impacts of human activities and natural processes on sulfate in water bodies. It is of great significance for the environmental protection and the sustainable utilization of water resources in the basin. By identifying sulfate sources, contributions and migration-transformation processes via hydrochemical-isotopic approaches, this research provides a scientific basis for water pollution prevention and management in the basin.

2. Materials and Methods

2.1. Study Area

The Liuyang River Basin is situated in eastern Hunan Province, China. Its trunk stream (the Liuyang River) serves as a primary tributary of the Xiangjiang River. Originating from the northern foothills of Dawei Mountain within the Luoxiao Mountain Range, the Liuyang River extends 234.8 km in total length and drains a catchment area of 4665 km^2. The basin exhibits diverse geomorphological features, characterized by a northeast to southwest topographic gradient. The upstream, spanning from Shuangjiangkou to eastern of Daweishan Town, features mountainous terrain. The midstream, extending from Shuangjiangkou to Zhentou Town, transitions to hilly landscapes, while the downstream (west of Zhentou Town) comprises extensive alluvial plains. There are also larger tributaries, such as Xiaoxi River and Jianjiang River developed within the basin. Climatically classified as a subtropical monsoon zone, the basin receives a multi-year average precipitation of 1345.7 mm, 47.7% of which occurs from May to July.

The Liuyang River Basin exposes relatively complete stratigraphy, spanning from the Pre-Sinian to Quaternary. The oldest units comprise the Pre-Sinian Banxi and Lengjiaxi Groups, consisting of metamorphic rocks that extensively outcrop in the study area. Clastic rocks of Devonian, Cretaceous, and Paleogene are predominantly distributed in the downstream. Carboniferous-Permian carbonate rocks occur along the Xiaoxi River, and Liuyang River upstream and downstream. Quaternary unconsolidated deposits concentrate along the river valley in the upstream and downstream. Limited exposures of Yanshanian magmatic rocks, primarily acidic and intermediate-acidic lithology, are localized in the headwater of Liuyang River and Xiaoxi River.

Within the basin, groundwater is classified into four aquifer types by aquifer media: weathered fissure water in metamorphic rocks, weathered fissure water in clastic rocks, karst water in carbonate rocks, and pore water in Quaternary unconsolidated deposits. weathered fissure water is widely distributed in the study area. Quaternary pore water is distributed along the surface water, while karst water predominantly emerges in the upstream and midstream. Groundwater recharge modes include atmospheric precipitation infiltration, recharge from surface water percolation, and lateral inflow from adjacent aquifers. Regionally, groundwater flows along a NE-SW direction (Figure 1).

2.2. Sample Collection and Analysis

Surface water and groundwater samples were collected in the study area in November 2023 (Figure 1). Surface water sampling adhered to the following principles: sampling points were evenly distributed along the river course with additional sites at tributaries-trunk stream confluences; all surface water samples (n = 28) were collected from mid-channel locations, including 21 from the trunk stream, 3 from Xiaoxi River, 3 from Jianjiang River, and 1 from Unnamed River. Groundwater samples were sourced from wells (n = 17) proximate to surface water sites, extracted from shallow aquifers. During sampling, a multiparameter water quality probe (Eureka Manta+30, Austin, TX, USA) recorded in situ measurements of pH, dissolved oxygen (DO), electrical conductivity (EC), and oxidation-reduction potential (ORP).

Anion, cation, and water stable isotope (δD_{H2O}, $\delta^{18}O_{H2O}$) samples were stored in 50 mL amber high-density polyethylene (HDPE) bottles. Samples for the analysis of $\delta^{34}S_{SO4}$ and $\delta^{18}O_{SO4}$ were collected in 1.5 L amber HDPE bottles. All samples were filtered through 0.45 μm cellulose acetate membranes prior to preservation. Cation samples were acidified with 2–4 mL ultrapure HNO_3 to pH < 2, anion samples were sealed unpreserved and refrigerated at 4 °C. Sulfate isotope samples were acidified with 1 mL ultrapure HCl to pH <2 in the laboratory, followed by supersaturated $BaCl_2$ addition to convert dissolved

SO_4^{2-} to $BaSO_4$. The precipitate was rinsed with deionized water to remove Cl^-, filtered through 0.22 μm membranes, and dried at 850 °C for 2 h in a muffle furnace.

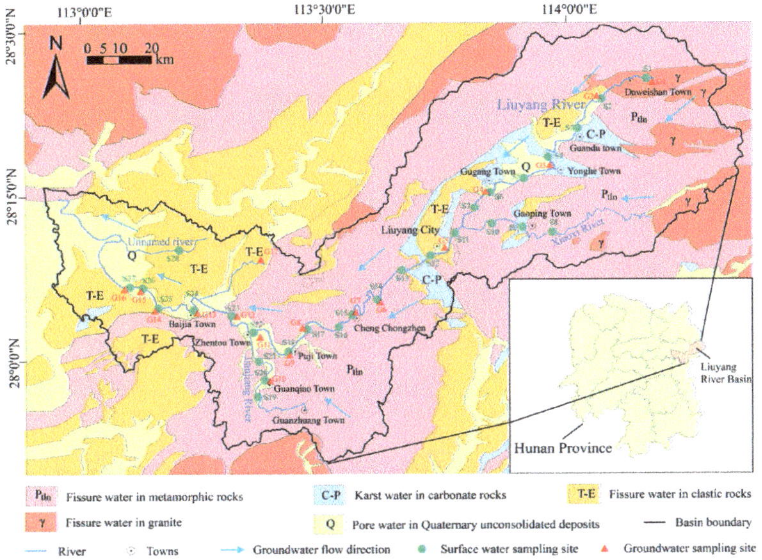

Figure 1. Hydrogeological conditions and distribution of sampling sites in the study area.

The concentrations of HCO_3^- were determined via titration, while other anions were analyzed using ion chromatography (Dionex ICS-3000, Sunnyvale, CA, USA). Cation concentrations were quantified through inductively coupled plasma optical emission spectrometry (Thermo ICAP-6300, Waltham, MA, USA), achieving measurement precision better than 2%. Total dissolved solids (TDS) were measured gravimetrically by drying and weighing. δD_{H2O} and $\delta^{18}O_{H2O}$ were analyzed using a laser-based liquid water isotope analyzer (IWA-35EP, LICA United Technology Co. Ltd., Beijing, China), with analytical precisions surpassing 0.1‰ and 0.5‰, respectively. $\delta^{34}S_{SO4}$ and $\delta^{18}O_{SO4}$ were determined via gas-source isotope ratio mass spectrometry (Thermo Delta V Advantage, Waltham, MA, USA and MAT253, Jacksonville, FL, USA), attaining precisions better than 0.2‰ and 0.5‰. All analyses were conducted at the laboratory in Wuhan Center of China Geological Survey.

2.3. The Stable Isotope Mixing Models in R (Simmr)

Simmr, a Bayesian mixing framework based on Dirichlet distributions, employs probabilistic source apportionment through R implementation to quantify proportional contributions of multiple sulfate sources [29]. Unlike conventional linear mixing models requiring fixed end-member mean values, Simmr inputs end-member ranges, thereby enhancing analytical capability for complex source mixtures. The model structure is formally expressed as

$$X_{ij} = \sum_{k=1}^{k} P_k\left(S_{jk} + C_{jk}\right) + \varepsilon_{ij} \tag{1}$$

$$S_{jk} \sim N\left(\mu_{jk}, \omega_{jk}^2\right) \tag{2}$$

$$C_{jk} \sim N\left(\lambda_{jk}, \tau_{jk}^2\right) \tag{3}$$

$$\varepsilon_{ij} \sim N\left(0, \sigma_j^2\right) \quad (4)$$

where X_{ij} is the isotopic value j of sample i ($j = \delta^{34}S_{SO4}$ and $\delta^{18}O_{SO4}$ in this study). S_{jk} is the isotopic value j of source k, characterized by its mean μ_{jk} and standard deviation $\omega^2{}_{jk}$. P_k quantifies the proportional contribution of source k (k is potential source of sulfate). C_{jk}, the isotopic fractionation factor for the isotope j in source k, is defined by its mean λ_{jk} and standard deviation $\tau^2{}_{jk}$, this term may be disregarded when significant isotopic fractionation is absent. The residual error ε_{ij}, with a mean of 0 and standard deviation of $\sigma^2{}_j$, accounts for unquantified variability in sample i.

3. Results and Discussion

3.1. Hydrochemical Characteristics

3.1.1. Hydrochemical Compositions and Types

The TDS values in surface water ranged from 48.70 to 262.00 mg·L^{-1}, with an average of 134.45 mg·L^{-1}. The EC values varied between 75.00 and 518.20 μS·cm^{-1}, averaging 281.47 μS·cm^{-1}. Groundwater exhibited higher TDS (186.07 mg·L^{-1}) and EC (351.79 μS·cm^{-1}) compared to surface water, suggesting more intensive water-rock interactions (Table 1). The maximum, minimum, and average concentrations of cations and anions are shown in Table 1. In surface water, average concentrations of cations were manifested as $Ca^{2+} > Na^+ > Mg^{2+} > K^+$, while anions followed the concentration gradient: $HCO_3^- > SO_4^{2-} > Cl^- > NO_3^-$. In groundwater, Ca^{2+} and HCO_3^- exhibit the highest average concentrations, followed by Na^+, Mg^{2+}, SO_4^{2-}, and NO_3^-, K^+, and Cl^- having the lowest concentrations. Along the flow path, surface water showed a gradually increase trend in both cation and anion concentrations. In the upstream, Ca^{2+}, HCO_3^-, and SO_4^{2-} of groundwater exhibited higher concentrations than the midstream and downstream. K^+ concentrations remained relatively stable spatially. Na^+, Mg^{2+}, Cl^-, and NO_3^- concentrations showed an initial increase followed by a gradual decrease.

Table 1. Statistics of the main hydrochemistry indexes in water bodies of the study area.

	Statistic	TDS	EC	K^+	Na^+	Ca^{2+}	Mg^{2+}	HCO_3^-	SO_4^{2-}	Cl^-	NO_3^-
		mg·L^{-1}	μS·cm^{-1}	mg·L^{-1}							
Surface water	Max	262	518.20	7.01	22.30	55.90	8.83	111	76.70	36	21.80
	Min	48.70	75	1.85	3.01	5.30	1.75	34	4.42	1.87	1.58
	Mean	134.45	281.47	4.34	11.99	22.99	5.13	76.23	19.06	17.15	8.70
Groundwater	Max	413	759.90	7.34	22.60	103	24.80	312	58.50	28.50	89.20
	Min	52.90	86.80	0.74	4.73	2.74	1.90	19.60	0.81	3.92	0.66
	Mean	186.07	351.79	2.40	10.86	34.47	9.79	114.70	18.75	14.07	26.45

Ions in the surface water and groundwater were dominated by $HCO_3^-$and Ca^{2+}. Hydrochemical types of surface water in the upstream were primarily HCO_3-Ca, transitioning to mixed HCO_3·Cl-Ca·Na and HCO_3·SO_4-Ca·Mg in the midstream, and finally to HCO_3·Cl-Ca·Na in the downstream. Hydrochemical types in groundwater were more complex. The upstream included HCO_3-Ca and HCO_3·SO_4-Ca types; the midstream and downstream were dominated by HCO_3-Ca·Mg type with minor HCO_3·SO_4-Ca type, some samples showed elevated Na^+, SO_4^{2-}, and Cl^-. Specifically, SO_4^2 was the dominant anion in sample S28. Samples S5, G3 (located downstream of mine) and G5, S12 (near Liuyang city) displayed a HCO_3·SO_4-Ca type, indicating the effluents from mining activities and industrial sewage (Figure 2).

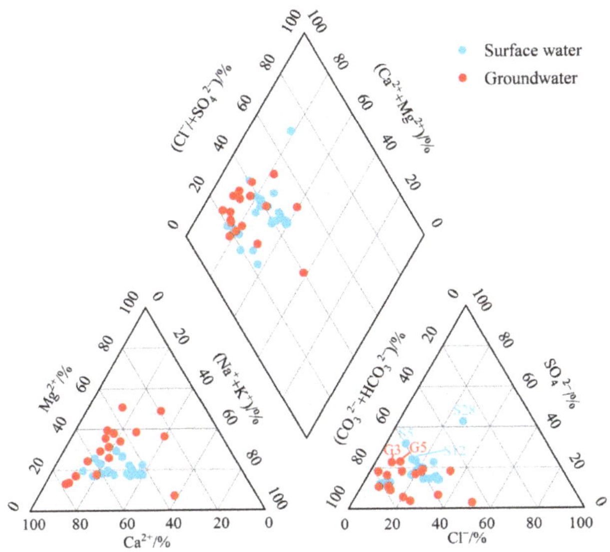

Figure 2. Piper triplex map in water bodies of the study area.

3.1.2. Sulfate Characteristics

The sulfate concentrations in surface water ranged from 4.41 to 76.70 mg·L^{-1}, with an average of 19.06 mg·L^{-1}. Historical data from three key monitoring sections in the Liuyang River Basin (Huanghuadong, Xiaoxi, and Sanshuichang) showed average sulfate concentrations of 5.98 mg·L^{-1}, 7.58 mg·L^{-1}, and 2.50 mg·L^{-1}, respectively [30]. It indicates a rising trend in recent years, likely linked to intensified human activities. Spatially, surface water in the upstream exhibited the widest sulfate variability (4.41–37.20 mg·L^{-1}), followed by midstream (17.20–32.60 mg·L^{-1}) and downstream (17.90–19.30 mg·L^{-1}). Notably, sulfate levels in Xiaoxi River (5.86 mg·L^{-1}) and Jianjiang River (12.93 mg·L^{-1}) were significantly lower than in the trunk stream, while the unnamed river showed an exceptionally high concentration of 76.70 mg·L^{-1}. Pronounced variation in the upstream suggests multiple sulfate sources, whereas the minimal fluctuation of downstream implies dominance by a single source. Groundwater sulfate concentrations (0.81–58.50 mg·L^{-1}) were slightly lower than those in the surface water. In the upstream, groundwater exhibited higher sulfate levels (23.29 mg·L^{-1}) compared to midstream (19.48 mg·L^{-1}) and downstream (14.86 mg·L^{-1}), with a gradual decline along the flow path (Figure 3).

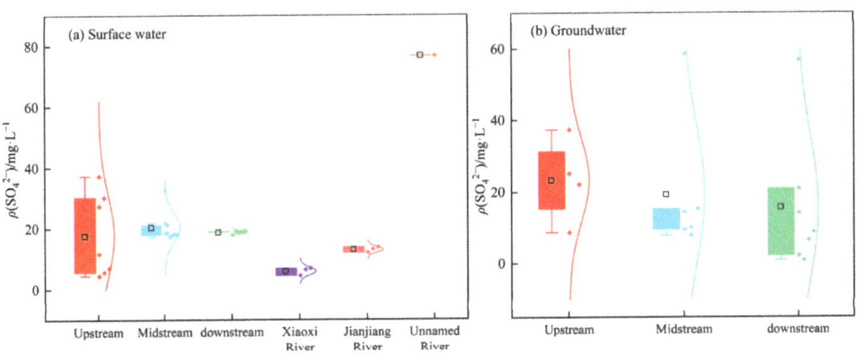

Figure 3. Box diagram of SO_4^{2-} concentration in water bodies of the study area.

For surface water, the minimum sulfate concentration (4.41 $mg \cdot L^{-1}$) was recorded at the headwater site (S1), attributable to its location in a granite-dominated area with minimal human activity and limited sulfate inputs. In contrast, the maximum sulfate concentration (76.70 $mg \cdot L^{-1}$) was found in an unnamed river (S28), likely due to industrial sewage discharge with high sulfate concentrations in the upstream of this river. Along the flow path, sulfate concentrations initially increased, then slightly decreased and stabilized. This pattern is linked to mining drainage in the upstream and sewage discharge in the mid-stream urban areas [31,32]. In groundwater, the lowest sulfate concentration (0.81 $mg \cdot L^{-1}$) occurred northwest of Zhentou Town (G12), potentially resulting from dilution or bacterial sulfate reduction (BSR) [33]. The highest concentration (58.50 $mg \cdot L^{-1}$) was detected in Liuyang city (G5), likely influenced by urban industrial effluents and mining drainage. Compared to surface water, groundwater sulfate concentrations exhibited a broader variation range, reflecting the control of complex lithology and hydrogeological conditions (Figure 4).

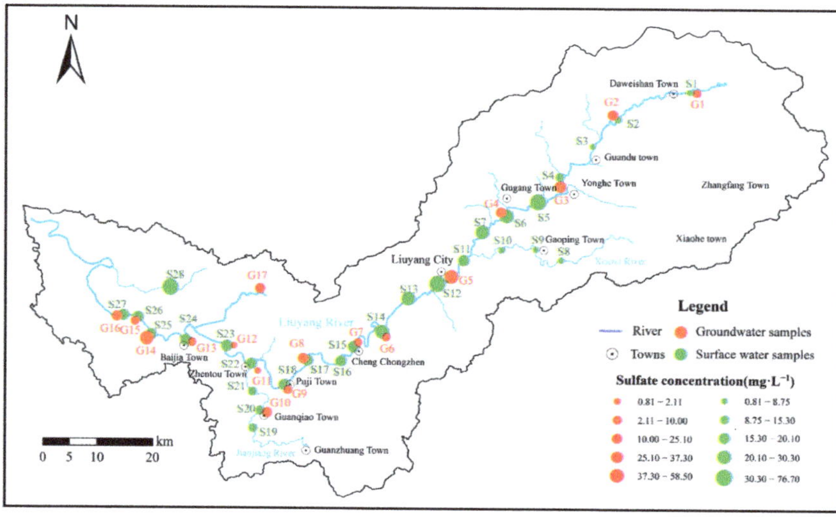

Figure 4. Spatial distribution of SO_4^{2-} concentration in water bodies of the study area.

3.2. Hydrochemical Analysis of Sulfate Sources

Gibbs diagram is a tool for preliminary analysis of hydrochemical controls by rock weathering, precipitation, and evaporation [34]. All surface water samples in the study area fall within the rock-dominated field, distant from the evaporation and precipitation dominance fields. This indicates that the hydrochemical composition of surface water is primarily governed by rock weathering processes. For groundwater, most samples cluster in the rock-dominated domain, with a few approaching the precipitation dominance field. This distribution suggests that water–rock interactions dominate groundwater chemistry, and some groundwater is affected by atmospheric precipitation (Figure 5).

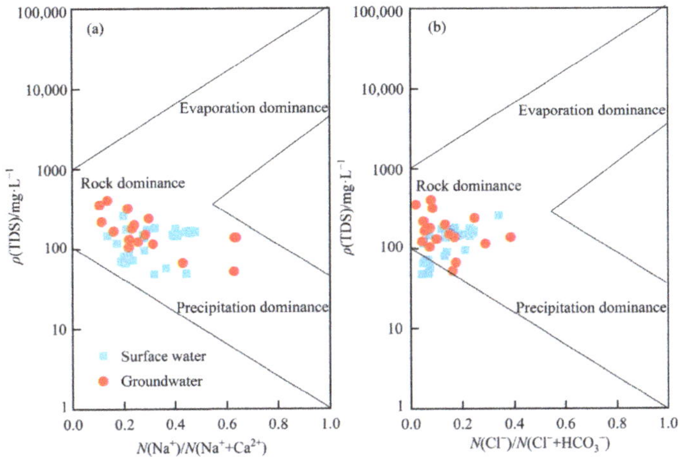

Figure 5. (**a**,**b**) Gibbs diagrams in water bodies of the study area.

To further distinguish different types of water–rock interactions, Gaillardet et al. delineated typical HCO_3^-/Na^+ and Ca^{2+}/Na^+ ratio ranges for carbonate rocks, silicate rocks, and evaporites [35]. In the study area, surface water samples plot between the silicate and carbonate rocks end members, positioned closer to the silicate domain. Groundwater samples also fall within the silicate–carbonate transition zone but exhibit stronger influence from carbonate dissolution. Evaporite dissolution shows a negligible impact on the hydrochemical characteristics in the study area (Figure 6). Along the flow path, samples S1–S2 and G1–G2 plot near the silicate dissolution end member, while S3–S12 and G3–G6 occupy intermediate positions between silicate and carbonate domains, reflecting increased carbonate dissolution effects. Subsequently, both surface water and groundwater gradually shift back toward silicate-dominated signatures, consistent with the exposed lithology of the basin.

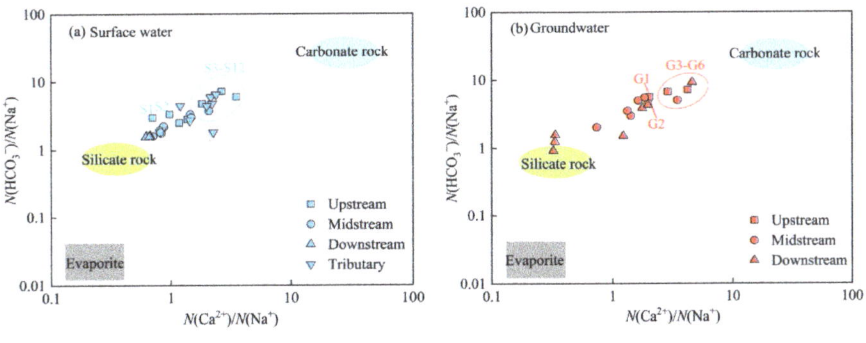

Figure 6. The relationship between molar ratios of HCO_3^-/Na^+ and Ca^{2+}/Na^+ in water bodies of the study area.

Water–rock interactions serve as a significant source of sulfate in water bodies. Analyzing ionic ratios associated with these processes enables identification of sulfate sources [36]. Water–rock interactions associated with sulfate primarily involve the dissolution of evaporite such as gypsum ($CaSO_4{\cdot}2H_2O$) and mirabilite ($Na_2SO_4{\cdot}10H_2O$), and the oxidation of sulfide including pyrite (FeS_2) [37]. As observed in Figure 7a, all surface water and groundwater samples plot above the gypsum dissolution line, with groundwater samples positioned farther from this line than surface water. This indicates limited Ca^{2+} and SO_4^{2-}

contributions from gypsum dissolution, and there are other Ca^{2+} and SO_4^{2-} sources in the water bodies. Most water samples in the study area plot between the calcite ($CaCO_3$) and dolomite ($CaMg(CO_3)_2$) dissolution lines (Figure 7b), implying Ca^{2+} and HCO_3^- predominantly originate from the dissolution of calcite and dolomite. Individual samples plotting above the calcite dissolution line may be attributed to gypsum dissolution, sulfuric acid-induced carbonate dissolution, or cation exchange processes. The milligram equivalent ratio of $[SO_4^{2-}] + [HCO_3^-]$ and $[Ca^{2+}] + [Mg^{2+}]$ provide preliminary evidence for sulfuric acid involvement in rock weathering [38]. A 1:1 ratio indicates the dissolution of carbonate rocks by sulfuric acid. Most surface water and groundwater samples align along the 1:1 line (Figure 7c), suggesting combined weathering effects from both carbonic acid and sulfuric acid. Groundwater samples in the middle and lower reaches plot below this line are due to reverse cation exchange processes or Ca-Mg silicate minerals dissolution. Figure 7d reveals a scattered distribution of samples across the mirabilite dissolution line, with no significant correlation between Na^+ and SO_4^{2-} concentrations, only individual surface water samples in the tributary plot near the mirabilite dissolution line. These findings suggest negligible contributions from mirabilite dissolution to sulfate in the basin.

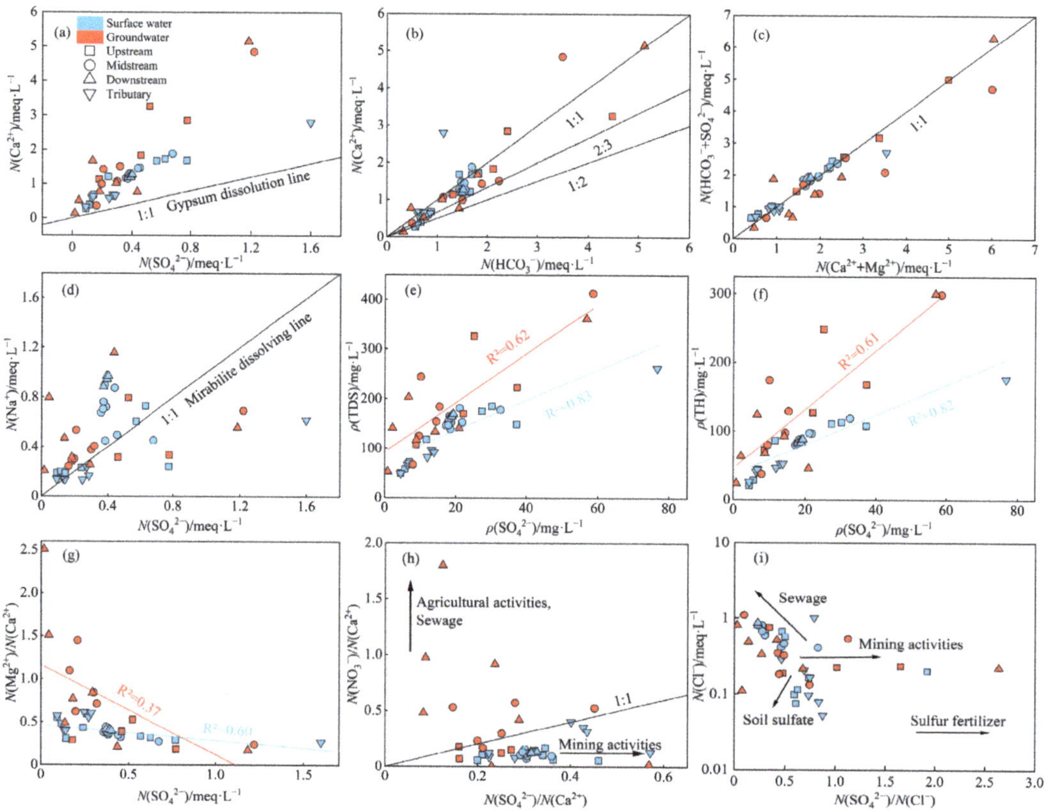

Figure 7. The ratio relationships of **(a)** Ca^{2+} and SO_4^{2-}; **(b)** Ca^{2+} and HCO_3^-; **(c)** $[Ca^{2+}] + [Mg^{2+}]$ and $[SO_4^{2-}] + [HCO_3^-]$; **(d)** Na^+ and SO_4^{2-}; **(e)** TDS and SO_4^{2-}; **(f)** TH and SO_4^{2-}; **(g)** Mg^{2+}/Ca^{2+} and SO_4^{2-}; **(h)** NO_3^-/Ca^{2+} and SO_4^{2-}/Ca^{2+}; **(i)** SO_4^{2-}/Cl^- and Cl^- in water bodies of the study area.

The study area's stratigraphy contains abundant pyrite. Mining activities expose pyrite to the atmosphere, generating sulfate and H^+ through oxidation processes. Water polluted by pyrite leachate shows elevated SO_4^{2-}, TDS, total hardness (TH) and Fe^{2+} coupled with depressed pH [39]. Sulfuric acid derived from sulfide oxidation further accelerates carbonate weathering, establishing correlations between SO_4^{2-} and Mg^{2+}/Ca^{2+} ratios. Surface water exhibits strong positive correlations between SO_4^{2-} and both TDS ($R^2 = 0.83$) and TH ($R^2 = 0.82$) ($p < 0.01$), while weaker associations are observed in groundwater (Figure 7e,f). Sample S4, affected by mine drainage in Qibaoshan mining area, exhibited abrupt increases in SO_4^{2-}, TDS, and TH compared to sample S3 in the upstream. As the Jianjiang River (S19–S21) receives discharge from Guanzhuang mining area, its hydrochemical characteristics showed depressed pH values (mean 6.9) and elevated Fe^{2+} concentrations (0.06 mg·L^{-1}) compared to other samples. Furthermore, surface water samples exhibit strong negative correlations between SO_4^{2-} and Mg^{2+}/Ca^{2+} ratios ($R^2 = 0.76$), contrasted with moderate correlations in groundwater ($R^2 = 0.41$) (Figure 7g). These findings indicate sulfide oxidation as the predominant source of sulfate, exerting more significant impact on surface water than groundwater.

Anthropogenic activities significantly influence sulfate concentrations in both surface water and groundwater. Primary anthropogenic sources include industrial and domestic sewage, agricultural activities, and mining drainage. The SO_4^{2-}/Ca^{2+} and NO_3^-/Ca^{2+} ratios serve as effective tracers for distinguishing impacts from diverse human activities on hydrochemical characteristics [40,41]. As shown in Figure 7h, all surface water samples plot below the 1:1 line, with SO_4^{2-}/Ca^{2+} ratios exceeding NO_3^-/Ca^{2+} ratios, indicating that mining activities have a significant impact on sulfate concentrations in surface water. This effect was particularly pronounced at sites S5–S7 and S19–S21 due to direct drainage inputs from Qibaoshan and Guanzhuang mining areas. Conversely, groundwater in the mid-lower reaches predominantly clusters above the 1:1 line. This indicates the influences of agricultural activities and sewage, related to the dense town and farmland in the mid-lower reaches. Some groundwater samples (G2–G4) are located below the 1:1 line, as is related to their proximity to the mining area.

The analysis of SO_4^{2-}/Cl^- ratios and Cl^- concentrations provides evidence for anthropogenic sulfate source identification. Sulfur fertilizers application elevates SO_4^{2-}/Cl^- ratios with depressed Cl^- levels, whereas mining activities induce high SO_4^{2-}/Cl^- ratios without significant Cl^- variation. Sewage inputs typically manifest as Cl^- enrichment with reduced SO_4^{2-}/Cl^- ratios, while low values in SO_4^{2-}/Cl^- ratios and Cl^- concentrations indicate natural source like soil or precipitation [42]. Figure 7i reveals concurrent influences from soil sulfate, mining activities, and sewage, with sulfur fertilizers demonstrating negligible impact. Surface water in the upper reaches, Xiaoxi River, and Jianjiang River exhibited dual control by soil sulfate and mining-derived inputs. Soil sulfate and sewage influences increased in the middle and lower reaches. Groundwater was predominantly influenced by sewage infiltration and soil, with mining-impacted exceptions at samples G3–G4.

3.3. Isotopic Analysis of Sulfate Sources

3.3.1. Analysis of Hydrogen and Oxygen Isotopes

δD and $\delta^{18}O$ isotopes in water bodies serve as effective tracers for identifying recharge sources and hydraulic connectivity between surface water and groundwater, while also indicating atmospheric precipitation influences on sulfate [43,44]. Surface water in the study area exhibited δD and $\delta^{18}O$ ranges of −47.86‰ to −31.60‰ and −7.44‰ to −5.16‰, averaging −36.25‰ and −5.77‰, respectively. Spatially, upper, middle, and lower reaches along with tributaries demonstrated diverse isotopic compositions, with average δD values of −34.63‰, −34.88‰, −33.59‰, and −41.99‰, and corresponding $\delta^{18}O$ values of −5.74‰, −5.51‰, −5.26‰ and −6.47‰. The δD values of groundwater samples ranged from −47.86‰ to −31.60‰, with an average of −32.03‰, while the $\delta^{18}O$ values ranged from −7.44‰ to −5.16‰, averaging −5.52‰. Spatially, upstream, midstream, and downstream exhibited average δD values of −31.39‰, −32.41‰ and −32.31‰, with corresponding $\delta^{18}O$ average of −5.59‰, −5.49‰ and −5.47‰, respectively (Table 2).

Table 2. Isotopic compositions of δD and $\delta^{18}O$ in water bodies of the study area.

	Statistic	Surface Water				Groundwater		
		Upstream	Midstream	Downstream	Tributaries	Upstream	Midstream	Downstream
δD (%)	Max	−32.05	−31.60	−33.28	−35.85	−29.55	−28.09	−29.39
	Min	−37.02	−38.55	−34.07	−47.86	−33.13	−38.02	−34.52
	Mean	−34.63	−34.88	−33.59	−41.99	−31.39	−32.41	−32.31
$\delta^{18}O$ (%)	Max	−5.20	−5.16	−5.19	−5.91	−5.38	−4.67	−4.32
	Min	−6.27	−6.08	−5.35	−7.44	−5.91	−6.57	−6.50
	Mean	−5.74	−5.51	−5.26	−6.47	−5.59	−5.49	−5.47

Due to the lack of local precipitation isotope values, Changsha's meteoric water line (LMWL: $\delta D = 8.44\delta^{18}O + 15.01$ [45]) was adopted as the regional reference. The global meteoric water line (GMWL) is defined as $\delta D = 8\delta^{18}O + 10$. Surface water and groundwater samples exhibit significant linear correlations between δD and $\delta^{18}O$. Surface water predominantly plot along the GMWL with minor deviations toward the lower right of the LMWL, implying atmospheric precipitation recharge and limited evaporation. Most groundwater samples were near the LMWL, indicating the recharge from atmospheric precipitation. Samples in the downstream show a deviation relative to the LMWL, and a $\delta^{18}O$ drift phenomenon exists, suggesting the influence of water–rock interactions and evaporation (Figure 8a). The relationship of $\delta^{18}O$ and Cl^- can indicate the intensity of evaporation [46]. Figure 8b reveals weak positive correlations between Cl^- concentrations and $\delta^{18}O$ values, indicating that evaporation effects were limited.

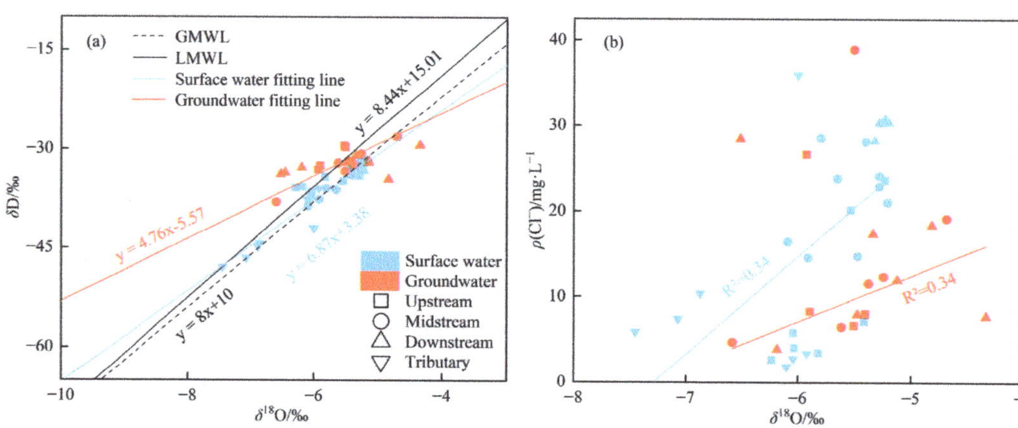

Figure 8. The relationships of (**a**) $\delta^{18}O$ and δD and (**b**) $\delta^{18}O$ and Cl^- in water bodies of the study area.

Hydraulic connectivity exists between surface water and groundwater in the upper, middle, and lower reaches. Groundwater in the downstream exhibited significantly enriched δD and $\delta^{18}O$ values relative to mid-upper reaches, suggesting limited hydraulic connectivity with midstream and upstream. Surface water in the upstream exhibited more depleted isotopic signatures than adjacent groundwater, with progressive enrichment along the runoff path. This indicates that groundwater recharges to surface water in the upstream.

Sample G5 in the midstream demonstrated marked isotopic enrichment relative to nearby samples S11–S13, revealing poor hydraulic connection. Conversely, isotopic similarities between samples S14–S18 and G6–G10 expressed active surface water-groundwater exchange. In the downstream, surface water exhibited minimal isotopic variation, while groundwater displayed progressive δD and $\delta^{18}O$ enrichment, reflecting weak hydraulic connectivity (Figure 8a).

3.3.2. Sulfur and Oxygen Isotopic Composition of Sulfate

The $\delta^{34}S_{SO4}$ and $\delta^{18}O_{SO4}$ values of surface water ranged from 1.70‰ to 7.43‰ and 1.89‰ to 7.92‰, with average values of 5.28‰ and 4.12‰, respectively. In the upper reaches, $\delta^{34}S_{SO4}$ and $\delta^{18}O_{SO4}$ values spanned 2.82–7.43‰ and 2.78–6.49‰, averaging 4.84‰ and 4.65‰. Midstream samples exhibited $\delta^{34}S_{SO4}$ values of 1.70–7.01‰ (mean: 5.14‰) and $\delta^{18}O_{SO4}$ values of 1.89–5.66‰ (mean: 3.65‰). Downstream water showed $\delta^{34}S_{SO4}$ values of 5.57–6.52‰ and $\delta^{18}O_{SO4}$ values of 3.27–7.92‰, with averages of 6.32‰ and 4.18‰. Spatially, $\delta^{34}S_{SO4}$ exhibited an overall increasing trend during the runoff process, contrasting with the decreasing pattern observed for $\delta^{18}O_{SO4}$.

Groundwater exhibited higher $\delta^{34}S_{SO4}$ (2.84–9.01‰) and $\delta^{18}O_{SO4}$ (2.30–8.60‰) values compared to surface water. Upstream samples showed $\delta^{34}S_{SO4}$ values of 6.19–8.38‰ and $\delta^{18}O_{SO4}$ values of 2.30–7.86‰, with averages of 6.97‰ and 5.24‰. In the middle reaches, $\delta^{34}S_{SO4}$ and $\delta^{18}O_{SO4}$ values spanned 6.59–8.66‰ and 5.02–8.60‰, averaging 7.27‰ and 6.45‰. The $\delta^{34}S_{SO4}$ and $\delta^{18}O_{SO4}$ values of downstream samples ranged from 2.84‰ to 9.01‰ and 3.62‰ to 8.28‰, with average values of 6.24‰ and 6.79‰, respectively. Spatially, $\delta^{34}S_{SO4}$ displayed a gradual decrease contrasting with the increasing $\delta^{18}O_{SO4}$ trend, demonstrating inverse variation pattern relative to surface water (Figure 9).

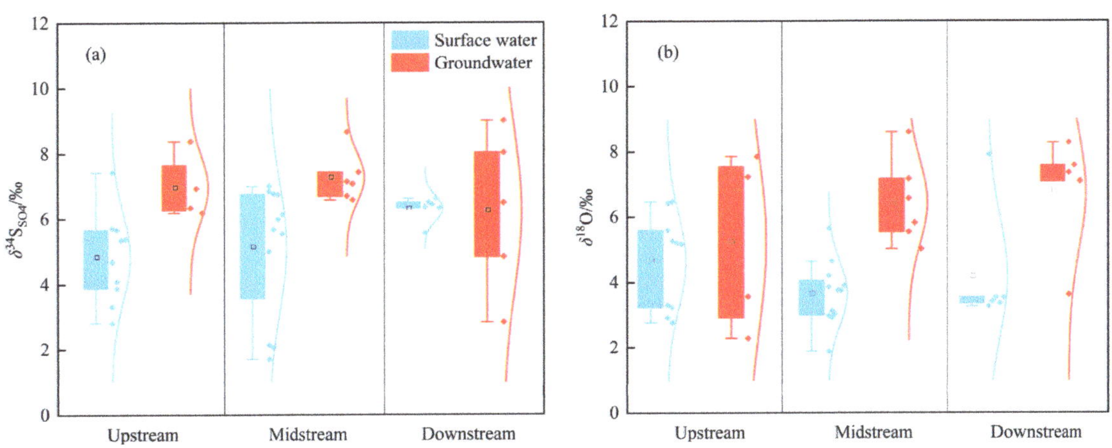

Figure 9. Spatial distributions of (**a**) $\delta^{34}S_{SO4}$ and (**b**) $\delta^{18}O_{SO4}$ in water bodies of the study area.

3.3.3. Analysis of Sulfur and Oxygen Isotopes

Sulfate undergoes significant isotopic fractionation predominantly during BSR process, which decreases sulfate concentrations while enriching $\delta^{34}S_{SO4}$ and $\delta^{18}O_{SO4}$ values [47]. Other biogeochemical pathways exhibit minimal fractionation effects. Consequently, sulfate from distinct sources demonstrates characteristic $\delta^{34}S_{SO4}$ and $\delta^{18}O_{SO4}$ signature ranges (Table 3).

Table 3. Characteristics of $\delta^{34}S_{SO4}$ and $\delta^{18}O_{SO4}$ from different potential sources.

Sources	$\delta^{34}S_{SO4}$ (‰)		$\delta^{18}O_{SO4}$ (‰)		Reference
	Average	Standard Deviation	Average	Standard Deviation	
Atmospheric precipitation	5.90	1.80	9.40	1.64	[48]
Soil sulfate	5.48	1.99	3.66	2.78	[48]
Sulfide oxidation	−0.10	4.90	−0.50	3.03	[32]
Sewage	8.70	2.40	6.86	1.60	[47]
Evaporite	22.50	7.50	14.00	1.00	[49]
Chemical fertilizers	2.30	7.70	16.10	4.00	[50]

Surface water and groundwater in the study area show no significant correlation between $\delta^{34}S_{SO4}$ and $1/SO_4^{2-}$, with most samples outside the BSR-dominated field. This indicates minimal influence of BSR process on $\delta^{34}S_{SO4}$ and $\delta^{18}O_{SO4}$ values, and suggests multiple sources contributing to sulfate. Most samples predominantly fall within soil sulfate and sewage end members, with some falling into sulfide oxidation and atmospheric precipitation domains. This reveals soil sulfate and sewage as primary sulfate sources, alongside sulfide oxidation and atmospheric inputs. The absence of samples in the evaporite range shows that few sulfate inputs derive from evaporite dissolution. No water samples fall within the fertilizer end member, as is related to the predominant use of nitrogen fertilizers in the basin (Figure 10a,b).

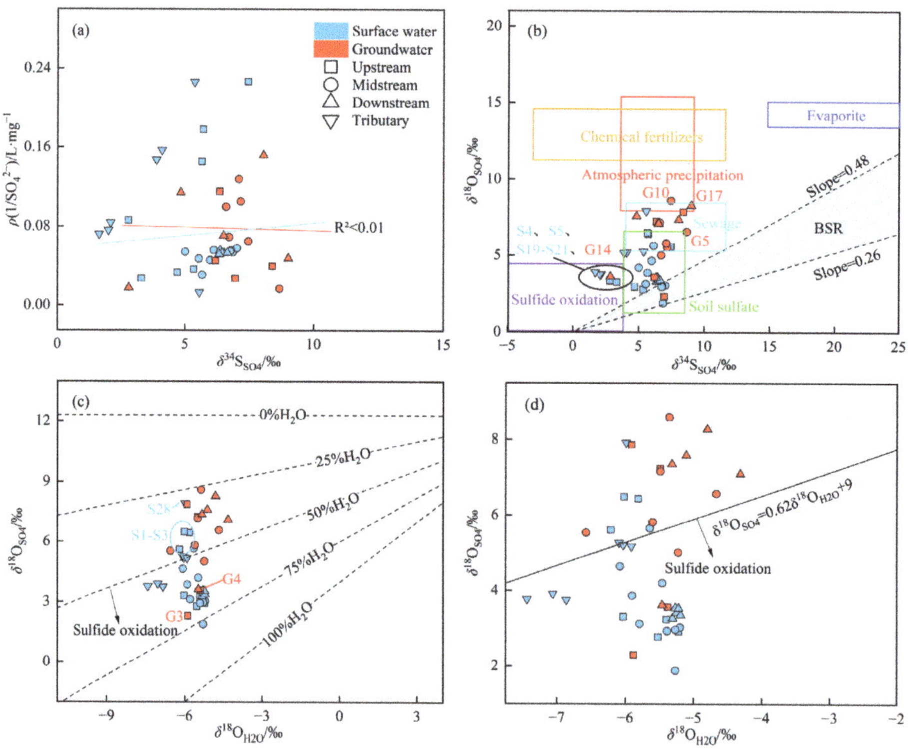

Figure 10. The relationships of (**a**) $\delta^{34}S_{SO4}$ vs. $1/SO_4^{2-}$, (**b**) $\delta^{34}S_{SO4}$ vs. $\delta^{18}O_{SO4}$ and (**c**,**d**) $\delta^{18}O_{H2O}$ vs. $\delta^{18}O_{SO4}$ in water bodies of the study area.

Sulfate in surface water was primarily derived from multiple sources, including soil, sulfide oxidation, sewage, and atmospheric precipitation (Figure 10b). Upstream samples (S4, S5) and Jianjiang River samples (S19–S21) plot in the upper-right quadrant of the sulfide oxidation end member. This indicates that sulfide oxidation dominates sulfate contributions in these water samples, accompanied by soil sulfate or sewage inputs. Sample S28, located in an unnamed river, exhibited the highest sulfate concentration (76 mg·L^{-1}) and $\delta^{18}O_{SO4}$ value (7.92‰) among surface water. This suggests sewage as the primary sulfate source, with additional contributions from precipitation and soil. Samples in the downstream showed minimal variation in $\delta^{34}S_{SO4}$ and $\delta^{18}O_{SO4}$ values, with isotopic compositions closely matching soil sulfate, confirming soil as the predominant sulfate source in these samples.

Groundwater shared similar sulfate sources with surface water. Its higher $\delta^{34}S_{SO4}$ and $\delta^{18}O_{SO4}$ values relative to surface water indicate reduced sulfide oxidation and soil but increased sewage and precipitation inputs on sulfate concentrations (Figure 10b). Specifically, sample G5 exhibited high sulfate concentration, $\delta^{34}S_{SO4}$ and $\delta^{18}O_{SO4}$ values, attributable to municipal sewage inputs within Liuyang city. Samples G10 and G17 from open dug wells showed significantly enriched $\delta^{18}O_{SO4}$, indicating inputs from direct atmospheric precipitation. Sample G14 displayed abrupt sulfate concentration increases with depleted $\delta^{34}S_{SO4}$ and $\delta^{18}O_{SO4}$ values relative to adjacent samples, suggesting anthropogenic sulfide inputs like coal combustion.

The relationship between $\delta^{18}O_{H2O}$ and $\delta^{18}O_{SO4}$ provides further validation of sulfide oxidation impacts on sulfate in water bodies [51]. Existing studies demonstrated that >50% of oxygen in sulfate derived from sulfide oxidation originates from water [52,53]. Most surface water samples in the study area plot below the 50%H_2O reference line, indicating >50% oxygen derivation from water and confirming sulfide oxidation as the dominant sulfate source (Figure 10c). Exceptions include samples S1–S3 and tributary sample S28, which plot above the 50%H_2O line, showing limited sulfide oxidation contributions due to absence of a mine drainage recharge. Groundwater samples predominantly plot above the 50%H_2O line, suggesting limited sulfide oxidation influence (Figure 10c). Affected by mine drainage, there were obvious contributions from sulfide oxidation in samples G3 and G4.

Balci et al. obtained the upper ($\delta^{18}O_{SO4} = 0.62\delta^{18}O_{H2O} + 9$) and lower ($\delta^{18}O_{SO4} = \delta^{18}O_{H2O}$) limit equations of sulfide oxidation through experiments [54]. Figure 10d shows that most surface water samples fall inside the sulfide oxidation domain, whereas most groundwater resided outside this range. This result is consistent with the analysis of Figure 10c. Furthermore, $\Delta\delta^{18}O$ ($\delta^{18}O_{SO4} - \delta^{18}O_{H2O}$) > 8‰ typically indicates active sulfide oxidation [55]. Surface water and groundwater in the study area exhibited average $\Delta\delta^{18}O$ values of 9.90‰ and 11.62‰, respectively, reaffirming sulfide oxidation as a major sulfate source in water bodies.

3.3.4. Sulfate Contributions from Different Sources

The preceding analysis of sulfate sources and biogeochemical processes showed that sulfate in the study area was derived from multiple sources. Soil sulfate, sewage discharge, sulfide oxidation, and atmospheric precipitation were the main sources, and minimal contributions originated from evaporites and fertilizers. Given the absence of pronounced BSR effects and negligible isotopic fractionation, the Simmr was employed to quantify different source contributions of sulfate in surface water and groundwater [56].

Quantitative analysis revealed marked spatial heterogeneity in sulfate contributions from different sources of surface water and groundwater (Figure 11). In surface water, sulfate was derived primarily from soil sulfate (35.70%), sulfide oxidation (26.56%), sewage (16.58%), and atmospheric precipitation (12.45%), with negligible contributions from evaporite dissolution (4.21%) and chemical fertilizers (4.50%). Groundwater sulfate originated predominantly from sewage (34.96%) and soil sulfate (28.09%), complemented by atmospheric precipitation (17.35%) and sulfide oxidation (12.25%), while evaporite (4.29%) and chemical fertilizers (3.06%) remain minor sources. Spatially, contributions from soil sulfate in surface water were manifested as downstream > midstream > upstream > tributaries, while sulfide oxidation showed an opposite law. Contributions from sewage and precipitation showed limited fluctuations, with only individual samples displaying significantly elevated contributions. In groundwater, contributions from sewage intensified in midstream and downstream regions, while contributions from soil sulfate diminished along the flow path. Sulfide oxidation and atmospheric precipitation displayed lower contributions except for individual samples.

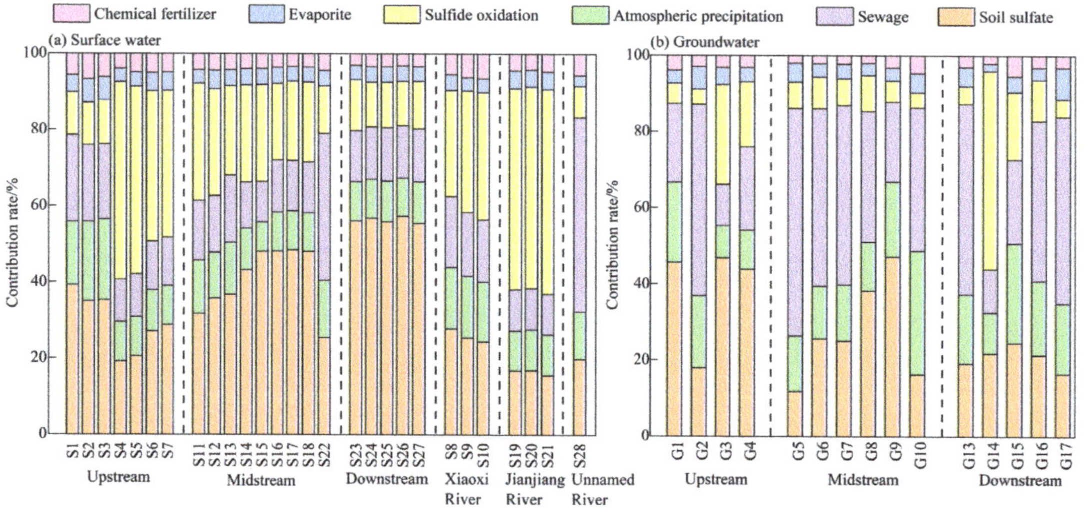

Figure 11. Sulfate contributions from different sources in water bodies of the study area.

3.4. Sulfate Migration and Transformation

The mid-upper reaches of the study area are surrounded by mountains on three sides with steep slopes. This topography facilitates soil sulfate transport to river through surface runoff (Figure 12a). Upstream areas expose extensive carbonate rocks, characterized by well-developed karst fissures and thin soil cover (Figure 1). These conditions enhance soil sulfate infiltration into groundwater. Downstream regions contain extensive farmland (Figure 12b). Intensive agricultural activities and high precipitation accelerate soil erosion [48]. However, the flat terrain and extensive impermeable surfaces restrict soil sulfate leaching into groundwater [57]. Thus, soil sulfate critically influences sulfate migration in both surface water and groundwater. The contribution of soil sulfate was significantly higher in surface water (35.70%) than groundwater (28.09%), and it was close to 60% when surface water flows to the downstream.

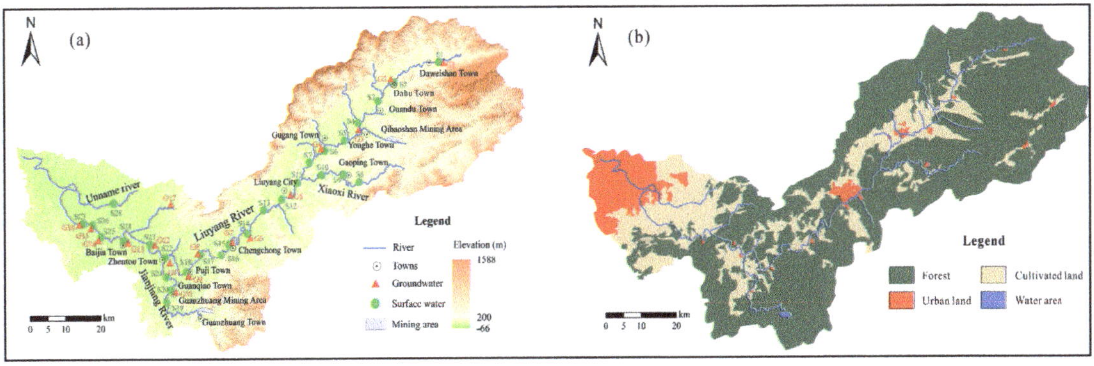

Figure 12. Topography (**a**) and land use types (**b**) of the study area.

The study area possesses abundant mineral resources. Sulfide oxidation during mining operations constitutes a critical biogeochemical process governing sulfate migration in surface water. Its contribution (26.52%) significantly exceeded that in groundwater (12.25%). This disparity arises because surface water directly receives mining drainage inputs, whereas transport of groundwater sulfate is constrained by limited flow systems and restricted runoff range. Specifically, after flowing through S4 and G3, sulfide-derived contributions increased significantly under the influence of the Qibaoshan mining area, while along the flow path, contributions from sulfide oxidation progressively decreased. Sulfate in Jianjiang River also primarily originated from sulfide oxidation influenced by Guanzhuang mining area. Additionally, sample G14 exhibited anomalously elevated sulfate concentrations and sulfide oxidation contribution. However, no sulfide deposits occur nearby, the contribution from sulfide oxidation may be linked to rural sulfur-rich coal combustion.

Sewage in the study area primarily comprises industrial discharges from chemical industry, paper plants and domestic sewage from rural settlements. It significantly influences sulfate in both surface water and groundwater. Surface water exhibited relatively low contributions from sewage (16.58%), attributable to dominant sulfate inputs from soil and sulfide oxidation. In contrast, sulfate in groundwater was significantly contributed by sewage (34.96%). This correlates with widespread rural wells featuring low water tables that enable sewage infiltration into aquifers via fractures [58]. For example, samples S22, G2 and G5–G7 were collected from the vicinity of towns and urban areas with a large number of industrial sewage discharge sources (Table 4). Sample S28 was collected from the vicinity of a factory. They received substantial industrial and domestic sewage, resulting in higher contributions from sewage (>40%).

Table 4. The quantity of industrial sewage discharge sources in the towns of the study area [59].

Towns	Quantity	Towns	Quantity	Towns	Quantity
Daweishan	17	Gaoping	28	Zhentou	35
Dahu	23	Gugang	33	Puji	18
Yonghe	20	Liuyang	199	Guanqiao	13
Guandu	14	Chengchong	44	Baijia	7

The study area receives abundant atmospheric precipitation, serving as the primary recharge source for different water bodies. Precipitation exhibited low and stable sulfate concentrations, contributing moderately to surface water (12.45%) and groundwater (17.26%). Precipitation contributions showed limited spatial variability, primarily controlled by recharge conditions and hydrogeological conditions [26]. For example, aquifer media in G1–G4 and G9–G17 predominantly consist of carbonate and clastic rocks with high permeability coefficient, and part of the wells were exposed to the air during sampling. They can receive more atmospheric precipitation recharge. Conversely, samples G5 and G8 (weathered fissure water in metamorphic rocks) exhibit limited precipitation recharge, resulting in comparatively minor atmospheric contributions to sulfate.

Evaporite and chemical fertilizers showed negligible contributions to sulfate without notable variation in the water bodies of the study area.

In summary, the migration and transformation of sulfate in surface water and groundwater within the study area results from the dual control of natural processes and anthropogenic activities. Soil, sulfide oxidation, sewage, atmospheric precipitation, evaporite, and chemical fertilizers govern sulfate migration and transformation through distinct geochemical processes and physical mixing processes (Figure 13).

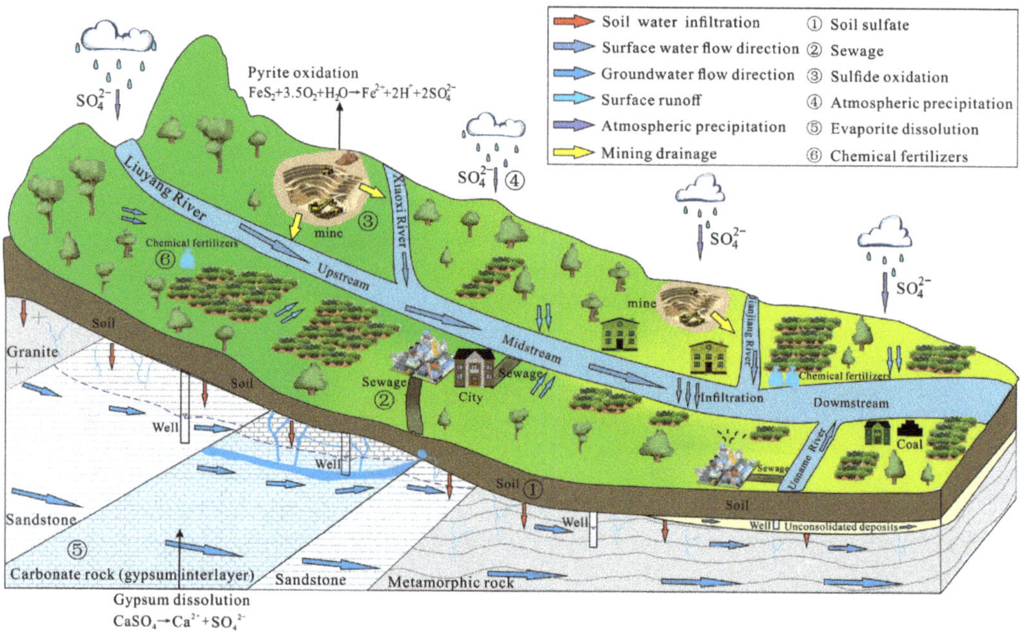

Figure 13. Pattern of sulfate migration and transformation in the water bodies of the study area.

4. Conclusions

Surface water and groundwater in the Liuyang River Basin exhibited hydrochemical types dominated by the HCO_3-Ca type. Anthropogenic impacts had significantly elevated SO_4^{2-} proportions in some samples, resulting in the emergence of $HCO_3 \cdot SO_4$-Ca·Mg and $HCO_3 \cdot SO_4$-Ca types of water. Sulfate in the water bodies was unevenly distributed in space, and the high value areas were distributed in areas with strong human activities such as mining areas and towns.

The analyses of hydrochemistry and isotopes showed that surface water and groundwater primarily underwent carbonate and silicate rock dissolution, with concurrent sulfide oxidation process. Evaporite dissolution (like gypsum) contributed minimally to sulfate. Both water bodies received atmospheric precipitation recharge and existed hydraulic connectivity, indicating precipitation as a major sulfate source. Mining and agricultural activities significantly influenced sulfate concentrations, where sulfide oxidation, soil sulfate and sewage constituted dominant contributors, while fertilizers inputs showed negligible impacts.

There were distinct variations in the contributions of different sulfate sources. Soil sulfate dominated sulfate inputs in surface water, while sewage constituted the primary contributor in groundwater. The calculation results of the Simmr model revealed that surface water sulfate was derived from soil sulfate (35.70%), sewage (16.58%), sulfide oxidation (26.56%), and atmospheric precipitation (12.45%), whereas groundwater sulfate originated from sewage (34.96%), soil sulfate (28.09%), atmospheric precipitation (17.35%), and sulfide oxidation (12.25%).

The migration and transformation of sulfate in surface water and groundwater were jointly influenced by natural processes and anthropogenic activities. Key controlling factors included geochemical processes (sulfide oxidation and evaporite dissolution) and physical mixing (soil, sewage, atmospheric precipitation, and fertilizers). In the absence of human influence, sulfate primarily originated from soil leaching and precipitation inputs. However, when water flowed through anthropogenically impacted zones like mining areas and towns, sulfide oxidation and sewage emerged as dominant factors for the sulfate migration and transformation.

However, this study only conducted sampling and analysis of surface water and groundwater in the Liuyang River Basin during the dry season, and it lacked comparative analysis between wet and dry seasons. Future research should strengthen investigations in the wet season to reveal temporal variations in sulfate distribution and sources in the basin. Additionally, future studies should perform direct measurements of isotope signatures from potential sulfate sources in the study area to enhance source identification accuracy.

Author Contributions: Conceptualization, L.W. and Y.Z.; methodology, L.W. and Y.L.; validation, L.W. and W.L.; formal analysis, L.W.; investigation, Y.L. and H.Z.; resources, Y.Z. and H.Z.; data curation, L.W.; writing—original draft preparation, L.W.; writing—review and editing, L.W. and Y.L.; visualization, Y.Z.; supervision, W.L.; project administration, Y.L.; funding acquisition, Y.Z. All authors have read and agreed to the published version of the manuscript.

Funding: This research was funded by the Natural Science Foundation of Hubei Province of China (No.2023AFD216) and Geological survey project of China Geological Survey (No.DD20230076).

Data Availability Statement: The data presented in this study are available on request from the corresponding author. The data are not publicly available due to privacy or ethical restrictions.

Conflicts of Interest: The authors declare no conflicts of interest.

References

1. Goldscheider, N.; Chen, Z.; Auler, A.S.; Bakalowicz, M.; Broda, S.; Drew, D.; Hartmann, J.; Jiang, J.; Moosdorf, N.; Stevanovic, Z.; et al. Global distribution of carbonate rocks and karst water resources. *Hydrogeol. J.* **2020**, *28*, 1661–1677. [CrossRef]
2. Li, X.; Pu, X.; Wang, W.; Dong, X.; Zhang, Y.; Wang, J.; Wang, Y.; Meng, M. Surface water environmental carrying capacity and surface water quality based on economy-society-environment nexus-Evidence from China. *Water-Energy Nexus* **2023**, *6*, 231–243. [CrossRef]
3. Wang, L.; Liu, S.; Ma, S.; Yang, Z.; Chen, Y.; Gao, W.; Liu, Q.; Zhang, Y. Quantile regression illuminates the heterogeneous effect of water quality on phytoplankton in Lake Taihu, China. *Water* **2024**, *16*, 2570. [CrossRef]
4. Man, K.; Ma, M.; Xu, J. Research on the mechanism of sulfate pollution of groundwater in Jiaozuo area. *Appl. Mech. Mater.* **2014**, *3517*, 436–439. [CrossRef]
5. Gani, A.; Hussain, A.; Pathak, S.; Ahmed, S.; Omar, P.J. Impact of pollutants on groundwater quality and health risk assessment of Quaternary aquifers in Northern India. *J. Hazard. Toxic Radioact. Waste* **2025**, *29*, 4024039. [CrossRef]
6. Filipović, M.; Terzić, J.; Reberski, J.L.; Vlahović, I. Utilizing a multi-tracer method to investigate sulphate contamination: Novel insights on hydrogeochemical characteristics of groundwater in intricate karst systems. *Groundw. Sustain. Dev.* **2024**, *27*, 101350. [CrossRef]

7. Chen, X.; Zheng, L.; Dong, X.; Jiang, C.; Wei, X. Sources and mixing of sulfate contamination in the water environment of a typical coal mining city, China: Evidence from stable isotope characteristics. *Environ. Geochem. Health* **2020**, *42*, 2865–2879. [CrossRef]
8. Kinnunen, P.; Kyllönen, H.; Kaartinen, T.; Mäkinen, J.; Heikkinen, J.; Miettinen, V. Sulphate removal from mine water with chemical, biological and membrane technologies. *Water Sci. Technol.* **2018**, *2017*, 194–205. [CrossRef]
9. Mao, H.; Wang, C.; Qu, S.; Liao, F.; Wang, G.; Shi, Z. Source and evolution of sulfate in the multi-layer groundwater system in an abandoned mine—Insight from stable isotopes and Bayesian isotope mixing model. *Sci. Total Environ.* **2023**, *859*, 160368. [CrossRef]
10. Klipsch, S.; Herwartz, D.; Voigt, C.; Münker, C.; Chong, G.; Böttcher, M.E.; Staubwasser, M. Sulfate sources, biologic cycling, and mobility in Atacama Desert soils revealed by isotope signatures. *Glob. Planet. Change* **2023**, *230*, 104290. [CrossRef]
11. Guan, H.; Xiao, H.; Xiao, H.; Xu, Y. The oxygen and sulfur isotopic compositions of soluble sulfate in the needles of *Pinus massoniana* Lamb.: Source discrimination and contribution estimation. *J. Geochem. Explor.* **2020**, *208*, 106402. [CrossRef]
12. Zhang, Q.; Wang, H.; Lu, C. Tracing sulfate origin and transformation in an area with multiple sources of pollution in northern China by using environmental isotopes and Bayesian isotope mixing model. *Environ. Pollut.* **2020**, *265*, 115105. [CrossRef]
13. Tostevin, R.; Turchyn, A.V.; Farquhar, J.; Johnston, D.T.; Eldridge, D.L.; Bishop, J.K.B.; McIlvin, M. Multiple sulfur isotope constraints on the modern sulfur cycle. *Earth Planet. Sci. Lett.* **2014**, *396*, 14–21. [CrossRef]
14. Hao, Y.; Pang, Z.; Gong, Q.; Li, N.; Liao, D.; Luo, Z. Oxygen and sulfur isotope systematics of dissolved sulfate in a nonvolcanic geothermal system: Sulfate source, evolution and impact on geothermometers. *Water* **2025**, *17*, 788. [CrossRef]
15. Mizutani, Y.; Rafter, T.A. Isotopic behaviour of sulphate oxygen in the bacterial reduction of sulphate. *Geochem. J.* **1973**, *6*, 183–191. [CrossRef]
16. Jiang, C.; Cheng, L.; Li, C.; Zheng, L. A hydrochemical and multi-isotopic study of groundwater sulfate origin and contribution in the coal mining area. *Ecotoxicol. Environ. Saf.* **2022**, *248*, 114286. [CrossRef] [PubMed]
17. Zheng, L.; Chen, X.; Dong, X.; Wei, X.; Jiang, C.; Tang, Q. Using $\delta^{34}S$-SO_4 and $\delta^{18}O$-SO_4 to trace the sources of sulfate in different types of surface water from the Linhuan coal-mining subsidence area of Huaibei, China. *Ecotoxicol. Environ. Saf.* **2019**, *181*, 231–240. [CrossRef]
18. Pu, J.; Wang, S.; Zhang, L. Combining hydrochemistry and isotopes to unravel sulfate sources in karst groundwater: A case study from southwest China. *Sci. Total Environ.* **2021**, *786*, 147312.
19. Han, D.; Song, X.; Currell, J.M. Identification of anthropogenic and natural inputs of sulfate into a karstic coastal groundwater system in northeast China: Evidence from major ions, $\delta^{13}C_{DIC}$ and $\delta^{34}S_{SO4}$. *Hydrol. Earth Syst. Sci.* **2016**, *20*, 1983–1999. [CrossRef]
20. Wang, H.; Zhang, Q. Research advances in identifying sulfate contamination sources of water environment by using stable isotopes. *Int. J. Environ. Res. Public Health* **2019**, *16*, 1914. [CrossRef]
21. Zhang, D.; Yang, J.; Huang, X.; Liu, S.; Zhang, Z. Sources of dissolved heavy metals in river water of the Yiluo River Basin based on sulfur isotope of sulfate. *China Environ. Sci.* **2019**, *39*, 2549–2559. [CrossRef]
22. Zhang, D.; Xue, T.; Xiao, J.; Chai, N.; Gong, S. Significant influence of water diversion and anthropogenic input on riverine sulfate based on sulfur and oxygen isotopes. *J. Hazard. Mater.* **2024**, *461*, 132622. [CrossRef]
23. Sun, J.; Takahashi, Y.; Strosnider, W.H.J.; Kogure, T.; Wu, P.; Cao, X. Tracing and quantifying contributions of end members to karst water at a coalfield in southwest China. *Chemosphere* **2019**, *234*, 777–788. [CrossRef]
24. Dogramaci, S.; McLean, L.; Skrzypek, G. Hydrochemical and stable isotope indicators of pyrite oxidation in carbonate-rich environment; the Hamersley Basin, Western Australia. *J. Hydrol.* **2017**, *545*, 288–298. [CrossRef]
25. Zak, D.; Hupfer, M.; Cabezas, A.; Jurasinski, G.; Audet, J.; Kleeberg, A.; McInnes, R.; Kristiansen, S.M.; Petersen, R.J.; Liu, H.; et al. Sulphate in freshwater ecosystems: A review of sources, biogeochemical cycles, ecotoxicological effects and bioremediation. *Earth-Sci. Rev.* **2021**, *212*, 103446. [CrossRef]
26. Cheng, L.; Jiang, C.; Li, C.; Zheng, L. Tracing sulfate source and transformation in the groundwater of the linhuan coal mining area, huaibei coalfield, China. *Int. J. Environ. Res. Public Health* **2022**, *19*, 14434. [CrossRef]
27. Bao, Y.; Pang, Z.; Huang, T.; Li, Y.; Tian, J.; Luo, J.; Qian, T. Chemical and isotopic evidences on evaporite dissolution as the origin of high sulfate water in a karst geothermal reservoir. *Appl. Geochem.* **2022**, *145*, 105419. [CrossRef]
28. Otero, N.; Canals, À.; Soler, A. Using dual-isotope data to trace the origin and processes of dissolved sulphate: A case study in Calders stream (Llobregat basin, Spain). *Aquat. Geochem.* **2007**, *13*, 109–126. [CrossRef]
29. Parnell, A.C.; Inger, R.; Bearhop, S.; Jackson, A.L. Source partitioning using stable isotopes: Coping with too much variation. *PLoS ONE* **2010**, *5*, e9672. [CrossRef] [PubMed]
30. Peng, Y.; Zou, L.; Tang, Y.; Chen, H.; Tang, Y.; Yu, L.; Li, H.; Liu, F. Assessment and temporospatial characteristics of water quality in Liuyang River based on homogeneity. *Hunan Agric. Sci.* **2021**, *10*, 50–57. [CrossRef]
31. Lin, Y.; Wang, Y.; Wu, Y.; Xu, B. Analyzing the source of sulfate in karst groundwater based on a Bayesian stable isotope mixing model: A case study of Xujiagou Spring Area, Northern China. *Water* **2025**, *17*, 794. [CrossRef]

32. Hu, J.; Chen, J.; Xu, D.; Wu, C.; Zhang, K.; Liu, J.; Liu, A.; Liu, C. Age and sources of the ore-forming material for the Qibaoshan Cu-polymetallic deposit in Hu'nan Province: Evidence from quartz vein Rb-Sr isotopic dating and S-Pb isotopes. *Geol. Bull. China* **2017**, *36*, 857–866. [CrossRef]
33. Zhang, D.; Liu, C. A preliminary study on sulfate reduction bacteria behaviors in groundwater by sulfur and carbon isotopes: A case study in Jiaozuo City, China. *Ecotoxicology* **2014**, *23*, 2014–2024. [CrossRef]
34. Gibbs, R.J. Mechanisms controlling world water chemistry. *Science* **1970**, *170*, 1088. [CrossRef]
35. Gaillardet, J.; Dupre, B.; Louvat, P.; Allègre, C.J. Global silicate weathering and CO_2 consumption rates deduced from the chemistry of large rivers. *Chem. Geol.* **1999**, *159*, 3–30. [CrossRef]
36. Olea-Olea, S.; Silva-Aguilera, R.A.; Alcocer, J.; Escolero, O.; Morales-Casique, E.; Florez-Peñaloza, J.R.; Almora-Fonseca, K.A.; Oseguera, L.A. Water–rock interaction processes in groundwater and flows in a maar lake in Central Mexico. *Water* **2024**, *16*, 715. [CrossRef]
37. Yang, Y.; Meng, Z.; Jiao, W. Hydrological and pollution processes in mining area of Fenhe river basin in China. *Environ. Pollut.* **2018**, *234*, 743–750. [CrossRef]
38. Spence, J.; Telmer, K. The role of sulfur in chemical weathering and atmospheric CO_2 fluxes: Evidence from major ions, δ^{13}C-DIC, and $\delta^{34}S\text{-}SO_4$ in rivers of the Canadian Cordillera. *Geochim. Cosmochim. Acta* **2005**, *69*, 5441–5458. [CrossRef]
39. Ren, K.; Zeng, J.; Liang, J.; Yuan, D.; Jiao, Y.; Peng, C.; Pan, X. Impacts of acid mine drainage on karst aquifers: Evidence from hydrogeochemistry, stable sulfur and oxygen isotopes. *Sci. Total Environ.* **2021**, *761*, 143223. [CrossRef]
40. Tang, L.; Yao, R.; Zhang, Y.; Ding, W.; Wang, J.; Kang, J.; Liu, G.; Zhang, W.; Li, X. Hydrochemical analysis and groundwater suitability for drinking and irrigation in an arid agricultural area of the northwest China. *J. Contam. Hydrol.* **2023**, *259*, 104256. [CrossRef]
41. Lu, T.; Li, R.; Ferrer, A.S.N.; Xiong, S.; Zou, P.; Peng, H. Hydrochemical characteristics and quality assessment of shallow groundwater in Yangtze River Delta of eastern China. *Environ. Sci. Pollut. Res.* **2022**, *29*, 57215–57231. [CrossRef]
42. Liu, C.; Lang, Y.; Satake, H.; Wu, J.; Li, S. Identification of anthropogenic and natural inputs of sulfate and chloride into the karstic groundwater of Guiyang, SW China: Combined δ^{37}Cl and δ^{34}S approach. *Environ. Sci. Technol.* **2008**, *42*, 5421–5427. [CrossRef] [PubMed]
43. Liu, Y.; Zhang, C.; Jiang, J.; Zhang, Y.; Wang, G.; Xu, L.; Qu, Z. Interaction between groundwater and surface water in the Qujiang River Basin in China: Evidence from chemical isotope measurements. *Water* **2023**, *15*, 3932. [CrossRef]
44. Zhang, C.; Li, X.; Wang, Z.; Hou, X.; Ma, J.; Gao, M.; Bai, Z.; Fu, C.; Gui, C.; Zuo, X. Combining stable isotope (^{2}H, ^{18}O, and $^{87}Sr/^{86}Sr$) and hydrochemistry to understand hydrogeochemical processes in karst groundwater system, Xin'an, northern China. *Environ. Earth Sci.* **2023**, *82*, 224. [CrossRef]
45. Xie, Y.; Zhang, X.; Yao, T.; Huang, H. Monitoring and analysis of stable isotopes of the near surface water vapor in Changsha. *Environ. Sci.* **2016**, *37*, 475–481. [CrossRef]
46. Sharma, M.K.; Kumar, P.; Prajapati, P.; Bhanot, K.; Wadhwa, U.; Tomar, G.; Goyal, R.; Prasad, B.; Sharma, B. Study of hydrochemical and geochemical characteristics and solute fluxes in Upper Ganga Basin, India. *J. Asian Earth Sci. X* **2022**, *8*, 100108. [CrossRef]
47. Torres-Martínez, J.A.; Mora, A.; Mahlknecht, J.; Kaown, D.; Barceló, D. Determining nitrate and sulfate pollution sources and transformations in a coastal aquifer impacted by seawater intrusion—A multi-isotopic approach combined with self-organizing maps and a Bayesian mixing model. *J. Hazard. Mater.* **2021**, *417*, 126103. [CrossRef]
48. Zhang, D.; Li, X.; Zhao, Z.; Liu, C. Using dual isotopic data to track the sources and behaviors of dissolved sulfate in the western North China plain. *Appl. Geochem.* **2015**, *52*, 43–56. [CrossRef]
49. Li, X.; Zhou, A.; Gan, Y.; Yu, T.; Wang, D.; Liu, Y. Controls on the δ^{34}S and δ^{18}O of dissolved sulfate in the Quaternary aquifers of the North China Plain. *J. Hydrol.* **2011**, *400*, 312–322. [CrossRef]
50. Hosono, T.; Tokunaga, T.; Tsushima, A.; Shimada, J. Combined use of δ^{13}C, δ^{15}N, and δ^{34}S tracers to study anaerobic bacterial processes in groundwater flow systems. *Water Res.* **2014**, *54*, 284–296. [CrossRef]
51. Saka, D.; Antwi, E.O.; Skrzypek, G.; Adu-Gyamfi, J.; Heng, L.; Attiogbe, F. Tracing sulfate sources in a tropical agricultural catchment with a stable isotope Bayesian mixing model. *Sci. Total Environ.* **2024**, *951*, 175502. [CrossRef]
52. Zhang, K.; Shi, Z.; Ding, X.; Ge, L.; Xiong, M.; Zhang, Q.; Lai, W.; Ge, L. Identification of anthropogenic and natural inputs of sulfate into river system of carbonate Zn-Pb mining area in southwest China: Evidence from hydrochemical composition, $\delta^{34}S_{SO4}$ and $\delta^{18}O_{SO4}$. *Water* **2024**, *16*, 2311. [CrossRef]
53. Cao, X.; Wu, P.; Zhou, S.; Sun, J.; Han, Z. Tracing the origin and geochemical processes of dissolved sulphate in a karst-dominated wetland catchment using stable isotope indicators. *J. Hydrol.* **2018**, *562*, 210–222. [CrossRef]
54. Balci, N.; Shanks, W.C.; Mayer, B.; Mandernack, K.W. Oxygen and sulfur isotope systematics of sulfate produced by bacterial and abiotic oxidation of pyrite. *Geochim. Cosmochim. Acta* **2007**, *71*, 3796–3811. [CrossRef]
55. Sun, J.; Kobayashi, T.; Strosnider, W.H.J.; Wu, P. Stable sulfur and oxygen isotopes as geochemical tracers of sulfate in karst waters. *J. Hydrol.* **2017**, *551*, 245–252. [CrossRef]

56. Kong, J.; Zhou, Z.; Xie, R.; Cao, W. Tracing the source and behaviour of sulphate in karst reservoirs, using stable isotopes and Bayesian isotopic-mixing models. *Sci. Total Environ.* **2024**, *958*, 177994. [CrossRef]
57. Zhang, J. Identification of Nitrate and Sulfate Sources and Their Increasing Mechanism in Karst Water System of Jinan Spring Catchment Based on Hydrochemistry and Multi-Isotopes. Doctor's Thesis, China University of Geosciences, Wuhan, China, 2022.
58. Torres-Martínez, J.A.; Mora, A.; Knappett, P.S.K.; Ornelas-Soto, N.; Mahlknecht, J. Tracking nitrate and sulfate sources in groundwater of an urbanized valley using a multi-tracer approach combined with a Bayesian isotope mixing model. *Water Res.* **2020**, *182*, 115962. [CrossRef]
59. Chen, Y. Present Situation of Environmental Quality and Control Measures in Liuyang River Section. Master's Thesis, Hunan Agricultural University, Changsha, China, 2017.

water

Article

A Robust Optimization Framework for Hydraulic Containment System Design Under Uncertain Hydraulic Conductivity Fields

Wenfeng Gao [1], Yawei Kou [1], Hao Dong [1], Haoran Liu [1] and Simin Jiang [2,*]

[1] Shandong Provincial Geo-Mineral Engineering Exploration Institute (No. 801 Hydrogeological and Engineering Exploration Brigade of Shandong Provincial Bureau of Geology and Mineral Exploration and Development), Jinan 250014, China
[2] College of Civil Engineering, Tongji University, Shanghai 200092, China
* Correspondence: jiangsimin@tongji.edu.cn

Abstract

Effective containment of contaminant plumes in heterogeneous aquifers is critically challenged by the inherent uncertainty in hydraulic conductivity (K). Conventional, deterministic optimization approaches for pump-and-treat (P&T) system design often fail when confronted with real-world geological variability. This study proposes a novel robust simulation-optimization framework to design reliable hydraulic containment systems that explicitly account for this subsurface uncertainty. The framework integrates the Karhunen–Loève Expansion (KLE) for efficient stochastic representation of heterogeneous K-fields with a Genetic Algorithm (GA) implemented via the pymoo library, coupled with the MODFLOW groundwater flow model for physics-based performance evaluation. The core innovation lies in a multi-scenario assessment process, where candidate well configurations (locations and pumping rates) are evaluated against an ensemble of K-field realizations generated by KLE. This approach shifts the design objective from optimality under a single scenario to robustness across a spectrum of plausible subsurface conditions. A structured three-step filtering method—based on mean performance, consistency (pass rate), and stability (low variability)—is employed to identify the most reliable solutions. The framework's effectiveness is demonstrated through a numerical case study. Results confirm that deterministic designs are highly sensitive to the specific K-field realization. In contrast, the robust framework successfully identifies well configurations that maintain a high and stable containment performance across diverse K-field scenarios, effectively mitigating the risk of failure associated with single-scenario designs. Furthermore, the analysis reveals how varying degrees of aquifer heterogeneity influence both the required operational cost and the attainable level of robustness. This systematic approach provides decision-makers with a practical and reliable strategy for designing cost-effective P&T systems that are resilient to geological uncertainty, offering significant advantages over traditional methods for contaminated site remediation.

Keywords: well placement optimization; groundwater management; robust optimization; MODFLOW; uncertainty quantification; contaminant plume containment

1. Introduction

Groundwater contamination has emerged as a global environmental issue, posing significant threats to both ecosystems and human health. Major sources of aquifer pollution include industrial activities, agricultural runoff, and improper waste disposal [1,2]. To effectively address this challenge, pump-and-treat (P&T) systems have been established as a mature and widely applied remediation technology, playing a crucial role in containing and removing contaminant plumes [3,4]. This technique involves installing extraction wells within the contaminated zone to pump out polluted groundwater for ex situ treatment, thereby containing the plume and restoring aquifer quality [5].

The migration of contaminant plumes in groundwater systems is a critical issue in environmental management, particularly in heterogeneous aquifers. A primary challenge in groundwater remediation is the effective placement of extraction wells and the optimization of pumping rates under conditions of subsurface uncertainty. P&T systems are highly sensitive to subsurface heterogeneity, especially the spatial distribution of hydraulic conductivity (K), a key parameter governing groundwater flow and contaminant transport [6–8]. Traditional modeling approaches often assume a homogeneous or simplified heterogeneous K-field, which can lead to significant inaccuracies in predicting groundwater flow and contaminant migration.

To address these limitations, geostatistical methods have become the mainstream approach for representing heterogeneous hydraulic conductivity fields. These methods treat K as a random spatial field characterized by statistical properties such as mean, variance, and spatial correlation structures, enabling a more realistic and continuous description of subsurface heterogeneity [9,10]. Common geostatistical techniques include variogram analysis and stochastic simulation, which generate multiple realizations of the K-field consistent with observed data and spatial statistics.

Despite their advantages, geostatistical representations typically lead to high-dimensional stochastic models, posing significant computational challenges for uncertainty quantification and optimization tasks. The dimensionality arises because the K-field is discretized over many spatial locations, each treated as a random variable. To overcome this, dimensionality reduction techniques are essential. Among these, the Karhunen–Loève Expansion (KLE), a form of Principal Component Analysis (PCA) for random fields, is widely adopted in groundwater modeling to reduce the complexity of the K-field while preserving its main statistical features [11–13]. KLE decomposes the random field into a series of orthogonal eigenfunctions weighted by uncorrelated random variables, allowing truncation after a limited number of terms that capture most of the variance. This reduction not only facilitates efficient stochastic simulations but also enables robust optimization frameworks that evaluate system performance across multiple realizations of the reduced-dimensional K-field.

To identify the optimal design and operation of P&T systems under these stochastic conditions, simulation-optimization (S-O) frameworks are widely employed. These frameworks integrate a groundwater simulation model, which predicts the fate and transport of contaminants for a given K-field realization, with an optimization algorithm that systematically searches for the best well configurations and pumping rates. Due to the complexity and non-linearity of the objective functions (e.g., minimizing cost while meeting cleanup targets), metaheuristic algorithms have gained prominence. Popular methods include Genetic Algorithms (GAs), Particle Swarm Optimization (PSO), and Simulated Annealing (SA), which are adept at navigating vast and rugged search spaces to find near-optimal solutions without requiring gradient information [14–16]. While powerful, these S-O approaches can be computationally intensive, and their application is often limited to a single or a few representative K-field realizations, which may not adequately account for the full spectrum of subsurface uncertainty [17,18].

Given the inherent uncertainty in the K-field and its profound impact on the performance of hydraulic containment systems, adopting a robust optimization (RO) approach becomes essential. Robust optimization aims to find solutions that perform well across a range of possible scenarios, rather than optimizing for a single realization of the uncertain parameter [19]. Instead of seeking a solution that is optimal under a single, deterministic condition, RO identifies a solution that exhibits the best robustness across a given range of uncertainty [6,20]. This means that the remediation system is designed to effectively contain the contaminant plume with a high degree of reliability, even if the actual hydraulic conductivity field deviates from the estimated one. In the context of optimizing well placement and pumping rates, this is typically achieved through a multi-scenario analysis framework, which involves generating multiple plausible realizations of the K-field to comprehensively evaluate candidate designs, thereby ensuring that the final optimized strategy maintains its effectiveness and reliability in the face of hydrogeological uncertainty.

In this study, we propose a novel robust simulation-optimization framework that integrates the efficiency of the Karhunen–Loève Expansion (KLE) for stochastic representation of hydraulic conductivity uncertainty with an advanced Genetic Algorithm to design a cost-effective and reliable pump-and-treat (P&T) system. The primary objective is to simultaneously optimize the locations and pumping rates of extraction wells to minimize the total operational cost (represented by total pumping rate) while ensuring a high and stable probability of containing the contaminant plume across a diverse ensemble of plausible heterogeneous aquifer conditions. This is achieved by evaluating candidate well configurations against multiple K-field realizations generated via KLE, thereby identifying solutions that are not only optimal for a single deterministic scenario but also robust and reliable under significant hydrogeological uncertainty.

The remainder of this paper is structured as follows. Section 2 details the methodology, including the numerical groundwater flow model, the optimization framework, the Karhunen–Loève Expansion technique for generating heterogeneous hydraulic conductivity fields, and the robustness assessment strategy. Section 3 describes the specific experimental design, outlining the case study setup and the multi-scenario evaluation process used to validate the proposed framework. Section 4 presents and discusses the results, analyzing the performance of optimized solutions under single-scenario and multi-scenario frameworks, evaluating their robustness across varying levels of aquifer heterogeneity, and examining the characteristics of the most stable well configurations. Finally, Section 5 offers concluding remarks, summarizing the key findings, discussing the implications for groundwater management, and identifying areas for future research.

2. Methodology

This study focuses on the optimization of well placement and pumping rates for a hydraulic containment system designed to prevent contaminant plume migration in heterogeneous aquifers. Given the subsurface uncertainty, particularly in the spatial distribution of hydraulic conductivity, we adopt a robust optimization framework that evaluates candidate solutions across multiple realizations of the K-field to identify configurations that maintain effective containment under varying K-field uncertainty. The methodology integrates four core components:

1. A numerical groundwater flow model based on MODFLOW for simulating hydraulic head responses under various well configurations;
2. An optimization model that defines decision variables, the objective function, and constraints related to well design and contaminant containment;
3. A Genetic Algorithm-based search strategy tailored for mixed-variable encoding and constrained optimization;
4. A stochastic modeling approach to quantify uncertainty in the K-field and assess solution robustness.

2.1. Groundwater Flow Modeling

The groundwater flow dynamics in the study area are governed by the transient saturated flow equation based on Darcy's law and mass conservation principles. The governing partial differential equation is expressed as

$$\nabla \cdot (K\nabla h) + Q = 0 \quad (1)$$

where h is the hydraulic head [L]; K is the spatially variable hydraulic conductivity tensor [LT^{-1}]; and Q is the source–sink term [T^{-1}].

This equation is subject to appropriate initial and boundary conditions that define the physical behavior of the aquifer system. In this study, the model domain is discretized into a regular grid using finite difference approximations, and the resulting system of equations is solved numerically using the widely used modular finite-difference groundwater flow model MODFLOW-2005 [21].

The model was implemented via the Python package FloPy 3.9.1, which enables efficient setup, execution, and post-processing of MODFLOW simulations within a scriptable environment [22]. Each optimization iteration involves dynamically updating well locations and pumping rates, followed by re-running the MODFLOW simulation to compute the resulting hydraulic head distribution. These results are then used to evaluate the satisfaction of head-based constraints designed to contain the contaminant plume.

2.2. Optimization Model for Hydraulic Containment

The optimization problem aims to determine the optimal configuration of a set of extraction wells that minimizes total pumping effort while ensuring effective hydraulic containment of the contaminant plume.

2.2.1. Decision Variables

The decision variables in this problem consist of the spatial coordinates (row and column indices) and the corresponding pumping rates of n extraction wells, resulting in a total of $3n$ decision variables. Specifically, each well is represented by three variables: x_{3i} (row index: integer variable), x_{3i+1} (column index: integer variable), and x_{3i+2} (pumping rate: real number variable), where $i = 0, 1, \ldots, n$ corresponds to n extraction wells.

This mixed-variable encoding allows for simultaneous optimization of both discrete well locations and continuous pumping rates.

2.2.2. Objective Function

The objective function aims to minimize the total pumping rate across all wells:

$$\text{Minimize } f(x) = \sum_{i=0}^{n-1} |x_{3i+2}| + P(x) \quad (2)$$

where $P(x)$ denotes a linear penalty term applied when any of the predefined constraints are violated. This ensures that infeasible solutions are gradually guided toward feasibility during the evolutionary process.

2.2.3. Hydraulic Head Constraints

To achieve effective containment of the contaminant plume, the optimization incorporates m hydraulic head constraints designed to maintain favorable flow conditions that restrict pollutant migration. These constraints are defined between predefined pairs of monitoring points:

$$h\left(\mathrm{loc}_j^{\mathrm{outer}}\right) - h\left(\mathrm{loc}_j^{\mathrm{inner}}\right) \geq \Delta h_{min},\ j = 1,2,\ldots,m \tag{3}$$

where j denotes the index for the m predefined pairs of monitoring points; Δh_{min} denotes a minimum head difference; and $\mathrm{loc}_j^{\mathrm{outer}}$ and $\mathrm{loc}_j^{\mathrm{inner}}$ are the locations of the outer and inner monitoring points in the j-th pair, respectively. This formulation enforces a controlled flow field that prevents downstream movement of contaminants by maintaining higher heads at outer locations relative to inner reference points.

2.2.4. Variable Bounds and Feasible Search Space

To define the feasible search space for well placement and pumping rate optimization, practical and hydrogeological considerations were incorporated. The spatial coordinates of each extraction well were constrained within a predefined subdomain of the aquifer model grid:

(1) Row indices of well locations were bounded between R_{min} and R_{max},

$$x_{3\mathrm{i}} \in [R_{\mathrm{min}}, R_{\mathrm{max}}],\ \mathrm{i} = 0,1,\ldots,m$$

(2) Column indices were restricted within C_{min} and C_{max},

$$x_{3\mathrm{i}+1} \in [C_{\mathrm{min}}, C_{\mathrm{max}}],\ \mathrm{i} = 0,1,\ldots,m$$

(3) Pumping rates were limited to a range $[Q_{\mathrm{min}}, Q_{\mathrm{max}}]$, where $Q_{\mathrm{min}} < 0$ denotes the maximum allowable extraction rate (negative value indicating withdrawal), and $Q_{\mathrm{max}} = 0$ reflects the physical constraint that no injection is permitted:

$$x_{3\mathrm{i}+2} \in [Q_{\mathrm{min}}, Q_{\mathrm{max}}],\ \mathrm{i} = 0,1,\ldots,m$$

These bounds ensure that all generated solutions are physically meaningful and operationally viable.

2.3. *Optimization Framework*

To solve the high-dimensional and constrained optimization problem, we employ a hybrid-variable Genetic Algorithm implemented using the pymoo optimization library [23]. Key components of the algorithm include the following:

(1) Sampling: A mixed-variable sampling strategy that combines integer random sampling for well coordinates and uniform real-value sampling for pumping rates.
(2) Crossover: Simulated Binary Crossover (SBX) is used for real-valued variables.
(3) Mutation: Polynomial Mutation (PM) is applied to real variables, while random integer mutation is used for coordinate variables.
(4) Selection: Tournament selection with a crowding distance is employed to maintain diversity in the population.

Through preliminary testing, multiple combinations of population sizes and generation limits were systematically evaluated to identify a configuration that optimally balances computational efficiency with solution quality. Although specific configurations (in Table A1) demonstrated effectiveness in the case study presented in Section 3, these parameters may be modified based on the specific requirements of the optimization problem and available computational capabilities.

Each candidate solution is rigorously evaluated through full MODFLOW simulations utilizing the FloPy interface. The simulation process dynamically updates well locations and pumping rates, computes the corresponding hydraulic head distribution, and systematically evaluates constraint satisfaction. This tightly integrated coupling between the optimization engine and the groundwater flow simulator enables the identification of well configurations.

2.4. Uncertainty Modeling

Subsurface heterogeneity, particularly the spatial variability of hydraulic conductivity, introduces significant uncertainty in predicting groundwater flow patterns and contaminant transport dynamics. To account for this uncertainty and enhance the robustness of the optimized well configurations, a stochastic modeling framework is employed to generate multiple realizations of the K-field. These realizations represent plausible subsurface structures with varying degrees of spatial correlation and heterogeneity.

2.4.1. Stochastic Generation of Hydraulic Conductivity Fields

To account for the spatial variability and uncertainty in hydraulic conductivity, a stochastic modeling framework is employed to generate multiple realizations of the K-field using the Karhunen–Loève Expansion (KLE) method [11]. This approach allows for the representation of heterogeneous subsurface properties while maintaining statistical consistency across different realizations.

The log-conductivity field, denoted as $Y(x,\omega) = lnK(x,\omega)$, where $x \in D$ and $\omega \in \Omega$ (a probability space), is modeled as a random process. The mean component of $Y(x,\omega)$ over all possible realizations is represented by $\langle Y(x,\omega)\rangle$. A covariance function $C(x,y)$, which is bounded, symmetric, and positive definite, is required to construct the KLE. This covariance function can be decomposed into eigenvalues τ_i and eigenfunctions $f_i(x)$ according to the second kind of the homogeneous Fredholm integral equation:

$$C(x,y) = \sum_{i=1}^{\infty} \tau_i f_i(x) f_i(y), \tag{4}$$

The random process $Y(x,\omega)$ can be expanded as

$$Y(x,\omega) = \langle Y(x,\omega)\rangle + \sum_{i=1}^{\infty} \xi_i \sqrt{\tau_i} f_i(x,\omega), \tag{5}$$

where ξ_i are independent standard Gaussian random variables. In practice, the infinite series is truncated to a finite number N_{KL} of terms to obtain a computationally feasible approximation:

$$Y(x,\omega) \approx \langle Y(x,\omega)\rangle + \sum_{i=1}^{N_{KL}} \xi_i \sqrt{\tau_i} f_i(x,\omega), \tag{6}$$

This finite-dimensional approximation ensures that the generated K-fields maintain the desired statistical properties (mean, variance, and spatial correlation) while being computationally manageable. The truncation parameter N_{KL} determines the level of detail in representing the heterogeneity, with higher values capturing more fine-scale variability but at increased computational cost.

Each realization captures different spatial patterns of conductivity while maintaining overall statistical consistency. This approach ensures that the optimization process considers a diverse set of subsurface scenarios, thereby improving the generalizability and reliability of the resulting well designs.

2.4.2. Robustness Evaluation Across Multiple Realizations

To assess the performance stability of each candidate solution under uncertain subsurface conditions, all identified well configurations are evaluated across a total of N independently generated K-fields. For each realization, the same MODFLOW-based simulation is executed to compute the resulting hydraulic head distribution and evaluate constraint satisfaction. Key metrics used to quantify robustness include the following:

(1) Mean Constraint Satisfaction Ratio (CSR_{mean}): the average proportion of constraints satisfied across all K-fields.
(2) Minimum CSR (CSR_{min}): the lowest constraint satisfaction observed among all realizations, indicating the worst-case performance.
(3) Standard Deviation of CSR (CSR_{std}): a measure of performance variability across different K-fields.
(4) Pass Rate (PR): the proportion of K-fields in which the solution meets a predefined threshold for constraint satisfaction.

These metrics enable the identification of well configurations that consistently perform well across a wide range of subsurface conditions. The integration of uncertainty quantification into the optimization workflow supports risk-informed decision-making by explicitly considering the impact of subsurface heterogeneity on system performance. This approach facilitates the selection of well designs that balance optimality and robustness, providing valuable insights for long-term planning and management of contaminated sites where subsurface information is inherently limited.

3. Case Study

3.1. Problem Description

Figure 1 illustrates the study area, which is impacted by a groundwater contaminant plume resulting from long-term leakage at a heavily polluting industrial facility located in the northwest portion of the study area. The contamination source originates from a continuous pollutant leak that has persisted for 20 years. As shown in Figure 1, the resulting contaminant plume demonstrates the spatial distribution of concentrations exceeding 5 mg/L, illustrating the extent of contamination after two decades of leakage.

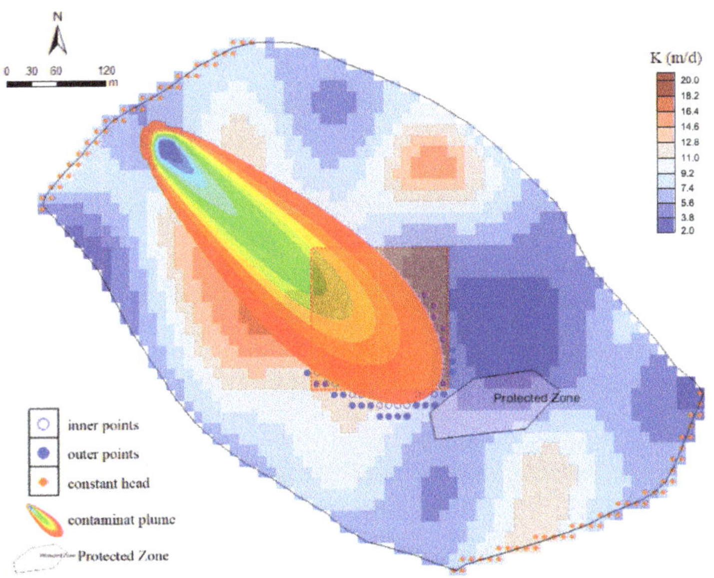

Figure 1. Study site layout with key monitoring locations.

To protect a designated protected zone located downstream, a hydraulic containment system must be implemented to prevent the further migration of the plume. This case study employs a two-dimensional steady-state groundwater flow model to investigate the application of a hydraulic containment system for the long-term containment of the contaminant plume. The top and bottom boundaries of the aquifer are considered impermeable. The lateral boundary conditions are defined as constant-head boundaries on the northwest and southeast sides, and as no-flow boundaries on the northeast and southwest sides.

As shown in Figure 1, the aquifer system is assumed to be heterogeneous and isotropic, with hydraulic conductivity following a log-normal distribution. The spatial correlation structure of the log-conductivity is characterized by the following exponential covariance function.

$$\gamma(l,l')_{\ln K} = \sigma^2_{\ln K}\exp\left(-\sqrt{\left(\frac{l_x - l'_x}{\lambda_x}\right)^2 + \left(\frac{l_y - l'_y}{\lambda_y}\right)^2}\right), \tag{7}$$

where $l = (l_x, l_y)$ and $l' = \left(l'_x, l'_y\right)$ denote two arbitrary spatial locations, $\sigma^2_{\ln K}$ is the variance, and λ_x and λ_y are the correlation lengths along x and y directions, respectively. Length scales of $\lambda_x/L_x = 0.30$ and $\lambda_y/L_y = 0.40$ are considered in this study, where L_x and L_y are the sizes of flow domain along the x and y directions, respectively. The mean of the $\ln K$ field is 2.30, while the variances are set to $\sigma^2_{\ln K} = 0.50$. In this study, the KLE method is adopted for generating multiple realizations of the K-field, and the truncation criterion for the KLE was based on capturing a target percentage (90% in this study) of the total variance of the log-conductivity field. The specific truncation parameter N_{KL} is automatically determined by the program based on the specified covariance model and grid. This approach was consistently applied across all scenarios, including those with varying heterogeneity levels.

To effectively contain the contaminant plume and prevent its migration toward the downstream protected zone, a series of extraction wells are proposed to be installed around the plume, forming a closed hydraulic containment system. The red dashed line in Figure 1 indicates the candidate range for well placement. The inner points (inner points) and outer points (outer points) shown in Figure 1 are used to implement the hydraulic head constraints described in Section 2.2. These points create a hydraulic gradient directed toward the center of the contaminant plume, thereby controlling its migration range and preventing further diffusion.

The primary objective of this study is to design an optimal well configuration for a hydraulic containment system that can effectively contain this plume under conditions of subsurface heterogeneity. The optimization aims to minimize the total pumping rate while ensuring that the hydraulic head constraints necessary for containment are satisfied across the domain.

3.2. Experiment Design

The experimental framework is designed to systematically evaluate the performance and robustness of optimized well configurations under significant subsurface uncertainty. The workflow is structured into three main components to enable a comprehensive assessment of solution optimality, generalizability, and resilience to geological heterogeneity.

Component 1: Single-Scenario Optimization and Cross-Validation

The first component establishes a baseline for comparison by performing deterministic optimizations under individual K-field realizations. A total of 50 independent hydraulic conductivity fields ($K_1, K_2, \ldots, K_{50}$) are generated using the KLE method. The choice of 50 realizations represents a practical balance between computational feasibility and the need for a sufficiently large ensemble to capture the range of possible subsurface conditions and provide statistical robustness in the subsequent multi-scenario assessment (requiring 50 solutions $\times$ 50 K-fields = 2500 simulations). This number was also found sufficient in preliminary evaluations to observe stable trends in the distribution of solution performance metrics (e.g., mean satisfaction ratio, standard deviation) without incurring excessive computational cost, and it is consistent with the ensemble sizes used in similar stochastic groundwater modeling studies. For each K_i, a separate optimization run is conducted as described in Section 2.3. This process yields 50 candidate solutions ($S_1, S_2, \ldots, S_{50}$), where S_i is the optimal well configuration identified under the K_i field.

To assess the transferability of these solutions, a cross-validation is performed. One of the 50 K-fields is selected as the true field (K_{true}), which represents the actual subsurface condition. Each of the 50 solutions is then evaluated on this K_{true} field to compute its Constraint Satisfaction Ratio (CSR). This component highlights the limitations of single-scenario optimization by demonstrating the performance gap between an estimated field (which may be obtained using geostatistics or groundwater inverse problem) and the true field.

Component 2: Multi-Scenario Robustness Assessment

To move beyond deterministic optimization and evaluate solution robustness, a comprehensive multi-scenario assessment is conducted. This comprehensive evaluation assesses the performance of each of the 50 solutions (S_i) under all 50 K-field realizations. For each (S_i, K_j), the Constraint Satisfaction Ratio is computed, resulting in a total of 2500 individual performance assessments (50 solutions × 50 K-fields). The core robustness assessment employs a three-step filtering process:

(1) Defining the Efficacy Threshold: An efficacy threshold for CSR_{mean} is established based on the statistical distribution of performance across all solutions and K-fields.
(2) Screening for Robustness: Solutions are screened based on their ability to meet or exceed this efficacy threshold in a high percentage (e.g., 75%) of the K-field realizations.
(3) Selecting the Most Stable Solutions: Among the solutions passing the robustness screen, the final subset is selected based on the lowest CSR_{std} across all K-fields, identifying the most stable performers.

Component 3: Application under Varying Heterogeneity Levels

This component applies the multi-scenario robustness assessment framework (Component 2) to evaluate how the level of aquifer heterogeneity influences the performance and selection of robust solutions. Three distinct levels of heterogeneity are considered, defined by the variance (σ^2) of the log-conductivity field: low ($\sigma^2 = 0.25$), moderate ($\sigma^2 = 0.50$), and high ($\sigma^2 = 1.00$).

For each heterogeneity level, a new set of K-field realizations is generated using the KLE method with the corresponding variance. The entire optimization and multi-scenario evaluation process is then repeated. This allows for a comparative analysis of solution robustness and the characteristics of the identified stable solutions under conditions of increasing subsurface uncertainty.

4. Results and Discussion

4.1. Single-Scenario Optimization and Cross-Validation

In the first component, the pymoo algorithm was independently applied to optimize well configurations under 50 randomly generated K-field realizations. Each optimization run yielded a candidate solution, representing a specific combination of well locations and pumping rates. These solutions were then evaluated on a selected true K-field—one of the 50 realizations, specifically the same K-field used as the background in Figure 1—to assess their performance in satisfying the hydraulic head constraints designed to contain the contaminant plume.

As illustrated in Figure 2, there is a strong positive correlation (correlation coefficient = 0.8541) between pumping rate and satisfaction. This quantitative relationship confirms that higher pumping rates generally lead to better constraint satisfaction, but at the potential cost of increased operational expenses. Statistical analysis reveals key characteristics: the first quartile (Q1), median (Q2), and third quartile (Q3) for pumping rates are 59.2, 84.5, and 117.8 m^3/d, respectively, while for satisfaction levels, they are 0.2941, 0.3824, and 0.7500, respectively. These metrics indicate that while most solutions have moderate pumping rates, satisfaction levels show greater variability, with the majority falling below 0.75. The 95% confidence interval around the trend line further illustrates the uncertainty in the relationship, suggesting that individual solution performance can deviate significantly from the general trend due to subsurface heterogeneity.

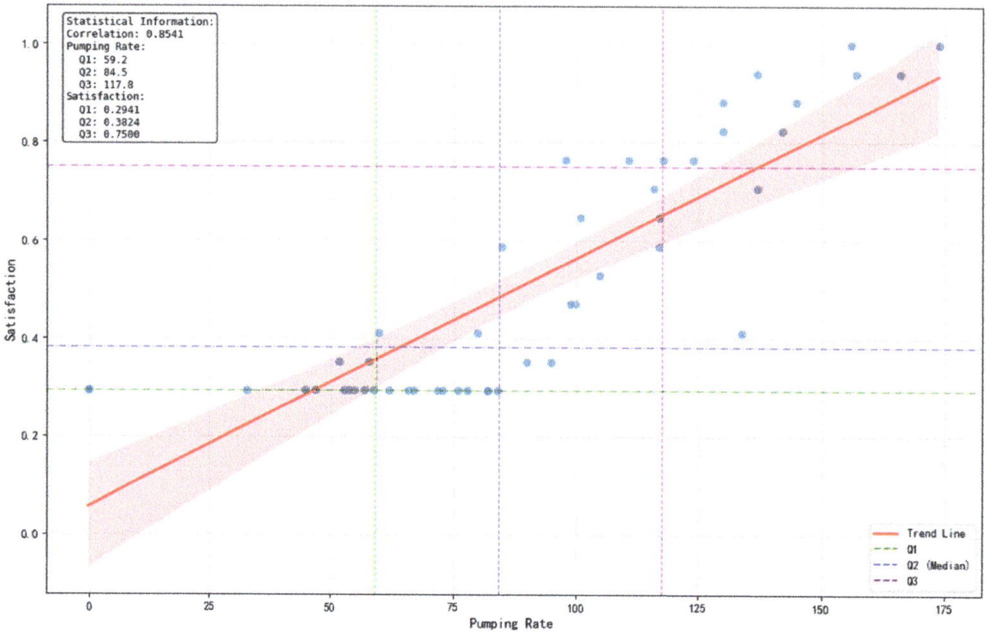

Figure 2. Relationship between total pumping rate and satisfaction ratio for solutions optimized under individual K-fields and evaluated on the true K-field (the red line indicates the linear regression trend, and the shaded area represents the 95% confidence interval).

To further evaluate the robustness of the optimized solutions, we conducted backward tracking using MODPATH to visualize the capture zones of the pumping wells under different K-field realizations. The blue pathlines represent the capture zone of the optimized pumping wells, providing insights into how effectively the contaminant plume can be contained.

Figure 3 illustrates the capture zone morphologies, highlighting the variability in performance due to subsurface heterogeneity. Figure 3a shows the optimal solution specifically for the true K-field. It achieved a pumping rate of 118 m^3/d and a satisfaction level of 0.76, demonstrating effective containment of the contaminant plume. In contrast, Figure 3b presents a randomly selected solution with a pumping rate of 90 m^3/d and a satisfaction level of 0.35. Its capture zone is inadequate, leading to failure in containing the contaminant plume. Another randomly selected solution (Figure 3c) exhibited a pumping rate of 98 m^3/d and a satisfaction level of 0.72, showing a capture zone similar to the optimal solution, indicating robust performance. Finally, a third randomly selected solution (Figure 3d) demonstrated a high pumping rate of 160 m^3/d and a satisfaction level of 0.94, achieving excellent contaminant containment but at the cost of substantial engineering expenses.

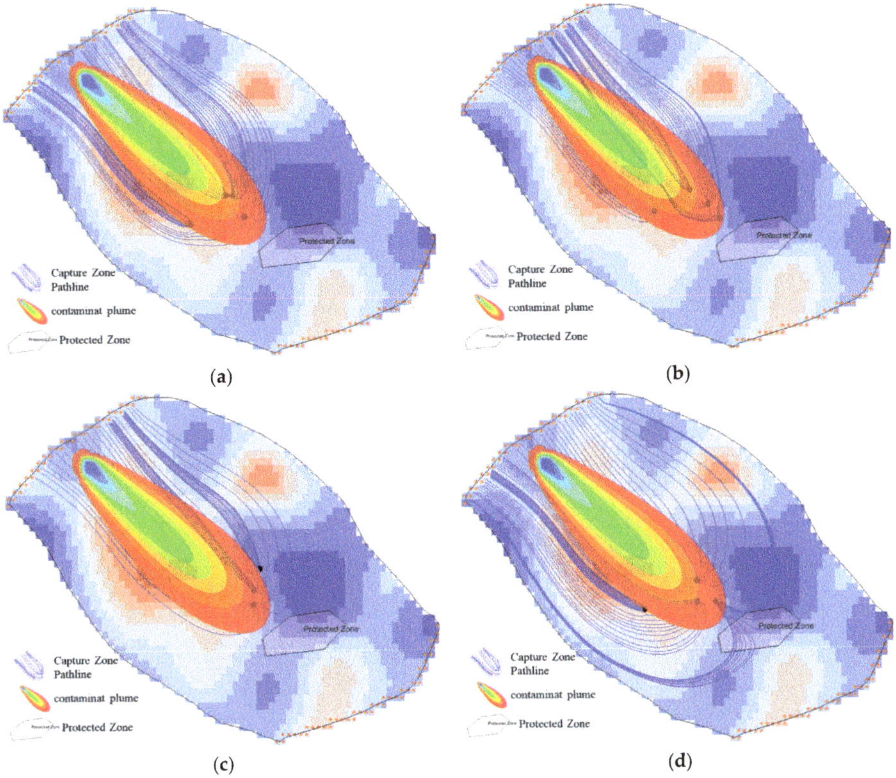

Figure 3. Capture zone morphologies of four selected solutions under the true K-field. (**a**) Optimal solution for the true K-field. (**b**) Randomly selected solution 1. (**c**) Randomly selected solution 2. (**d**) Randomly selected solution 3.

The results indicate that while some solutions perform well under their respective K-fields, their effectiveness varies significantly when applied to the true K-field. This highlights the importance of evaluating candidate solutions under different subsurface conditions rather than relying solely on deterministic optimization. The significant performance variability observed across different K-field realizations demonstrates a critical limitation of single-scenario approaches and underscores the need for a more comprehensive, multi-scenario evaluation to identify well designs that are genuinely robust and reliable under uncertain subsurface conditions.

4.2. Multi-Scenario Robustness Assessment

The single-scenario optimization results presented in Section 4.1 highlighted the significant vulnerability of solutions to geological uncertainty, despite the observed correlation between pumping rate and satisfaction. To identify well designs that are resilient to subsurface variability, the comprehensive multi-scenario evaluation described in Section 3.2 (Component 2) was conducted. This involved assessing the performance of each of the 50 optimized solutions (S*i*) under all 50 K-field realizations. For each solution-K-field pair (S_i, K_j), the satisfaction ratio (CSR) was computed, resulting in a total of 2500 individual performance assessments. This extensive evaluation captures the constraint satisfaction behavior of each solution under different realizations of the same geostatistical model.

The analysis is structured around the three-step filtering process outlined in Section 3.2 (Component 2) to identify solutions that are both effective and stable across diverse subsurface conditions.

Step 1: Defining the Efficacy Threshold

An efficacy threshold for the mean satisfaction ratio (CSR_{mean}) is established based on the statistical distribution of performance across all solutions and K-fields (as described in Section 2.4 for metric definitions). This threshold balances two criteria: (1) a sufficiently high level of constraint satisfaction, and (2) placement within the top quartile of all solutions in terms of performance.

A solution is considered effective only if its CSR_{mean} across all K-field realizations is at least 75%—a level deemed necessary for reliable containment. Meanwhile, this 75% satisfaction requirement aligns closely with the third quartile (Q3 = 72%) of the CSR_{mean} distribution shown in Figure 4, indicating that solutions meeting this threshold outperform 75% of all other solutions. Therefore, an efficacy threshold of CSR_{mean} = 0.75 is adopted, ensuring that only solutions with both high performance and a strong statistical standing are considered for further evaluation.

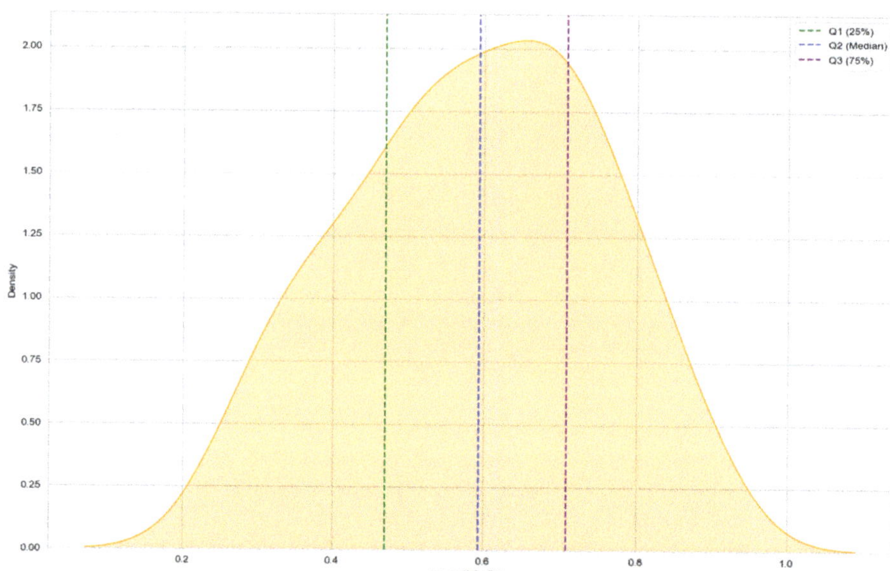

Figure 4. Kernel density estimate of CSR_{mean} across K-field realizations for all solutions.

Step 2: Screening for Robustness

Building upon the efficacy threshold defined in Step 1, the second step focuses on identifying robust solutions—those that maintain their efficacy across a wide range of K-field realizations. A solution is deemed robust if its satisfaction ratio is greater than or equal to the efficacy threshold (0.75) in at least 75% of the 50 K-fields (i.e., in at least 38 out of 50 realizations). This majority rule criterion ensures that the selected solutions are not only effective on average but also consistently reliable, as their performance is not contingent on a few favorable K-field realizations.

Step 3: Selecting the Most Stable Solutions

Among the solutions that pass the robustness screening in Step 2, the final step is to select those with the highest stability and the least sensitivity to geological variations. This is achieved by calculating the standard deviation of CSR (CSR_{std}) for each robust solution across all 50 K-fields. For this study, the five solutions with the smallest CSR_{std} are selected as the most stable solutions. This number (5) is chosen arbitrarily for demonstration purposes and can be adjusted based on the specific requirements of the analysis or the total number of solutions passing the previous screening steps. This selection minimizes the risk of encountering a scenario where the solution's performance drops significantly below its average, providing a high degree of confidence in its long-term reliability.

The results of this three-step filtering process are presented in Table 1, which summarizes the key performance metrics for the five most stable solutions. The analysis reveals that these solutions achieve a high level of average performance, with CSR_{mean} ranging from 79.88% to 86.82%. The top solution (S_{29}) achieves an average CSR of 86.82%, demonstrating exceptional effectiveness across the ensemble of K-field realizations. The CSR_{std} ranges from 19.40% to 24.36% for these solutions. S_{29} exhibits the highest stability with the lowest standard deviation of 19.40%, indicating that its performance is the least sensitive to geological variations. This low variability is visually confirmed in Figure 5, which presents a kernel density estimate (KDE) comparison of their satisfaction ratio distributions. The tightly clustered and right-skewed KDE curves for all five solutions provide strong quantitative evidence of their robustness. The high peak and narrow width of the curves confirm that these solutions spend the majority of their time in high-performance states, with minimal excursions into low-performance regions.

Table 1. Key statistics of the most stable well configurations under geological uncertainty.

Solution_Id	Total Pumping (m^3/d)	CSR_{mean}	CSR_{std}	PR (CSR < 50%)
S_{29}	174	86.82%	19.40%	10%
S_{19}	166	86.00%	20.30%	10%
S_{18}	156	84.47%	21.87%	12%
S_{28}	157	85.29%	22.53%	12%
S_{42}	142	79.88%	24.36%	20%

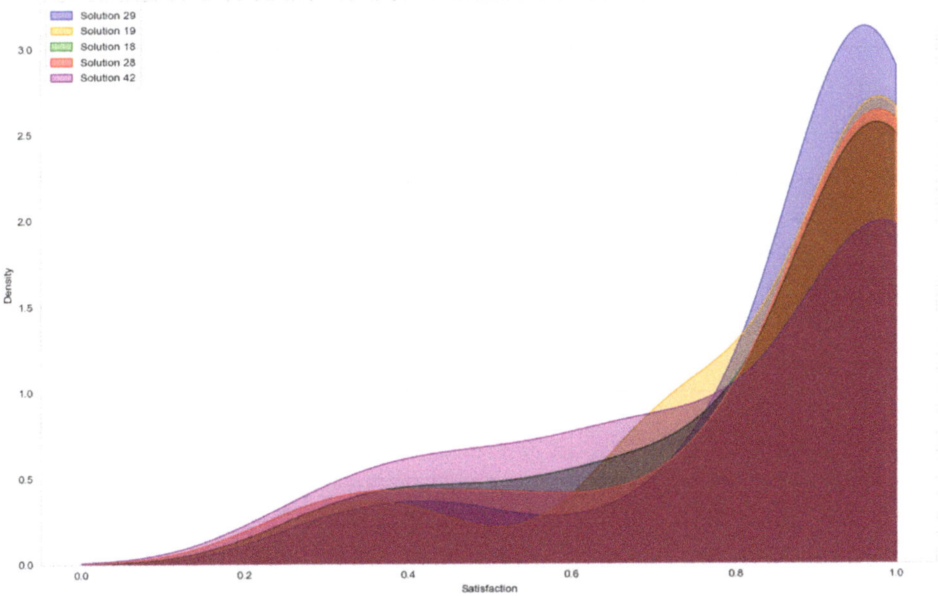

Figure 5. Kernel density estimate (KDE) comparison of satisfaction ratios for the top five most stable solutions.

Figure 6 illustrates the capture zone morphologies of the most robust solution S_{29} under four different randomly selected K-fields, illustrating performance variability due to subsurface heterogeneity. Figure 6a shows the capture effectiveness of S_{29} in its corresponding hydraulic conductivity field K_{29}, where it achieved a satisfaction level of 0.75, demonstrating effective containment of the contaminant plume. In contrast, Figure 6b,d presents S29's capture performance in randomly selected conductivity fields K_{15} and K_{42}, respectively, with improved results where the CSR reached 0.86 and 1.00, respectively. However, Figure 6c demonstrates a failure scenario, where S29's capture zone in K_{21} is inadequate, with a CSR of only 0.53. These visualizations clearly demonstrate that even the most robust solution identified through the multi-scenario assessment (S_{29}) exhibits significant performance variability across different K-field realizations. While it achieves effective containment (CSR $\geq$ 0.75) in most scenarios (e.g., K_{15}, K_{29}, K_{42}), its performance can drop below the desired threshold in others (e.g., K_{21}, CSR = 0.53), highlighting that no single solution can guarantee universal effectiveness across all possible subsurface conditions.

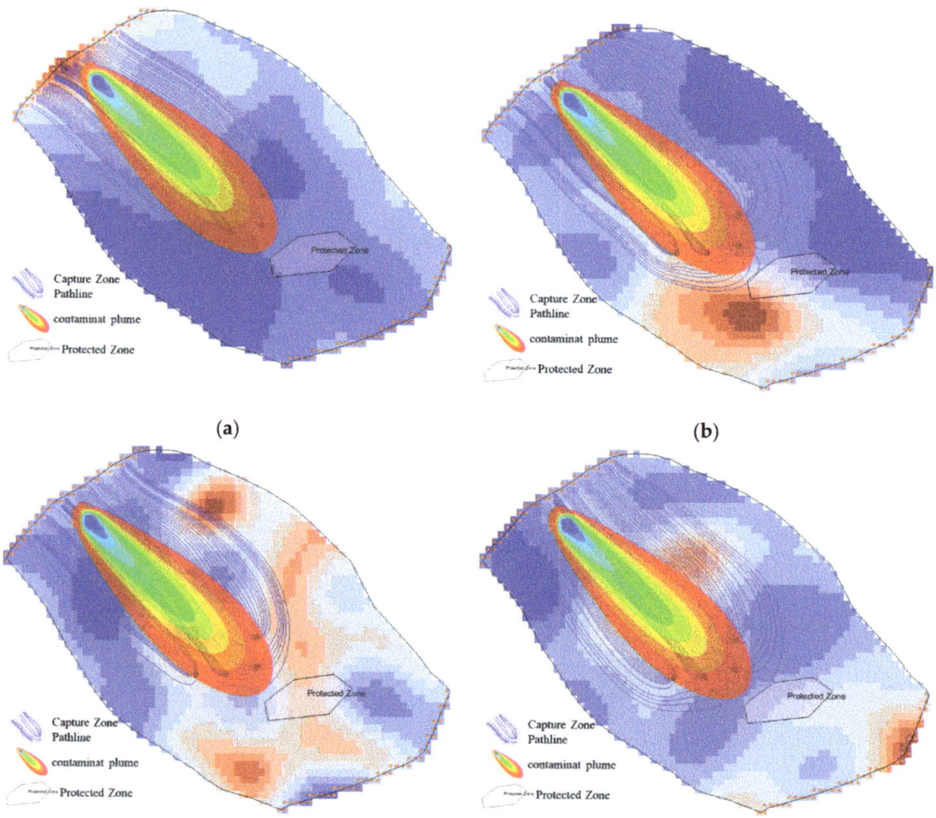

Figure 6. Capture zone morphologies of the most robust solution S_{29} under four different randomly selected K-fields. (**a**) K_{29} (CSR = 0.75). (**b**) K_{15} (CSR = 0.86). (**c**) K_{21} (CSR = 0.53). (**d**) K_{42} (CSR = 1.00).

However, as shown in Table 1, even the most robust solutions still exhibit a non-negligible percentage of K-field realizations (ranging from 10% to 20%) where performance drops below the desired threshold (CSR < 50%), indicating that perfect universality cannot be guaranteed.

To better understand these failure scenarios, Figure 7 presents the satisfaction ratios of the top five most stable solutions across all K-fields where at least one solution's performance drops below the CSR of 50%. This comprehensive view reveals that within this set of challenging K-fields, there are five specific K-fields (K_{22}, K_{23}, K_{30}, K_{35}, K_{43}) where the performance degradation is particularly severe. In these five K-fields, all five of the top solutions (S29, S28, S18, S28, S42) simultaneously exhibit satisfaction ratios below 50%, indicating a systemic failure of the hydraulic containment strategy.

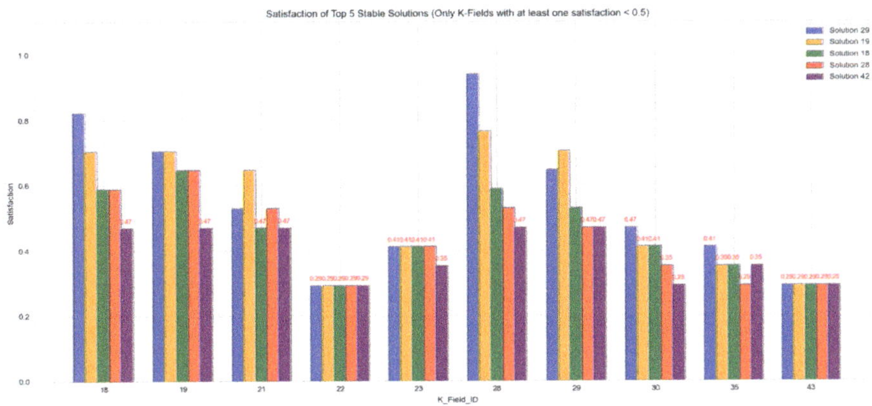

Figure 7. Satisfaction ratios of the top 5 most stable solutions across K-fields (only K-fields with at least one satisfaction < 0.5).

4.3. Evaluation of Solution Robustness Under Varying Heterogeneity Levels

The multi-scenario robustness assessment framework presented in Section 4.2 (Component 2) demonstrated its effectiveness in identifying well configurations that are resilient to subsurface uncertainty under a specific level of heterogeneity ($\sigma^2 = 0.50$). To further evaluate the generalizability and adaptability of this approach, Component 3 of the experimental framework (Section 3.2) applies the same multi-scenario evaluation process to scenarios with varying degrees of aquifer heterogeneity.

This component aims to investigate how the level of subsurface heterogeneity influences the performance characteristics of the identified robust solutions and the overall effectiveness of the optimization framework. Three distinct levels of heterogeneity were considered, defined by the variance (σ^2) of the log-conductivity field: low ($\sigma^2 = 0.25$), moderate ($\sigma^2 = 0.50$), and high ($\sigma^2 = 1.00$). For each heterogeneity level, a new set of 50 independent K-field realizations was generated using the KLE method with the corresponding variance. This systematic approach allows for a direct comparison of solution performance, robustness metrics, and the characteristics of the identified stable solutions across the spectrum of subsurface uncertainty, as summarized in Table 2.

Table 2. Performance summary of the most stable solution identified under each heterogeneity level.

Heterogeneity Level	Best Solution	Total Pumping (m^3/d)	CSR_{mean}	CSR_{std}	PR(CSR < 50%)
Low ($\sigma^2 = 0.25$)	S_{29}	202	94.82%	11.86%	2%
Medium ($\sigma^2 = 0.50$)	S_{29}	174	86.82%	19.40%	10%
High ($\sigma^2 = 1.00$)	S_{40}	207	77.06%	21.95%	12%

A key finding is that a consistent trend emerges regarding solution robustness and reliability. As heterogeneity increases from low ($\sigma^2 = 0.25$) to high ($\sigma^2 = 1.00$), the CSR_{mean}

of the most stable solution decreases (from 94.82% to 77.06%). Concurrently, the performance stability, measured by CSR_{std}, deteriorates from 11.86% to 21.95%, and the risk of encountering a failure scenario (CSR < 50%) increases from 2% to 12%. These findings underscore the complex interplay between aquifer heterogeneity, containment effectiveness, and system reliability, highlighting the critical need to explicitly account for the specific characteristics of geological uncertainty during the design phase.

However, the relationship between aquifer heterogeneity and the total pumping rate required for effective containment is not monotonic. The most stable solution under moderate heterogeneity ($\sigma^2 = 0.50$) requires the lowest total pumping rate (174 m^3/d), while higher rates are needed under both low (202 m^3/d) and high (207 m^3/d) heterogeneity levels. This non-monotonic behavior suggests that the optimal operational cost is not simply dictated by the degree of geological variability.

A comparative analysis of the capture zone morphologies across different heterogeneity levels (Figures 8 and 9) more clearly illustrates the impact of heterogeneity intensity. Under low heterogeneity ($\sigma^2 = 0.25$), the capture zones for solution S_{29} are highly stable and effective in most cases (e.g., K_8, K_{16}, CSR = 0.94, 1.00), with only a few instances showing slightly worse performance (e.g., K_{37}, CSR = 0.47). This is consistent with the extremely low performance standard deviation (11.86%) and only 2% failure rate in Table 2 for this scenario. However, under high heterogeneity ($\sigma^2 = 1.00$), the capture zones for the best-performing solution (S_{40}) exhibit significant instability, showing a stark contrast from good performance in K_{18} (CSR = 0.94) to clear failure in K_5 (CSR = 0.41). This directly corresponds to the highest performance standard deviation (21.95%) and 12% failure rate for this scenario in Table 2. Overall, as heterogeneity increases from low to high, the capture zone morphology of the robust solution transitions from highly consistent and reliable to significantly variable and prone to failure.

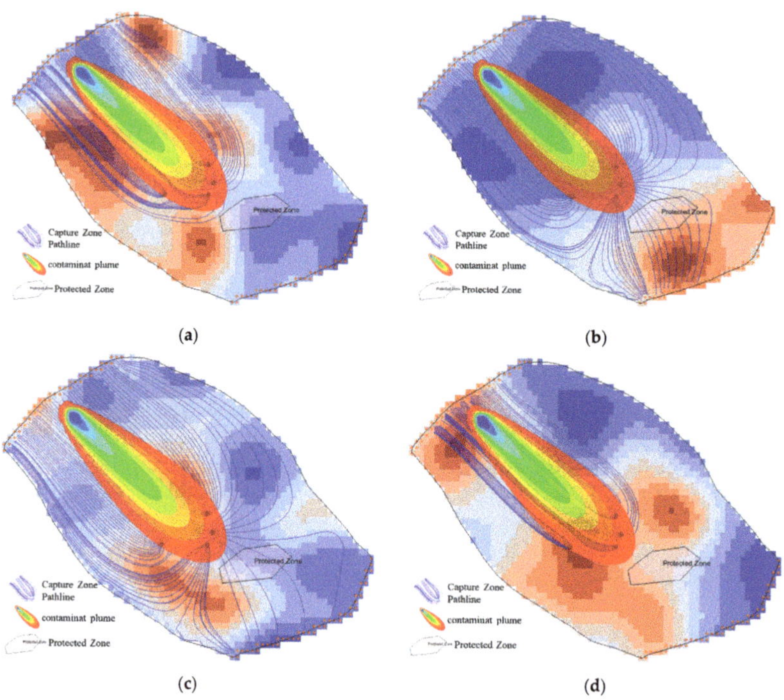

Figure 8. Capture zone morphologies of the most robust solution S_{29} under low heterogeneity. (**a**) K_{29} (CSR = 0.88). (**b**) K_8 (CSR = 0.94). (**c**) K_{16} (CSR = 1.00). (**d**) K_{37} (CSR = 0.47).

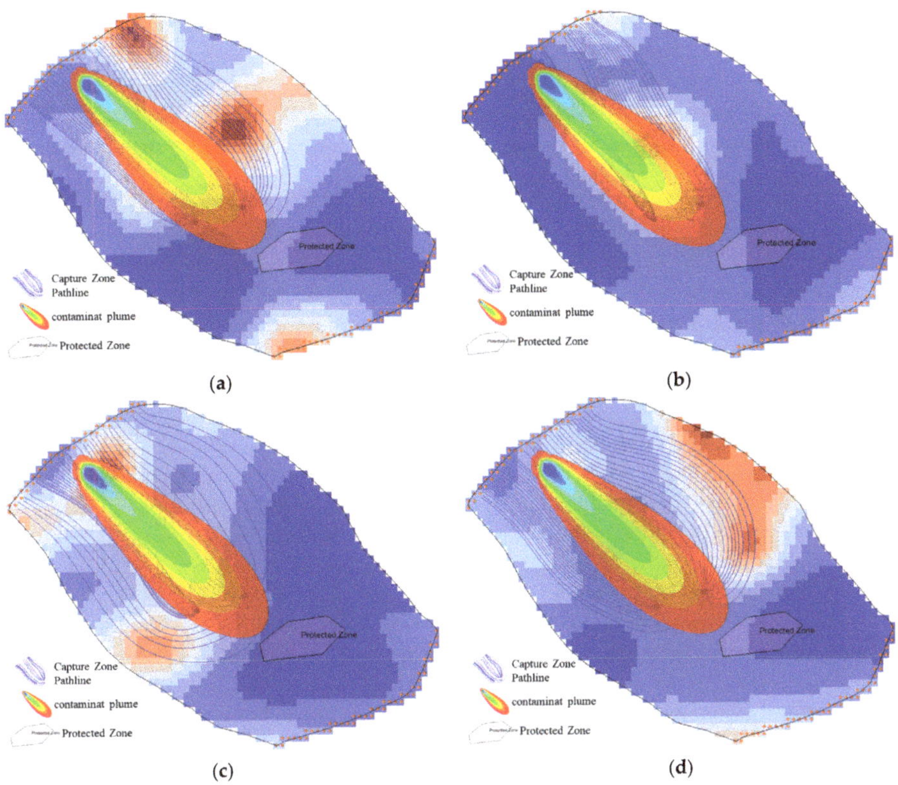

Figure 9. Capture zone morphologies of the most robust solution S_{40} under high heterogeneity. (**a**) K_{40} (CSR = 0.65). (**b**) K_5 (CSR = 0.41). (**c**) K_{18} (CSR = 0.94). (**d**) K_{37} (CSR = 0.71).

5. Conclusions

(1) This study presents a robust optimization framework for the design of hydraulic containment systems in heterogeneous aquifers, specifically addressing the challenge of contaminant plume migration. The framework integrates an optimization algorithm (pymoo) with a numerical groundwater flow model (MODFLOW) to simultaneously optimize well placement and pumping rates. Its key innovation lies in the explicit consideration and mitigation of subsurface uncertainty through a multi-scenario evaluation approach. Rather than relying on a single deterministic K-field, the method assesses candidate solutions across an ensemble of K-field realizations generated via the Karhunen–Loève Expansion (KLE) method. This allows for the identification of well configurations that are not only optimal under a specific scenario but also robust and reliable across a spectrum of plausible subsurface conditions.

(2) The effectiveness of the proposed framework was demonstrated through a numerical case study. The results confirm that traditional single-scenario optimization is highly sensitive to the specific realization of the K-field used for training, often leading to significant performance degradation when applied to different subsurface conditions. In contrast, the multi-scenario robustness assessment successfully identified a small subset of stable solutions (e.g., S29, S28, S18, S28, S42) that maintained a high CSR_{mean} (e.g., 79.88% to 86.82%) and low performance variability (e.g., CSR_{std}: 19.40% to 24.36%) across diverse K-field scenarios. These robust solutions effectively balanced containment performance with operational cost.

(3) Moreover, the application of the framework under varying heterogeneity levels revealed valuable insights, such as the consistent trend of decreasing robustness (lower mean satisfaction, higher variability) with increasing subsurface uncertainty, and the non-monotonic relationship between heterogeneity and required pumping rates. This underscores the framework's capability to provide nuanced guidance for risk-informed decision-making in groundwater management.

Despite its demonstrated effectiveness, this study has several limitations.

First, the computational cost is significant. The requirement for numerous, independent MODFLOW simulations (2500 in the main assessment) represents a substantial computational burden. This brute-force approach, while effective for demonstrating the framework's conceptual feasibility, may limit its application to larger domains, finer grids, or more complex, transient, three-dimensional models without access to significant computational resources. To address this critical limitation, future research will prioritize the integration of surrogate modeling techniques. Methods such as Gaussian Process Regression (Kriging), Polynomial Chaos Expansion, or advanced neural networks (e.g., Physics-Informed Neural Networks) could be employed to approximate the expensive MODFLOW simulations. These surrogate models, once trained on a subset of simulations, can dramatically reduce the number of required full physics-based model runs (potentially from thousands to hundreds or tens), thereby making the robust optimization framework computationally feasible for real-world, large-scale applications.

Second, the current framework focuses on a two-dimensional, steady-state flow model and a conservative contaminant. Extending the approach to three-dimensional, transient flow and transport models, including reactive transport, would be necessary for broader applicability to real-world sites with complex hydrogeology and diverse pollutants.

Third, this study assumes that the statistical properties of the K-field (mean, variance, correlation structure) are known or can be accurately estimated, which may not always be the case in practical scenarios with limited data.

Future research will focus on enhancing computational efficiency through surrogate modeling or adaptive sampling, incorporating more complex physical processes (3D, transient, reactive transport), exploring multi-objective optimization frameworks (e.g., balancing cost, time, and robustness), investigating methods for robust design under imprecise probabilistic information or incomplete probabilistic information, and applying the framework to field data or benchmark sites to validate its operational effectiveness.

Author Contributions: Conceptualization, W.G. and S.J.; Formal Analysis, W.G. and Y.K.; Methodology, S.J.; Project Administration, W.G. and S.J.; Resources, W.G.; Software, H.L. and S.J.; Supervision, S.J.; Validation, H.D. and H.L.; Writing—Original Draft, W.G. and Y.K.; Writing—Review and Editing, S.J. All authors have read and agreed to the published version of the manuscript.

Funding: This research was supported by a Geological Exploration Project funded by a Scientific and Technological Innovation Team Project of the Shandong Province Geo-mineral Engineering Exploration Institute (2022JBGS801-17), and by the Department of Natural Resources of Shandong Province (Lu Di Huan 2018-06).

Data Availability Statement: The code supporting this study are currently undergoing processing for archiving and will be made publicly available upon completion.

Conflicts of Interest: The authors declare that this research was conducted in the absence of any commercial or financial relationships that could be construed as potential conflicts of interest.

Appendix A

Table A1. Hyperparameters of the Genetic Algorithm (GA) used in pymoo.

Hyperparameter Category	Parameter Name	Symbol	Value	Description
Algorithm Settings	Population Size	pop_size	50	Number of individuals in each generation.
	Number of Generations	n_gen	100	Maximum number of generations for the optimization run.
Sampling	Sampling Strategy	-	Mixed-Variable Sampling	Custom strategy combining integer (for well coordinates) and real-value (for pumping rates) sampling.
Crossover	Crossover Operator	-	Simulated Binary Crossover (SBX)	Crossover operator used for real-valued variables (pumping rates).
	Crossover Probability	prob	0.9	Probability of crossover occurring between two parents.
	Distribution Index (SBX)	eta	15	Distribution index for SBX, controlling the spread of offspring around parents.
Mutation	Mutation Operator	-	Mixed-Variable Mutation	Custom mutation operator.
	Mutation Probability	prob	0.1	Probability of mutation occurring for each variable.
	Distribution Index (PM)	eta	20	Distribution index for Polynomial Mutation (PM), used for real variables.
Selection	Selection Strategy	-	Tournament Selection	Selection method (implicitly used by pymoo's GA).
	Tournament Size	-	2	Number of individuals participating in each tournament (default for pymoo GA).
Diversity	Eliminate Duplicates	-	True	Flag to enable the removal of duplicate solutions within the population.

References

1. Masoner, J.R.; Cozzarelli, I.M. Spatial and Temporal Migration of a Landfill Leachate Plume in Alluvium. *Water Air Soil Pollut.* **2015**, *226*, 18. [CrossRef]
2. Samlani, N.; Pino, D.S.; Bertolo, R.; Pak, T. A comprehensive dataset of environmentally contaminated sites in the state of São Paulo in Brazil. *Sci. Data* **2024**, *11*, 263. [CrossRef]
3. Baú, D.A.; Mayer, A.S. Optimal design of pump-and-treat systems under uncertain hydraulic conductivity and plume distribution. *J. Contam. Hydrol.* **2008**, *100*, 30–46. [CrossRef] [PubMed]
4. Huang, C.; Mayer, A.S. Pump-and-treat optimization using well locations and pumping rates as decision variables. *Water Resour. Res.* **1997**, *33*, 1001–1012. [CrossRef]
5. Sbai, M.A. Well Rate and Placement for Optimal Groundwater Remediation Design with A Surrogate Model. *Water* **2019**, *11*, 2233. [CrossRef]
6. Ricciardi, K.L.; Pinder, G.F.; Karatzas, G.P. Efficient Groundwater Remediation System Design Subject to Uncertainty Using Robust Optimization. *J. Water Resour. Plan. Manag.* **2007**, *133*, 253–263. [CrossRef]
7. Gernez, S.; Bouchedda, A.; Gloaguen, E.; Paradis, D. Comparison Between Hydraulic Conductivity Anisotropy and Electrical Resistivity Anisotropy from Tomography Inverse Modeling. *Front. Environ. Sci.* **2019**, *7*, 67. [CrossRef]
8. Aissat, R.; Pryet, A.; Saltel, M.; Dupuy, A. Comparison of different pilot point parameterization strategies when measurements are unevenly distributed in space. *Hydrogeol. J.* **2023**, *31*, 2381–2400. [CrossRef]
9. Kitanidis, P.K. *Introduction to Geostatistics: Applications in Hydrogeology*; Cambridge University Press: Cambridge, UK, 1997.
10. Deutsch, C.V.; Journel, A.G. *GSLIB: Geostatistical Software Library and User's Guide*, 2nd ed.; Oxford University Press: Oxford, UK, 1998.

11. Zhang, D.; Lu, Z. An efficient, high-order perturbation approach for flow in random porous media via Karhunen–Loève and polynomial expansions. *J. Comput. Phys.* **2004**, *194*, 773–794. [CrossRef]
12. Yang, H.-Q.; Zhang, L.; Xue, J.; Zhang, J.; Li, X. Unsaturated soil slope characterization with Karhunen–Loève and polynomial chaos via Bayesian approach. *Eng. Comput.* **2019**, *35*, 337–350. [CrossRef]
13. Zhu, B.; Hiraish, T. A decomposed Karhunen–Loève expansion scheme for the discretization of multidimensional random fields in geotechnical variability analysis. *Stoch. Environ. Res. Risk Assess.* **2024**, *38*, 1215–1233. [CrossRef]
14. Ahlfeld, D.; Sawyer, C. Well Location in Capture Zone Design Using Simulation and Optimization Techniques. *Groundwater* **1990**, *28*, 507–512. [CrossRef]
15. Nicklow, J.; Reed, P.; Savic, D.; Dessalegne, T.; Harrell, L.; Chan-Hilton, A.; Karamouz, M.; Minsker, B.; Ostfeld, A.; Singh, A.; et al. State of the Art for Genetic Algorithms and Beyond in Water Resources Planning and Management. *J. Water Resour. Plan. Manag.* **2009**, *136*, 413–432. [CrossRef]
16. Fienen, M.N.; Corson-Dosch, N.T.; Jahn, K.L.; White, J.T. Comparing single and multiple objective constrained optimization algorithms for tuning a groundwater remediation system. *Environ. Model. Softw.* **2024**, *173*, 105952. [CrossRef]
17. McKinney, D.; Lin, M. Genetic algorithm solution of groundwater management models. *Water Resour. Res.* **1994**, *30*, 1897–1906. [CrossRef]
18. Luo, Q.; Wu, J.-F.; Yang, Y.; Qian, J.; Wu, J.-C. Multi-objective optimization of long-term groundwater monitoring network design using a probabilistic Pareto genetic algorithm under uncertainty. *J. Hydrol.* **2016**, *534*, 352–363. [CrossRef]
19. Luo, Q.; Wu, J.-F.; Yang, Y.; Qian, J.; Wu, J.-C. Optimal design of groundwater remediation system using a probabilistic multi-objective fast harmony search algorithm under uncertainty. *J. Hydrol.* **2014**, *519*, 3305–3315. [CrossRef]
20. Wagner, B. Evaluating Data Worth for Ground-Water Management under Uncertainty. *J. Water Resour. Plan. Manag.* **1999**, *225*, 281–288. [CrossRef]
21. Harbaugh, A.W.; Banta, E.R.; Hill, M.C.; McDonald, M.G. *MODFLOW-2000, the U.S. Geological Surveymodular Ground-Water Flow Model-User Guide to Modularization Concepts and the Groundwater Flow Process*; Open-File Report 00-92; U.S. Geological Survey: Reston, VA, USA, 2000.
22. Bakker, M.; Post, V.; Langevin, C.D.; Hughes, J.D.; White, J.T.; Starn, J.J.; Fienen, M.N. Scripting MODFLOW Model Development Using Python and FloPy. *Groundwater* **2016**, *54*, 733–739. [CrossRef] [PubMed]
23. Blank, J.; Deb, K. Pymoo: Multi-Objective Optimization in Python. *IEEE Access* **2020**, *8*, 89497–89509. [CrossRef]

water

MDPI

Article

Integrated Assessment of the Hydrogeochemical and Human Risks of Fluoride and Nitrate in Groundwater Using the RS-GIS Tool: Case Study of the Marginal Ganga Alluvial Plain, India

Dev Sen Gupta [1], Ashwani Raju [2,*], Abhinav Patel [3], Surendra Kumar Chandniha [4], Vaishnavi Sahu [5], Ankit Kumar [2], Amit Kumar [6,*], Rupesh Kumar [7] and Samyah Salem Refadah [8]

1 Defence Geoinformatics Research Establishment, Defence Research & Development Organisation, Sector-37A, Chandigarh 160036, India
2 Remote Sensing & GIS Laboratory, Department of Geology, Institute of Science, Banaras Hindu University, Varanasi 221005, Uttar Pradesh, India; ankitgeo@bhu.ac.in
3 Hydrogeology Laboratory, Department of Geology, Institute of Science, Banaras Hindu University, Varanasi 221005, Uttar Pradesh, India; abhinav.geology17@bhu.ac.in
4 College of Agricultural Engineering and Technology & Research Station, Indira Gandhi Krishi Vishwavidyalaya, Raipur 495334, Chhattisgarh, India; surendra.chandniha@igkv.ac.in
5 Department of Botany, Institute of Science, Banaras Hindu University, Varanasi 221005, Uttar Pradesh, India; vaishusahu682@gmail.com
6 School of Hydrology and Water Resources, Nanjing University of Information Science and Technology, Nanjing 211544, China
7 Jindal Global Business School (JGBS), Jindal Global University, O. P, Sonipat 131029, Haryana, India; rupesh.kumar@jgu.edu.in
8 Department of Geography and GIS, Faculty of Arts and Humanities, King Abdulaziz University, Jeddah 21589, Saudi Arabia; srefadah@kau.edu.sa
* Correspondence: ashwani.geo@bhu.ac.in (A.R.); amitkdah@nuist.edu.cn (A.K.)

Abstract: Groundwater contamination with sub-lethal dissolved contaminants poses significant health risks globally, especially in rural India, where access to safe drinking water remains a critical challenge. This study explores the hydrogeochemical characterization and associated health risks of groundwater from shallow aquifers in the Marginal Ganga Alluvial Plain (MGAP) of northern India. The groundwater chemistry is dominated by Ca-Mg-CO_3 and Ca-Mg-Cl types, where there is dominance of silicate weathering and the ion-exchange processes are responsible for this solute composition in the groundwater. All the ionic species are within the permissible limits of the World Health Organization, except fluoride (F^-) and nitrate (NO_3^-). Geochemical analysis using bivariate relationships and saturation plots attributes the occurrence of F^- to geogenic sources, primarily the chemical weathering of granite-granodiorite, while NO_3^- contaminants are linked to anthropogenic inputs, such as nitrogen-rich fertilizers, in the absence of a large-scale urban environment. Multivariate statistical analyses, including hierarchical cluster analysis and factor analysis, confirm the predominance of geogenic controls, with NO_3^--enriched samples derived from anthropogenic factors. The spatial distribution and probability predictions of F^- and NO_3^- were generated using a non-parametric co-kriging technique approach, aiding in the delineation of contamination hotspots. The integration of the USEPA human health risk assessment methodology with the urbanization index has revealed critical findings, identifying approximately 23% of the study area as being at high risk. This comprehensive approach, which synergizes geospatial analysis and statistical methods, proves to be highly effective in delineating priority zones for health intervention. The results highlight the pressing need for targeted mitigation measures and the implementation of sustainable groundwater management practices at regional, national, and global levels.

Keywords: fluoride and nitrate; Ganga Alluvial Plain; kriging; USEPA approach; mitigation and adaptation

1. Introduction

Groundwater constitutes about 30% of global freshwater resources and is vital for agriculture, drinking water, and industry. India holds 4% of global freshwater but hosts 17% of the world's population, thereby facing severe water scarcity [1]. Over 25% of global groundwater use occurs in India, but excessive extraction has led to issues like depletion, pollution, and saltwater intrusion [2,3]. Per capita water availability, currently at 1486 m^3, is expected to drop below 1140 m^3 by 2050, exacerbating the crisis. Climate change further intensifies this challenge, with groundwater depletion rates projected to triple by 2041–2080, due to rising irrigation demands [4–6]. The contamination of groundwater with fluoride is one of the primary global issues. It has been previously estimated that in 25 nations, a huge percentage of the population, numbering around 200 million people, suffers from fluorosis globally, and the causes behind it have been attributed to the contamination of groundwater with fluoride, including in India [7,8]. All countries are more prone to fluoride exposure due to the consumption of fluoride-rich groundwater. Fluoride is the most electronegative element, which occurs as fluoride ions (F^-), and is commonly associated with natural processes for its release in the groundwater system [9–11]. Fluoride is a basic element in the growth of teeth and bones; the presence of fluoride ensures strength, but in excess, it causes serious health issues.

Fluoride enters the human body through both anthropogenic and geogenic methods, with geogenic contributions being more prominent in India [1,7]. Geogenic sources include the weathering process, leaching, fluoride-bearing mineral dissolution, the exchange of ions, rock–water interactions, and evaporative enrichment [10,11]. The anthropogenic sources, namely, industrial activities, coal burning, and phosphate fertilizers, increase the concentrations of fluoride in groundwater. Sulfate and nitrate are two other potentially hazardous ions found in fertilizer products. Fluoride keeps teeth healthy (the prevention of dental caries, dental enamel construction, and the establishment of bone minerals). Fluoride concentrations below 0.5 mg/L in drinking water can lead to dental issues such as cavities, diminished bone mineralization, and a lack of dental enamel production.

Conversely, concentrations exceeding 1.5 mg/L pose significant health risks, including dental fluorosis and skeletal fluorosis, which can severely impair bone and joint health [12–14]. The fluoride content greatly depends on the mineralogy of the host lithology, which dictates the level of its concentration in groundwater [2,5,15–23]. Fluoride contamination in aquifers arises due to rock erosion and exposure to minerals rich in fluoride. High fluoride concentrations in groundwater are expected in calcium-deficient sodium bicarbonate-type water. The alkaline etiquette of water promotes the conversion of fluorite (CaF_2) to fluoride. The main sources of Fluoride include rock minerals, namely, calcium fluoride [CaF_2], fluorapatite [$Ca_5(PO_4)3F$], and cryolite [Na_3AlF_6], whereas anthropogenic sources contain coal fly ash, petrol refinery waste, enhanced brick kilns dust, and extreme use of chemical fertilizers [23,24].

Fluoride and nitrate are significant ionic constituents in groundwater that influence its suitability for drinking. Fluoride primarily originates from geogenic sources, such as the dissolution of fluoride-rich minerals, while nitrate is often introduced through anthropogenic activities, including the use of agricultural fertilizers, domestic effluents, dairy operations, and nearby waste treatment plants [8,25–28]. According to USEPA, the recommended limit for nitrate in drinking water is 10 mg/L, while the BIS allows up to 45 mg/L, and the WHO sets the maximum at 50 mg/L. Recently, nitrate contamination has become a global issue, affecting most of those Asian and African countries where the uncontrolled use of fertilizers has been observed [29]. Nitrate contamination reduces the pH of water, making it acidic. Its hazardous effects cause methemoglobinemia and the creation of nitrosamine compounds in the human body. Methemoglobinemia disease is a critical condition that may sometimes prove carcinogenic and otherwise hazardous to health; it can be connected to cardiac dysrhythmias, circulatory failure, oxygen deprivation with cyanosis, and central nervous system (CNS)-related effects [27].

The Mahoba district in the Bundelkhand gneissic complex is characterized by granite rock geology, meaning that the weathering of granite rocks leads to the contamination of groundwater with fluoride. In this region, fluoride and nitrate are identified as the primary contaminants, posing substantial risks to human health. The human health risk assessment (HRA) model, as outlined by the USEPA (2014) [30], was employed to evaluate the health impacts of these contaminants via multiple exposure pathways. Geographic information systems (GIS), supported by statistics-based techniques like interpolation and regression analysis, were used to predict contaminant concentrations at unsampled locations. These methods draw upon established techniques [31–37] to generate spatially explicit risk maps. By integrating the predicted contaminant distribution with human census data, the study effectively visualized the spatial extent of health risks [12,13,38,39]. These vivid risk maps highlight hotspots, aiding in targeted mitigation and management strategies for groundwater quality improvement in the region. One tool of Arc GIS, named 'geostatistical analyst', is an extension module that has been used for exploratory data analysis, cross-validation purposes, the semivariogram model, and establishing the distribution pattern of groundwater contaminant concentrations. Kriging is a prediction tool based on statistical analysis that defines the stochastic theory for evaluating the spatial behavior of geographic data [22]. Ordinary kriging (OK) is used to create the spatial distribution of water quality and contaminants in the study area, whereas indicator kriging (IK) is used to calculate the probability of contaminants. The sub-lethal elements occurring at the local and/or regional scales put human health "at risk". HRA evaluates the potential exposure and adverse effects of toxic substances on the population through risk evaluation, exposure assessment, and hazard characterization (IPCS 2004). The globally adopted USEPA (2014) HRA model serves as a framework to assess the impact of contaminants on human health. Researchers worldwide have utilized these policies and guidelines to highlight the human health risks posed by environmental contaminants [40–48].

Fluoride and nitrate contamination in groundwater is a global concern, and the Bundelkhand region faces this issue, along with significant water depletion [33–35,49–51]. The majority of the population relies on shallow, unconfined aquifers for drinking water, exacerbating their vulnerability due to widespread poverty [14,52,53]. Exposure to nitrate and fluoride and their impacts on human health is important to understand in the present context. This research explores the hydrogeochemical evolution of the groundwater samples collected, with special reference to fluoride and nitrate contaminants. Different statistical approaches have been used to decipher the inter-correlation of hydrogeochemical parameters spatially, and a human health risk assessment was carried out using the USEPA approach. Research gaps on the Marginal Ganga Alluvial Plain include inadequate spatial and temporal assessments of groundwater contamination by fluoride and nitrate, particularly under the influence of agricultural and industrial activities. Yet, there has been limited exploration of contamination from combined geogenic and anthropogenic factors. The proper isotopic studies have not been carried out to understand source-related issues in further detail. Additionally, health risk assessments that consider local socio-economic and demographic conditions are scarce, hindering effective mitigation strategies. The present study seeks to develop a comprehensive framework by integrating hydrogeochemical analysis, multivariate statistical techniques, and the urbanization index to address groundwater contamination challenges in the sparsely populated regions of Bundelkhand. Furthermore, the application of non-parametric co-kriging for the spatial prediction of fluoride (F^-) and nitrate (NO_3^-) contamination represents a significant methodological advancement, enabling the precise delineation of high-risk hotspots and enhancing the effectiveness of targeted mitigation strategies. The research objectives are to: (i) investigate the chemical processes influencing groundwater quality, focusing on fluoride and nitrate pollution, (ii) elucidate the patterns of fluoride and nitrate distribution using multivariate statistical approaches and geostatistical modeling, and (iii) quantify exposure risks to the population through the Human Health Risk Assessment framework, applying probabilistic models to predict contaminant impacts effectively.

2. Materials and Methods

2.1. Study Area

Mahoba, a district of Bundelkhand region, Uttar Pradesh, is a part of the Marginal Ganga Alluvial Plain (MGAP), situated between the Indo-Gangetic Plain to the north and the Vindhyan Range to the south (Figure 1). The total geographic area of this district is about 2884 sq. km. Mahoba lies in the geographic location of N 25° 01′ 30″ to N 25° 39′ 40″ and E 79° 15′ 00″ to E 80° 10′ 30″. The normal decadal average rainfall in the monsoon season for Mahoba is 753 mm, whereas decadal non-monsoon rainfall has been recorded at 76 mm [52]. The four area blocks (Panwari, Charkhari, Kabrai, and Jaitpur) are drained by certain rivers, namely, the Dhasan, Urmil, Birma, and Arjun. May has the highest temperatures, shooting up to 50 °C, and January is the coldest month, with a temperature of less than 4 °C [54].

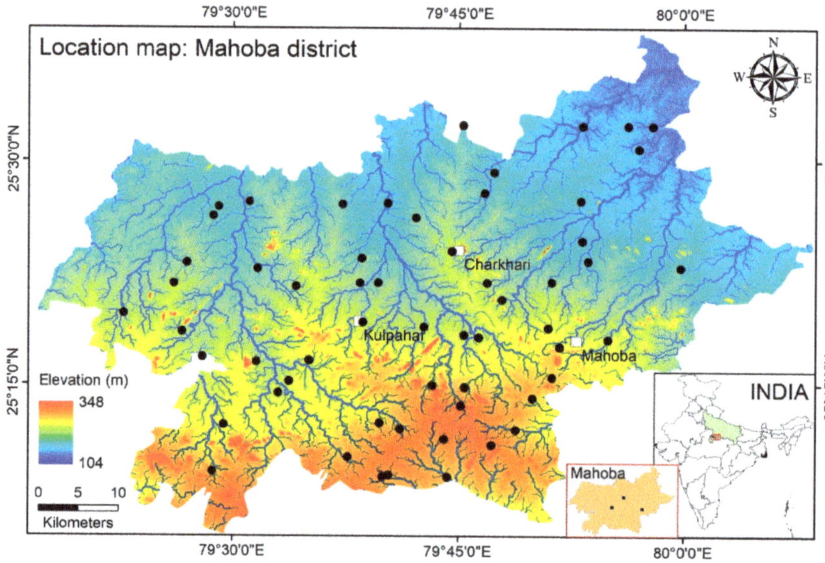

Figure 1. Study area map, with sampling locations (black circle) in the Marginal Ganga Alluvial Plain.

The Mahoba district in the Bundelkhand consists of granite, granodiorite, granitic gneisses, and dolerite, due to the presence of the Bundelkhand Massif Terrain (Figure 2). Granite rocks are a type of intrusive igneous rock composed mostly of quartz, mica, and alkali feldspar. The depth-to-water level in the Mahoba command area for the pre-monsoon season is 8.75 mbgl, while for the post-monsoon season, it is 6.29 mbgl [52]. The groundwater level is low in the Mahoba district because the district mainly comprises the hard rock formation of the Bundelkhand Massif. The rocks are massive; therefore, rainfall does not percolate the rocks [55]. However, some water can percolate through the cracks in the rocks, which helps to keep the water level stable. The occurrence of groundwater in this terrain is extremely uncertain, due to its hilly and rugged terrain. Fluoride, nitrate, sulfate, and calcium are the primary contaminants in the groundwater. Agriculture is the main source of economic activity because a large percentage of the population lives in rural areas. Irrigation practice depends on groundwater (60%) and surface water (40%) [54]. The land use/land cover (LULC) regulates the infiltration behavior of any region, and the LULC map of the study area shows the major cover of agriculture in the region (Figure 3). The shortage of water bodies and streams results in an insufficient amount of water for irrigation throughout the year. Water scarcity, the ingestion of contaminated water, unhygienic sanitation, and several health issues are common problems facing the people of the district.

Additionally, improper solid waste management and landfilling lead to the contamination of groundwater.

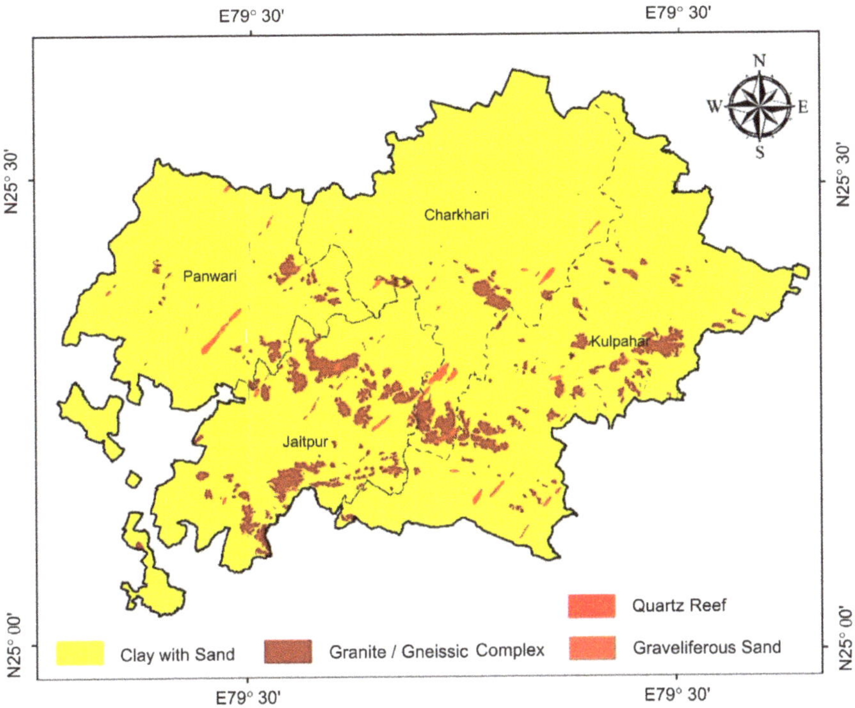

Figure 2. Lithological map of the study area, highlighting the major geology of the region.

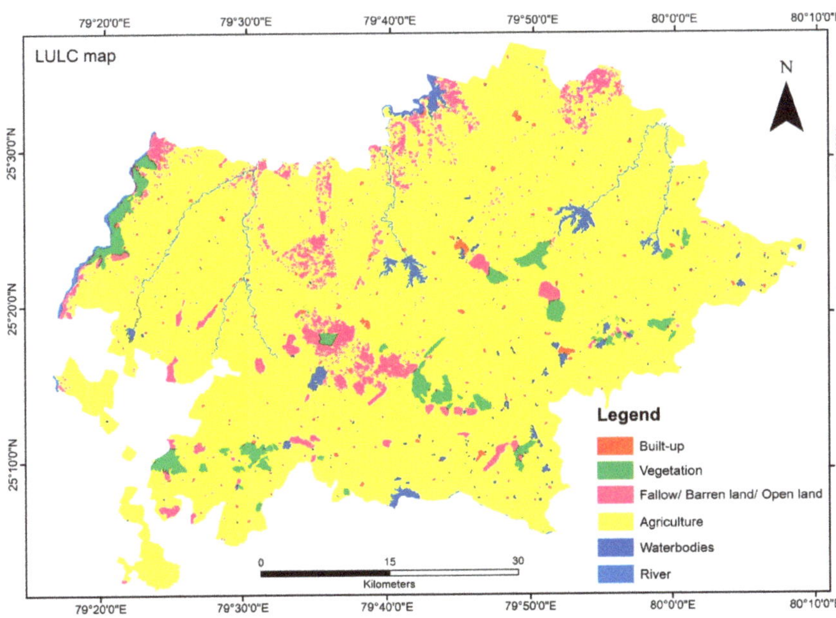

Figure 3. LULC map of the study area, highlighting the major land cover in the region.

2.2. Groundwater Sample Collection

Groundwater sampling ($n = 60$) was planned to be one sample from every 60 grids divided equally, but, in the field, this pattern was modified based on groundwater source availability (see the Supplementary Materials (SM: Table S1)). Two sets of groundwater samples taken from the India Mark-II samples after filtering (with 0.20 μm Millipore membrane syringe filters) were collected in pre-washed HDPE bottles for major ions and trace metals, respectively. To each 100 mL of collected samples, ultra-pure HNO_3 was added for the preservation of heavy metals. On-site parameters such as coordinates, electrical conductivity (μS/cm), and pH (Consort C831) were measured through portable meters [1]. The main cations (Ca^{2+}, K^+, Mg^{2+}, Na^+) and anions (NO_3^-, F^-, Cl^-, SO_4^{2-}) in the laboratory samples were evaluated in an ion chromatograph (model: Dual channel 930 Compact IC Flex ChS/PP) and via ICP-MS for heavy metals (Fe, Cr, Cu, As, Bi, Co, Al, and Pb). The instruments were verified for correct calibration, and the results were then cross-checked using the charge balance error (CBE) within ± 5%. The results showed that many samples exceeded the permissible limits prescribed by the WHO [56] and BIS [57].

2.3. Bivariate and Ternary Plots

To analyze the hydrogeochemical evolution of groundwater and the mechanisms driving the release of ionic species, various molar ratios and bivariate plots were utilized. Classical hydrogeochemical tools, such as Gibbs plots [58], were employed to identify the dominant processes controlling water chemistry, including evaporation, rock–water interactions, and precipitation. Chadha's (1999) [59] rectangular diagram was also applied to classify groundwater types and elucidate the prevailing weathering processes shaping the ionic composition in the study area. These analytical approaches provide insights into the geochemical behavior of groundwater and the factors influencing its quality.

2.4. Multivariate Statistical Approach

The study addresses the complexity of analyzing groundwater quality across multiple sampling sites with numerous hydrochemical parameters, which constitutes a multivariate problem [9,10,35,60]. To simplify this complexity, factor analysis (FA), which is a dimension reduction technique, and hierarchical cluster analysis (HCA) were employed to group those sampling sites with similar underlying hydrochemical characteristics, enabling the classification of groundwater samples based on shared underlying processes. These statistical approaches facilitate the systematic interpretation of hydrogeochemical data and the identification of dominant contamination sources [48,61,62]. Both FA and HCA were normalized and standardized to a z-score [11]. Barlett's test of sphericity and KMO were applied, followed by a varimax rotation for FA, which yielded 5 PCs (eigenvalues > 1). HCA is a very good correlation technique for understanding the factors affecting the hydrogeochemistry seen in this study area. In HCA, multivariate similarity or dissimilarity has been performed using Ward's (1963) [63] method. This technique helps in the clustering of samples, showing interrelation between all the samples using the variance approach [2].

2.5. Geostatistical Modeling Using 'Indication Kriging'

The dissolved elemental concentration of F^- and NO_3^- has been considered herein as a random variable [Z (x_i); iε {1. . .n}] for spatial interpolation. Each random variable has been used to prepare a regional database transformed into a smooth continuous surface using 'kriging'. The output is in the form of a thematic map, showing the spatial distribution of F^- and NO_3^-. Kriging, a geostatistical prediction model, explains the distribution of a random variable (z) in a geographical space (reference). Kriging adopts the regionalized variable theory to determine the spatial continuity of the random variable (z) in a framework of linear model expressed by the function of (a) the structural component spatial trend or mean, [m(x_i)], (b) a spatially autocorrelated random variable (regionalized, [ε'(x_i)]), and (c) random noise (stochastic variation, independent of a location, (ε) (Equation (1)).

$$z(x_i) = m(x_i) + \varepsilon' x_i + \epsilon \quad (1)$$

The spatial autocorrelation $[\varepsilon'(x_i)]$ is the random variable dependency, based on their spatial arrangement $[z(x_i)]$, using semivariance $[\gamma_i(h)]$. Semivariance is half the variance between all the possible random variables (average square difference between the variables, i.e.,$[z(x_i)]$ and $[z(x_i + h)]$, in a space separated by a lag distance (h) (Equation (2)):

$$\gamma(h) = \frac{1}{2n(h)} \sum_{i=1}^{n(h)} \{z(x_i) - z(x_i + h)\}^2 \quad (2)$$

where $\gamma(h)$ is semivariance, $[z(x_i)]$ is the value at the location (x_i) of a known variable, n is the number of pairs of sample point $z(x_i)$, and "h" represents the distance (lag distance) between a pair of points (x_i) and $(x_i + h)$.

Semivariance transforms the random variable into a continuous raster surface by exploring its spatial behavior when distributed in a regional space, using the parameters derived from the semivariogram plot. These parameters, i.e., nugget (C_o), sill (C_o + C_1), and range (C_2), were estimated from the best-fitting semivariogram model. Nugget defines the highest degree of autocorrelation; sill is the point where the best-fit curves level off, and range is the lag distance over which random variables are spatially dependent. The semivariance $[\gamma(h)]$ derived from the best-suited semivariogram model is later substituted into Equation (3) to derive the weights (λ_i).

$$\sum_{i=1}^{n} \lambda_i \gamma(x_i) + \varepsilon = \gamma(x_i) \quad (3)$$

Finally, spatial prediction (z') at an unknown location is estimated from the known variable $[z(x_i)]$ by introducing a weight (λ_i) derived from Equations (3) and (4).

$$z'(x_0) = \sum_{i=1}^{n} \lambda_i z(x_i) \quad (4)$$

However, to obtain probability-based predictions from continuous or categorical data, a more advanced approach called indicator kriging (IK) is used. It transforms the data into binary indicator codes based on the desirable thresholds, thereby minimizing the influence of outliers in the data. Indicator kriging uses a desirable threshold to transform the random variables $[z(x_i)]$ into the indicator codes $[I(x_i) = 0\ or\ 1]$ for spatial autocorrelation. The indicator codes were considered to be 0 if the concentration of the random variable was below the desirable threshold: $[z(x_i) \geq z_{th}]$, otherwise 1. Later, spatial autocorrelation between the indicator codes used semivariance $[\gamma_i(h)]$, derived from the best-fitting semivariogram plot (Equation (5)).

$$\gamma_i(h) = \frac{1}{2n(h)} \sum_{i=1}^{n(h)} \{x(x_i; z_{th}) - I(x_i + h; z_{th})\}^2 \quad (5)$$

Finally, the derived semivariance $[\gamma_i(h)]$ is integrated with a Lagrange multiplier to derive the weights (λ_i), which are later used to generate predicted probability estimates varying in class from 0 to 1, using Equation (6).

$$I'(x_0; z_{th}) = \sum_{i=1}^{n} \lambda_i I(x_0; z_{th}) \quad (6)$$

The suitability of the selected semivariogram model is tested using a cross-validation plot generated between the predicted and measured probability values. In addition, the maximum percentage of indicator values is modeled by the semivariogram, which is calculated by Equation (7); here, $\gamma_i(h)$ is the total semivariance or sill, and $\varepsilon(x_i)$ is an error related to the nugget.

$$Maximum\ Semivariance\ \% = \frac{[\gamma_i(h)] - [\varepsilon(x_i)]}{[\gamma_i\ (h)]} \times 100 \quad (7)$$

2.6. Human Health Risk Assessment Using the USEPA Approach

The study area, which has been identified as water-deficient, faces challenges such as irregular monsoonal rainfall, reduced recharge capacity, declining water levels, and intensified water–rock interactions, leading to elevated contaminant levels in aquifers. Inadequate sanitation and water treatment render the available water unsuitable for drinking, necessitating an assessment of the health risks from exposure to dissolved ions. The human health risk assessment (HRA) evaluates the population risks associated with such exposures. To observe human health at risk, a widely known empirical model defined by USEPA has been used in this article. The empirical model explains the relationship between dose response and human exposure in a contaminated environment. Humans can be exposed to a particular contaminant via three different routes, i.e., the direct ingestion of contaminated drinking water, absorption via the skin, and inhalation. Since the present approach involves the assessment of human health at risk from exposure to dissolved ionic concentrations in the drinking groundwater resources of the Mahoba area, a risk assessment for the direct intake (ingestion) of drinking water has been considered here. As the observed nitrate and fluoride levels are above the prescribed permissible limit, only these two substances have been considered in this USEPA-based empirical model. The derived estimates of potential health risks from exposure to fluoride and nitrate were combined with the human population dataset to determine the spatial extent of the exposure risk in a geographical space. The sequential formulation for USEPA-based human health risk assessment is as follows (Equation (8)):

$$CDI_w = \frac{C_w\,.IR.EF.ED}{BW.AT} \quad (8)$$

where CDI is the exposure dose (mg/kg/day) of the contaminants, Cw is the contamination of groundwater in mg/L, IR is the ingestion groundwater ratio (L/day), EF is the exposure frequency (365 days per year), ED is the exposure length (30 years), BW is a body weight (in kg) of approximately 60 kg, and AT denotes the average period of exposure (365 days).

To calculate the hazard quotient (HQ) using Equation (9), the estimated CDIs for individual components have been combined with the reference dose of a particular contaminant (RD). The values for fluoride were collected from the integrated risk information system, USEPA 2012, and are 0.007 mg/kg/day. When the HQ value is greater than 1, it means that the non-carcinogenic risk is greater than is permissible.

$$HQ_w = \frac{CDI_w}{RfD_w} \quad (9)$$

The hazard index (HI) is used to assess the risk of a specific diversity of interest after exposure to a mixture of chemicals. The HI values of fluoride and nitrate were calculated using Equation (10).

$$HI = HQ_{F-} + HQ_{NO3-} \quad (10)$$

Since the health risk assessment, as derived from the USEPA approach, is closely linked with the human population consuming drinking water, a close approximation of health risk estimates that is integrated with accounts of urbanization would give a precise distribution of human health at risk. The level of urbanization in the study area was assessed by integrating the harmonized night-time light (NTL) data from DMSP-OLS and VIIRS with a gridded population density dataset from CIESIN [19,64]. The NTL data represent urban intensity through night-time light emissions, but features like streetlights, government workplaces, parks, etc., have no population during the night-time. To address this issue, the population density dataset is adjusted by assigning > 1,000,000 administrative unit counts to a unit pixel. However, this approach misrepresents low-population areas near urban centers by equating them with highly populated regions [6]. Thus, a combined analysis of NTL and CIESIN data using Equation (11) is employed for more accurate urban population estimates.

$$UI = \frac{(NTL_{rescale}) \times (PD_{rescale})}{Total\ Area} \tag{11}$$

To assess the human health risk in the studied region, the world gridded population datasets from the University of Columbia (census) were integrated with calculated HI values to create a risk map using Equation (12).

$$Human\ Health\ Risk\ Index\ (HHRI) = \frac{(UI) \times (HI)}{Total\ Area} \tag{12}$$

An overview of the methodology adopted in the present study is shown in Figure 4.

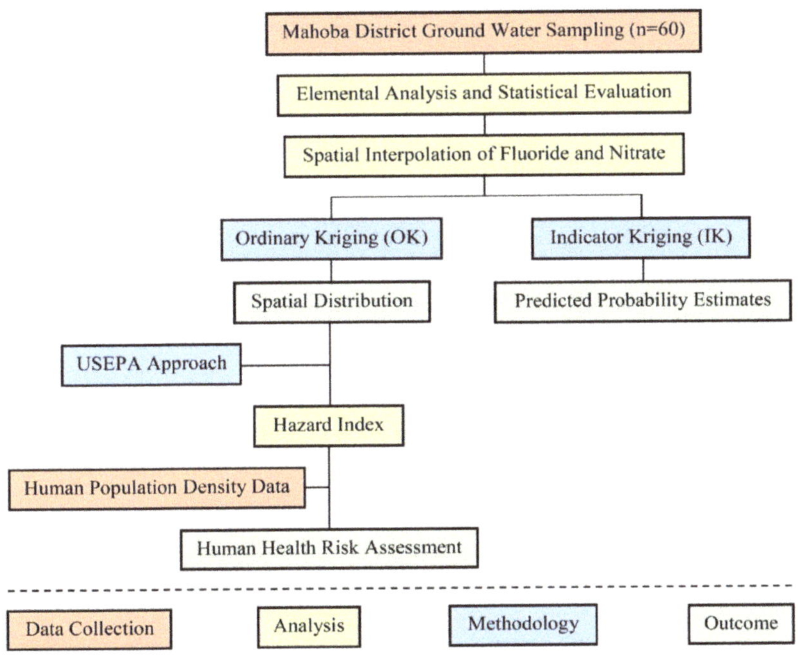

Figure 4. The methodology adopted for this study.

3. Results and Discussion

3.1. Hydro-Geochemistry and Water Type

Most of the groundwater samples appeared to be moderately mineralized and did not appear to vary much in terms of their major ion chemistry across spatial scales, as reflected in their skewness values in Table 1. However, when compared with WHO standard, it is evident that 5%, 26.7%, and 43.3% of the groundwater samples for pH, F^-, and NO_3^- exceeded the recommended levels in water for drinking purposes. The consumption of fluoride-rich groundwater in the long term may lead to various skeletal health ailments, while NO_3^- poisoning leads to blue baby syndrome, which has been reported around the globe. Therefore, it is vital to understand the release mechanism of these inorganic contaminants, which have proved to be fatal to human life and may lead to socioeconomic instabilities. The hydrogeochemical facies analysis of groundwater is an important technique using Chadha's (1999) rectangular plot, by which we understand the groundwater facies as being in four zones. Zone 1 corresponds to Ca-Mg-HCO_3 water types; this type of water facies generally results in recently recharged groundwater water. The most dominating water facies, as per the Chadha rectangular plot (Figure 5), is Ca-Mg-Cl, which describes groundwater with a considerably long residence time, thereby facilitating ion exchange processes in the aquifer system. The peculiar water type associated with the high F^- value is Na-HCO_3, which shows the relationship between the release of F^- and Na^+ under alkaline conditions from the silicate-bearing minerals found abundantly in the study area [65].

Table 1. Descriptive statistics of the various hydrochemical parameters and their comparison with the World Health Organization-recommended limit (WHO 2017).

Parameter	Mean	Min	Max	Std. Dev	Skewness	WHO (2017)	% Exceeding
pH	7.7	6.7	8.9	0.5	0.2	6.5–8.5	5.0
EC (μS/cm)	907.3	628.0	1247.0	131.8	0.0		
CO_3^{2-} (mg/L)	14.6	0.0	41.0	7.3	1.3		
HCO_3^- (mg/L)	206.7	107.0	380.0	58.2	0.3		
F^- (mg/L)	1.4	0.5	3.0	0.5	1.0	1.5	26.7
Cl^- (mg/L)	95.2	19.7	215.0	35.2	1.3		
SO_4^{2-} (mg/L)	59.7	10.0	131.0	15.6	1.0	200–400	
NO_3^- (mg/L)	55.6	20.1	118.1	27.2	0.9	50.0	43.3
Na^+ (mg/L)	70.6	37.5	97.8	16.5	−0.1		
K^+ (mg/L)	1.2	0.0	2.8	0.6	0.3	12.0	
Mg^{2+} (mg/L)	25.6	5.3	40.2	7.6	−0.5		
Ca^{2+} (mg/L)	68.2	41.1	89.9	11.8	−0.1		

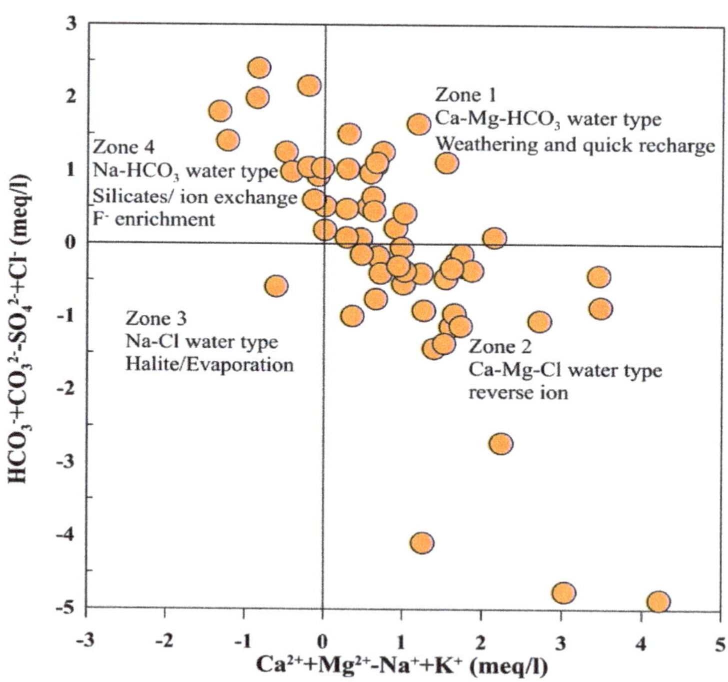

Figure 5. Chadha rectangular plot for various water types.

3.2. Solute Acquisition Processes

The source of solute in groundwater can generally be categorized as geogenic, anthropogenic, or natural-borne. The chemistry of groundwater evolves through various processes, including an interaction between the groundwater and aquifer matrix, human activities, or a combination of these factors [33]. When the alkalinity of groundwater is high, the hydroxyl ion readily replaces the fluoride ion in minerals such as mica, elite, and amphiboles, releasing F- into the groundwater [23]. Groundwater chemistry evolves due to the continuous interactions of recharging water and soil in the aquifer matrix, chemicals from anthropogenic activities, and other factors [33]. Furthermore, many sodium and bicarbonate minerals increase the concentration of fluoride in the groundwater. When calcium-rich groundwater transforms into sodium-rich groundwater, the probability of fluoride solubility increases [23]. Rock-dominant processes contribute the major ion contents to the groundwater samples for the present study. An increase in $Na^+ + K^+/Na^+K^+ + Ca^{2+}$ (Figure 6a,b) is the result of an increase in the concentration of monovalent cations due to the presence of silicate minerals in the aquifer matrix (Equations (13) and (14)).

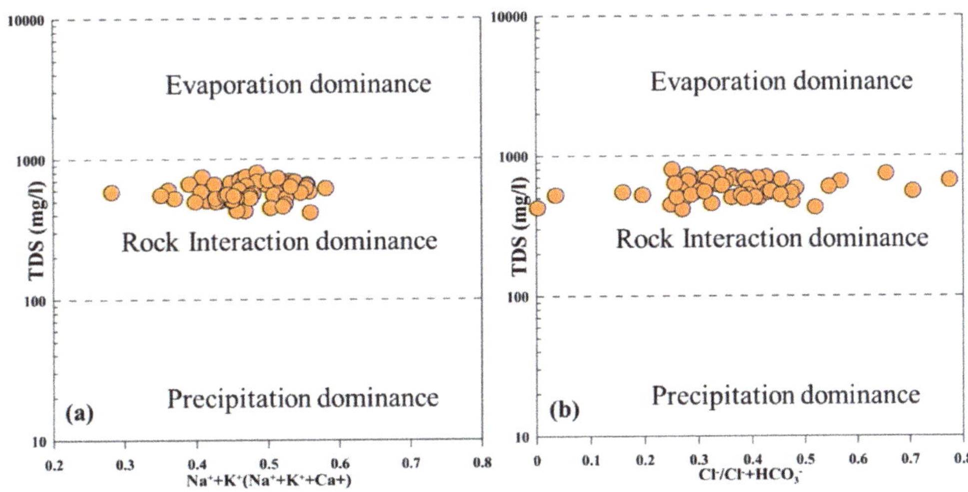

Figure 6. (**a**,**b**): Gibb's plot showing (**a**) the ratio between $(Na^+ + K^+)/(Na^+ + K^+ + Ca^{2+})$ as a function of TDS; (**b**) the ratio between $(Cl^- + NO^{3-})/(Cl^- + NO^{3-} + HCO^{3-})$ as a function of TDS.

The atmospheric input in the solute/salts dissolved in subsurface water can be understood by their ionic ratio to Cl^- [47]. In the present setting, the mean ratios of K^+/Cl^- and Na^+/Cl^- of groundwater are 1.68 and 0.03, respectively, which are far higher than marine aerosols ($Na^+/Cl^- = 0.85$ and $K^+/Cl^- = 0.017$), showing solute derived from host rock rather than atmospheric direct input [1,3,46]. The Na^+/Cl^- level for 86% of the groundwater samples is greater than 1, clearly demonstrating the presence of other sources of Na^+ than halite dissolution (Figure 7a). The molar ratio Ca^{2+}/Mg^{2+} for 28% of the groundwater samples of >2 depicts the importance of silicate weathering. However, the significant samples fall within the molar ratio $(Ca^{2+}/Mg^{2+}) > 1 < 2$, which might be due to the prevalence of Ca-rich silicate minerals or the ion exchange process in the substratum. The contribution of monovalent cations to the total cation can be understood through the $Na^+ + K^+$ vs. total cation process (Figure 7b). A significant amount of the groundwater samples are clustered around the $Na^+ + K^+ = 0.5 \times$ total cation (TZ^+) calculation, identifying the source of these ions ($Na^+ + K^+$) as silicate weathering [66].

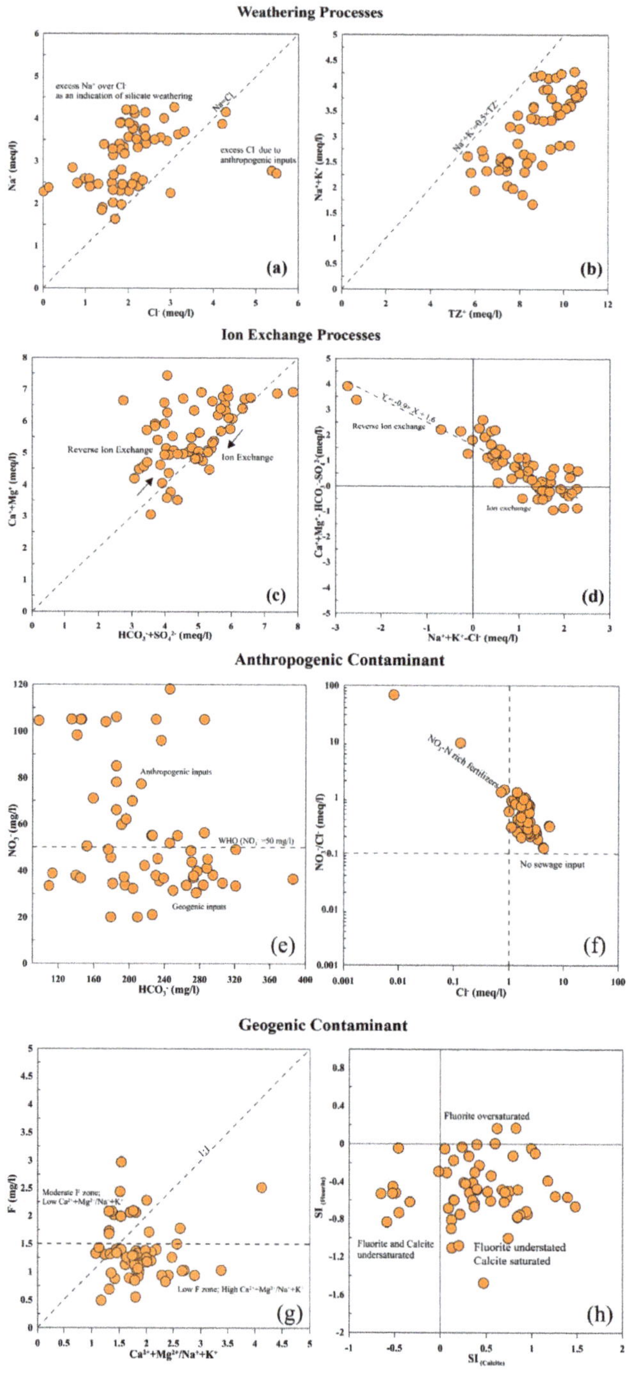

Figure 7. (**a–h**): Bivariate plots demonstrating the solute acquisition process: (**a**) Na^+ vs. Cl^-; (**b**) TZ^+ vs. $Na^+ + K^+$; (**c**) $SO_4{}^{2-}$ + $HCO_3{}^-$ vs. Mg^{2+} + Ca^{2+}; (**d**) Na^+ + K^+-Cl^- vs. Ca^{2+} + Mg^{2+}-HCO^{3-}-$SO_4{}^{2-}$; (**e**) $HCO_3{}^-$ vs. $NO_3{}^-$; (**f**) Cl^- vs. $NO_3{}^-/Cl^-$; (**g**) F^- vs. Ca^{2+} + Mg^{2+}/Na^+ + K^+; (**h**) the saturation index (calcite) vs. the saturation index (fluorite).

Groundwater samples falling below the 1:1 line in the bivariate plot of (Ca^{2+} + Mg^{2+}) vs. (HCO_3^- + SO_4^{2-}) indicate a contribution from sources of non-carbonate origin (Figure 7c), with the excess HCO_3^-/SO_4^{2-} balanced by alkali cations (Na^+ + K^+). Similarly, samples falling above the equi-line show an excess of cations that would be balanced out by Cl^- and other present anions. The molar ratio [(Ca^{2+} + Mg^{2+})/HCO_3^-] value was found to be less than 1, whereas the unbalanced HCO_3^- (left unbalanced by Ca^{2+} + Mg^{2+}) is supposed to be balanced by Na^+ + K^+, which is derived from silicate weathering. The lithology showing carbonate dominance has a higher Ca^{2+}/Na^+ ratio, Mg^{2+}/Na^+ ratio, and HCO3-/Na+ ratio of 50, 10, and 120, respectively, but these ratios abruptly decrease in the silicate end members [67]. In the study area, Ca^{2+}/Na^+ = 1.16, Mg^{2+}/Na^+ = 0.73, and HCO_3^-/Na^+ = 1.14 show a high probability of silicate weathering. Few of the established thermodynamics-equilibrated equations responsible for the elevated concentration of these ions in the groundwater can be understood through Equations (13)–(15).

$$NaAlSi_3O_8 + 20H_2O + 2CO_2(g) \rightarrow Na^+ + 2HCO_3^- + 4H_4SiO_4 + Al_2Si_2O_5(OH)_4 \quad (13)$$

$$KAlSi_3O_8 + 20H_2O + 2CO_2(g) \rightarrow K^+ + 2HCO_3^- + 4H_4SiO_4 + Al_2Si_2O_5(OH)_4 \quad (14)$$

$$Ca_2\left(Mg,Fe\right)_4Al_2Si_7O_{22}\left(OH\right)_2 + 20H_2O + 2CO_2(g) \rightarrow 2Ca^{2+} + 4\left(Mg,Fe\right)^{2+} + 2HCO_3^- + 4H_4SiO_4 + Al_2Si_2O_5(OH)_4 \quad (15)$$

Ion exchange processes involve the interaction of groundwater with clay minerals, which involves either Na+ from the clay going into the groundwater or in a reverse ion exchange, with Na+ from groundwater replacing Ca^{2+} and Mg^{2+} from the aquifer matrix (Figure 7d). The present study area portrays the ion exchange reactions contributing metal cations to the groundwater. Furthermore, Figure 7d shows a negative alignment with a slope of −0.9, confirming the role of exchange reactions. The result also found that 43.3% of the groundwater samples exceeded the recommended value for nitrate (Table 1). Figure 7e displays the groundwater samples rich in NO_3^- but usually low in HCO_3^- concentration, which might be due to being recently recharged through irrigation return. Furthermore, Figure 7f depicts elevated contents of NO_3^-/Cl^- in groundwater samples that have resulted from the use of nitrogen-based fertilizers contaminating the shallow groundwater at the local scale, although less prominent at the regional scale [11]. The F-rich groundwater is associated with low to moderate Ca^{2+} + Mg^{2+}/Na^+ + K^+ under alkaline conditions (Figure 7g). The bivariate plot of the saturation index of calcite $SI_{calcite}$ vs. $SI_{fluorite}$ (Figure 7h) shows most of the groundwater samples falling in the fourth quadrant of the calcite-saturated and fluorite-undersaturated zone. This explains the removal of Ca^{2+} through $CaCO_3$ precipitation and the further dissolution of CaF_2 (fluorite) that is present in the aquifer matrix through the following reaction (Equation (16)).

$$CaF_2 + 2HCO_3^- = CaCO_3 \text{ (precipitate)} + 2F^- + H_2O + CO_2 \quad (16)$$

3.3. Multivariate Statistical Approach

3.3.1. Factor Analysis

The factor analysis extracted five principal components with an eigenvalue > 1 for retention (Figure 8a), after varimax rotation with a total variance of 74% in the dataset (Table 2). The Kaiser–Meyer–Olkin sampling adequacy value for this study was 0.58, and Bartlett's test of sphericity test was < 0.05, which is well suited for the PCA. Factor 1 accounted for 32.3% of the total variance and was positively loaded with Na^+, Ca^{2+}, HCO_3^-, and EC, signifying the control of these ions in mineralization, and, for this region, this can be attributed to the geogenic factor. However, a few of the samples were also loaded with NO_3^-, derived through human-induced activities. Factor 2 contributed to 12.3% of the total variance and was in positive loading with Cl^-, Ca^{2+}, and K^+; a few groundwater samples were highly loaded with NO_3^- concentration for this PC2 (Figure 8b). Factor 3 accounted for 11.5% of the variance, negatively loaded with pH and positively loaded with

Ca^{2+} and SO_4^{2-}, while factors 4 and 5 accounted for 9.4% and 8.5% of the total variance, respectively. It has been noted that both geogenic and anthropogenic factors influence groundwater chemistry; however, based on the factor scores, geogenic causes seem to dominate in the present study area.

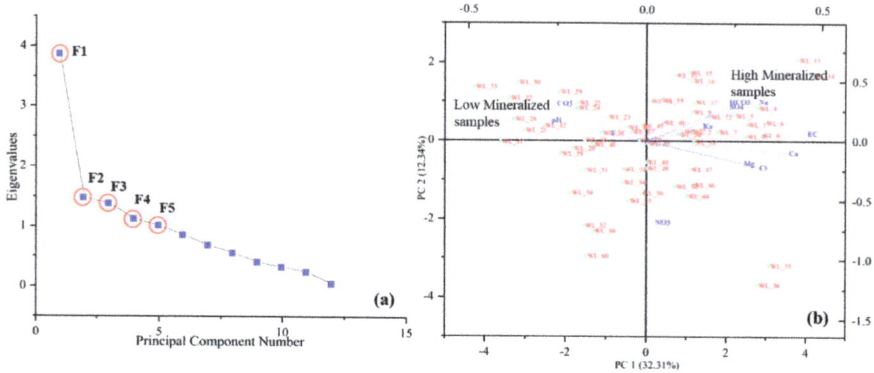

Figure 8. (**a**,**b**): Factor analysis results in (**a**) a scree plot with principal components (PCs) extracted for an eigenvalue greater than 1; (**b**) from the groundwater sample distribution in PC1 and PC2, it is evident that the groundwater samples falling in the right quadrant are comparatively more mineralized.

Table 2. Principal components (PCs) extracted through FA, with an eigenvalue of >1.

Parameter	1	2	3	4	5
pH	0.0	−0.1	−0.7	0.2	0.0
EC (μS/cm)	0.7	0.4	0.4	−0.1	0.3
CO_3^{2-}	−0.4	−0.3	0.0	0.6	0.2
HCO_3^-	0.8	−0.2	0.1	0.0	0.3
F^-	−0.1	−0.1	0.2	0.0	−0.7
Cl^-	0.2	0.9	0.1	−0.2	0.1
SO_4^{2-}	0.2	0.0	0.8	0.3	0.0
NO_3^-	−0.2	−0.1	0.1	−0.8	0.3
Na^+	0.8	0.3	0.1	0.1	−0.1
Ka^+	0.0	0.7	0.1	0.4	0.3
Mg^{2+}	0.1	0.2	0.3	−0.2	0.8
Ca^{2+}	0.4	0.5	0.6	−0.3	0.0
% Variance	32.3	12.3	11.5	9.4	8.5
Cumulative %	32.3	44.7	56.2	65.6	74.0

3.3.2. Hierarchical Cluster Analysis

Hierarchical clustering (HCA) was performed in both the row and column of the computed data matrix, and the data values were transferred to the color scale, which displays how the magnitude of the physicochemical parameters vary across each groundwater sample; HCA clustered them into two groups: cluster 1 and cluster 2. Since most of the variance in the dataset is attributed to Factors 1 and 2, which were then plotted with the samples to check the influence of these factors on the groundwater samples (Figure 8b), two clusters were retained for this study. Apart from these two samples (WL_35 and WL_36), the HCA result aligns with the result of the factor analysis (Figure 9). The row and column of the dendrogram correspond to physicochemical parameters and water samples, respectively. It is evident that the groundwater samples of cluster 1 are more mineralized concerning the HCO_3^-, Mg^{2+}, and Na^+ ions resulting from geogenic processes, as shown in the heatmap (Figure 9). However, a few of the groundwater samples also show the significant loading of NO_3^-, Cl^-, and SO_4^{2-} for cluster 1, resulting from human-induced activities. In the field study, the samples of cluster 1 mostly fall in village/agricultural lands that are affected by nitrate fertilizers. Cluster 2, on the other hand, is less mineralized compared to cluster 1, in terms of the major ions except NO_3^- (Table 3; Figure 9). The sample locations of cluster 2 are further away from agricultural lands as they fall in or around townships, so these groundwater samples reflect lower nitrate contamination than those from cluster 1 locations. It is evident that most of the groundwater samples influenced by Factor 1 (Figure 8b) fall into cluster 1 (Figure 9). Thus, cluster 1 groundwater samples are comparatively more mineralized than those in cluster 2.

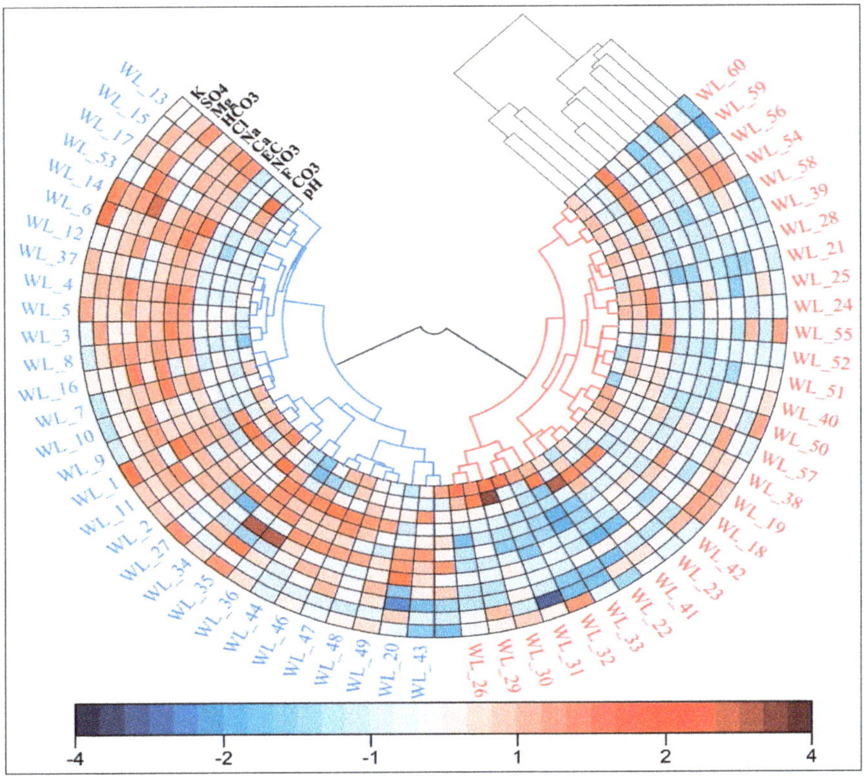

Figure 9. Dendrogram with a heatmap extracted using 12 physicochemical parameters; cluster 1 groundwater samples are more mineralized compared to the Cluster 2 groundwater samples.

Table 3. HCA resulted in two clusters, with their respective descriptive statistics.

	Cluster 1 (n = 31)				Cluster 2 (n = 29)			
Parameter	**Min**	**Max**	**Mean**	**Std Dev**	**Min**	**Max**	**Mean**	**Std Dev**
CO_3^{2-}	6.0	48.0	15.3	8.9	6.0	48.0	19.2	8.9
HCO_3^-	97.0	386.0	249.2	60.8	113.0	289.0	187.2	38.2
F^-	0.6	2.5	1.4	0.5	0.5	3.0	1.4	0.5
Cl^-	60.0	195.0	94.3	35.1	0.3	82.9	54.8	21.1
SO_4^{2-}	10.0	131.0	60.6	18.2	10.0	79.9	56.9	15.5
NO_3^-	30.6	118.1	55.2	27.3	20.1	106.1	55.9	27.4
Na^+	37.5	97.8	81.1	13.6	42.3	89.2	59.4	11.2
K^+	0.1	2.8	1.3	0.6	0.0	2.0	1.1	0.5

Table 3. *Cont.*

Parameter	Cluster 1 (n = 31)				Cluster 2 (n = 29)			
	Min	Max	Mean	Std Dev	Min	Max	Mean	Std Dev
Mg^{2+}	5.3	40.2	27.9	7.4	11.1	36.4	23.2	7.3
Ca^{2+}	56.1	89.9	76.5	8.6	41.1	72.6	59.4	7.4
pH	6.7	8.4	7.6	0.4	6.9	8.9	7.8	0.5
TDS	559.0	802.0	666.9	53.8	415.0	638.0	523.6	54.9
EC (μS/cm)	871.0	1247.0	1008.5	77.8	628.0	981.0	799.2	81.6

3.4. Spatial Distribution of F^- and NO_3^-

The concentrations of fluoride and nitrate have been spatially plotted, showing significant variations within the current study area. In the spatial interpolation process, the dissolved elemental concentrations of nitrate and fluoride were considered as random variables, and, using ordinary kriging, the data were transformed into a continuous surface. The thematic map shows the spatial distribution of fluoride, where the maximum and minimum concentrations in groundwater are 2.57 and 0.29, respectively, with an average value of 0.98 mg/L. The concentrations of fluoride in the various samples (~27%) exceed the permissible limits of the WHO, as has been identified in the Charkhari region. The alkaline nature of water is responsible for anionic exchanges, which may lead to the enhancement of fluoride concentrations in groundwater. The fluoride may be geogenic in origin, because of apatite weathering. The Bundelkhand gneissic complex has granitic rocks; the fluoride comes from the fluorapatite in granitic rocks and from the chemical dissolution of fluoride in groundwater by CaF_2. The hydroxide ion can replace the fluoride ion due to its negative charge and similarity in size. When the alkalinity of groundwater is high, hydroxyl ions readily replace the fluoride ions. About 70% of the total study area contains fluoride at its highest severity, which is in moderate concentrations (Figure 10a; Table 4). The Charkhari block has a relatively higher concentration of fluoride than other blocks like Panwari, Kabrai, and Jaitpur.

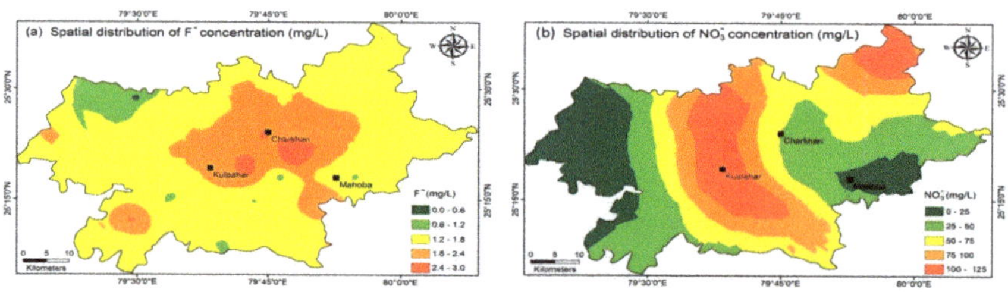

Figure 10. **(a,b)**: Spatial distribution map of F^- and NO_3^- concentration derived using ordinary kriging.

Table 4. Spatial distribution of F^- and NO_3^- from ordinary kriging.

Index	Class	F^-		Class	NO_3^-	
		Area (km²)	Area (%)		Area (km²)	Area (%)
Very Low Concentration	0–0.6	1.24	0.043	0–25	519.41	18.32
Low Concentration	0.6–1.2	161.86	5.613	25–50	793.07	27.97
Moderate Concentration	1.2–1.8	1992.71	69.092	50–75	629.98	22.22
High Concentration	1.8–2.4	648.25	22.48	75–100	471.35	16.63
Very High Concentration	2.4–3	79.94	2.772	100–125	421.22	14.86

The nitrate in groundwater primarily originates from agricultural fertilizers, with additional contributions from nearby waste, treatment plants, dairies, and domestic effluents. In the district, nitrate concentrations range from 20.08 to 118.09 mg/L, with an average of 55.57 mg/L. Spatially, higher concentrations are observed in the central region, following a north–south pattern, gradually decreasing towards the eastern and western boundaries (Figure 10b; Table 4). The Kulpahar and some parts of the Charkhari blocks are more affected by agricultural practices throughout the year. The fertilizer input is the major factor of nitrate contamination, but spatial maps of fluoride and nitrate are somehow oppositely correlated because fluoride is predominantly enhanced in an alkaline medium, whereas nitrate creates an acidic medium. However, the correlation cannot be established perfectly as fluoride is solely geogenically sourced, whereas nitrate is anthropogenically sourced, and their dominance primarily depends on the respective source concentration efficiency of contamination.

3.5. Semivariogram Models and the Predicted Probability Distribution of F^- and NO_3^-

As mentioned, of all the measured variables, the dissolved concentration of F^- and NO_3 was found to be above the WHO 2017 permissible limit. The concentration of F^- and NO_3^- have been considered as random variables and were used to determine the predicted probability of risk-quantified estimates at unsampled locales using indicator kriging (IK). IK is a non-parametric approach that uses a structural component (spatial trend), spatially autocorrelated random variable (auto-correlation), and random noise (stochastic variation) for interpolation. The random variables (F^- and NO_3^-) are transformed into indicator codes, and the spatial structure of the codes is best fitted to the experimental semivariogram model. The best-fit semivariogram model helps estimate the value of semivariogram parameters, i.e., nugget (Co), sill (C0 + C1), and range (C2, Lag distance), which later provides the value of semivariance ($\gamma_{(i)}$). Integrating the weights (λ_i) to the obtained semivariance value resulted in thematic maps depicting the spatial distribution of the projected probability of NO_3 and F. The predicted probability of F^- and NO_3^-, as determined from the semivariogram model, shows variable classes ranging between 0 and 1.

The best-fitting experimental semivariogram model for F^- and NO_3^- is exponential and indicates the nugget (C0) variance as 0. The zero-nugget variance suggests that the random variables (F^- and NO_3^-) that have been transformed to the indicator codes are significantly spatially auto-correlated within the range (C2) of 27.65 and 56.22 km, respectively (Figure 11a,b; Table 5). In addition, a small nugget value also suggests that the indicator kriging method chosen to determine the predicted probability of the random variable performs well. However, the range variation (the lag distance upon which spatially autocorrelated indicator codes are present) given by the exponential semivariogram models of F^- and NO_3^- indicate various sources of F^- and NO_3^-. The maximum value of the semivariance, i.e., the sill (C0 + C1), for a semivariogram model of F^- and NO_3^- are 24.58 and 20.74, respectively, which values are relatively higher than the nugget variance (Figure 11a,b; Table 5). The exponential semivariogram successfully models the random variables defined by the indicator codes, with no anisotropy seen. The prediction error and cross-validation for IK were achieved through the predicted mean error (PMR), root mean square predicted standardized error (RMS-PSE), root mean square error (RMS), and average standard error (ASE) (Figure 12a,b and Table 6). The predicted probability of F^- and NO_3^-, as determined from IK, have been classified into 0.0–0.2 (very low), 0.2–0.4 (low), 0.4–0.6 (moderate), 0.6–0.8 (high), and 0.8–1.0 (very high), based on their predicted concentration spatial distribution (Figure 13a,b) and (Table 7). The results indicated that most of the areas around Kulpahad and Charkhari, covering ~14–16% of central Mahoba district, fall within the high to very high probability class. The areas occupied by the high to very high probability classes of F^- and NO_3^- mostly fall within the rural sector and have negligible human populations, suggesting that the high predicted probability of NO_3^- can be exclusively attributed to the use of fertilizers on farms. The regional groundwater flow in the study area could be the cause of the probability distribution's steady reduction from the southwest to the northeast directions. As discussed earlier, nitrate concentration decreases gradually from the central to the eastern and western ends of the district.

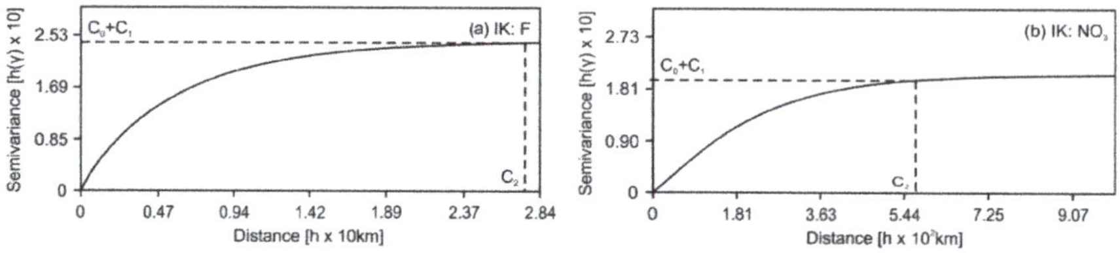

Figure 11. **(a,b)**: Best-fitted experiential semivariogram plot F^- and NO_3^- concentration.

Table 5. Semivariogram parameters (nugget, sill, and range) derived from the best-fitting theoretical model.

Method	Transformed Variables	Semivariogram Model	Threshold	Semivariogram Parameters			Anisotropy
				Nugget (C_0)	Sill ($C_0 + C_1$)	Range (C_2)	
IK	F	Exponential	1.5	0	24.58	27.65	No
IK	NO_3	Exponential	50	0	20.74	56.22	No

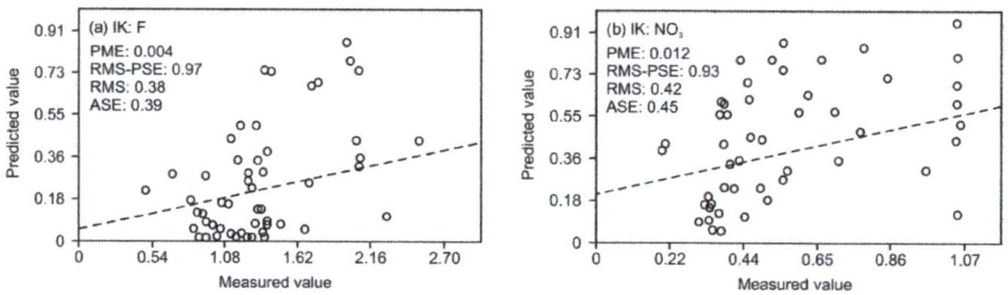

Figure 12. **(a,b)**: Cross-validation plots for IK showing the relationship between the measured and predicted indicator values of F^- and NO_3^- concentration.

Table 6. Prediction error and cross-validation results for IK.

Method	Transformed Variables	Semivariogram Model	Predicted Mean Error (PME)	Root Mean Square Predicted Standardized Error (RMS-PSE)	Root Mean Square Error (RMS)	Average Standard Error (ASE)
IK	F	Exponential	0.004	0.97	0.38	0.39
IK	NO_3	Exponential	0.012	0.93	0.42	0.45

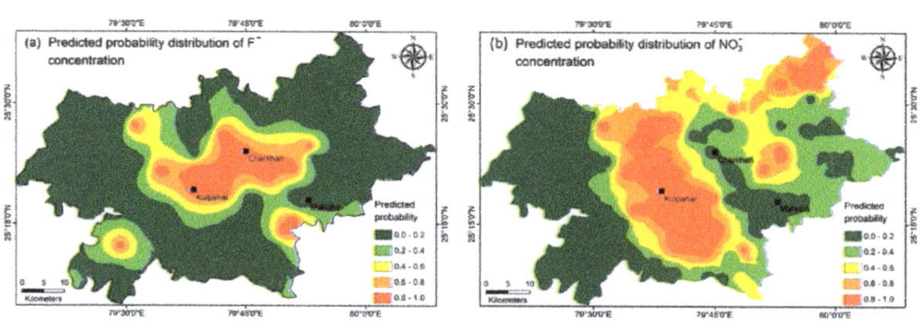

Figure 13. **(a,b)**: Predicted probability distribution maps of F^- and NO_3^- concentrations in drinking groundwater using indicator kriging.

Table 7. Predicted probability distribution of F^- and NO_3^- from IK.

Index	Class	F^-		NO_3^-	
		Area (km^2)	Area (%)	Area (km^2)	Area (%)
Very Low Probability	0–0.2	1811.38	63.89	1047.62	36.95
Low Probability	0.2–0.4	405.35	14.30	604.09	21.31
Moderate Probability	0.4–0.6	238.08	8.40	426.02	15.03
High Probability	0.6–0.8	184.7	6.51	304.38	10.74
Very High Probability	0.8–1.0	195.52	6.90	452.92	15.98

3.6. Health Risks from Exposure to F^- and NO_3^-

The hazard index (HI) denotes the sum fraction of the non-carcinogenic contaminants measured in drinking water and is used to estimate the risk of health effects [30,68]. Without any means of water treatment, people living in rural India are compelled to consume contaminated groundwater for potable use. HI from exposure to combined F^- and NO_3^- in drinking water can be calculated using the USEPA approach. This approach is empirical and combines standard body mass and exposure time, as used in the present study. The USEPA recommends estimating the health risks individually for males, females, and infants [69]. However, as no record of demographic data is available for the study area, the urbanization index (UI) derived from the human population data from the CIESIN and DSMP night-time light data has been used to estimate the human health risk from exposure to F^- and NO_3^- in the drinking water (Figure 14).

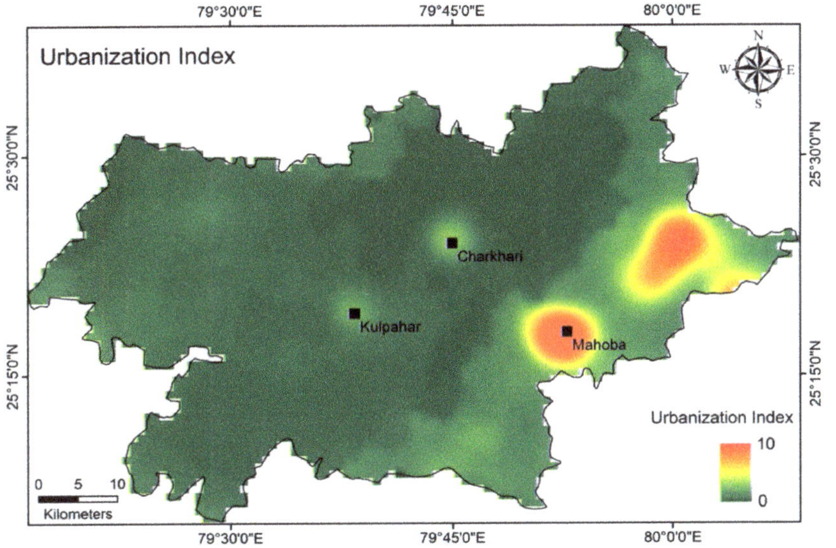

Figure 14. Urbanization index (UI) derived from the integration of DMSP-OLS, VIIRS, and the gridded population density dataset from CIESIN.

The derived HI (Figure 15a) indicates the spatial distribution of risk due to the consumption of F^- and NO_3^--contaminated water. The HI is classified into five classes, from very low to very high. The health risk is potentially alarming in the central parts around Kulpahar and Charkhari, indicating that ~23% of the total area is above high health risk (Figure 13; Table 8). The human health risk index classes correspond to the geographic distribution of the probability of threats to human health. The regional distribution of the HI, as predicted by Equation (10), has been associated with the urbanization index (UI), yielding a composite health risk map (Figure 15b). The health risk map calculates the impacted area and the urban population affected due to contamination by F^- and NO_3^- ions exceeding acceptable levels. The high fluoride and nitrate levels in groundwater have a deleterious influence on human health. The human health risk map shows the distribution of health risks to the human population from exposure to a prolonged intake of water contaminated with F^- and NO_3^-. The derived human health risk map has been classified into five classes from 0–15, representing very low to very high health risks, respectively (Figure 15b; Table 9). The results indicated that the areas lying towards the east and southeast show a high risk to human health from exposure to contaminated drinking water. Since HHRI is closely linked with the spatial distribution of urbanization (high population density in the east and southeast), the health risk from exposure to contaminated drinking water is also high in the east and southeast of the study area. The gradual decrease in the health risk from the south and southeast to the west is due to the negligible population density in most of the populated area.

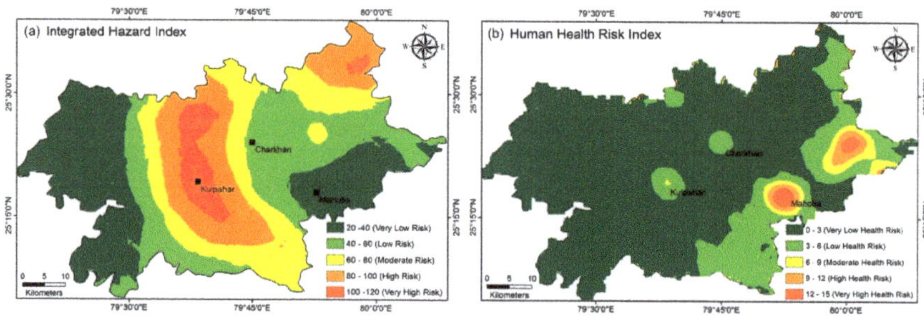

Figure 15. (**a**) The hazard index of F^- and NO_3^-; (**b**) the human health risk index, showing the spatial extent of varying degrees of risk due to F^- and NO_3^- concentrations in the groundwater. Note that the extent of high to very high health risk lies around the urban center of Mahoba.

Table 8. Integrated hazard index (HI), derived from the USEPA 2012 approach.

Class	Index	Area (km^2)	Area (%)
20–40	Very Low Risk	897.24	31.65
40–60	Low Risk	698.34	24.63
60–80	Moderate Risk	594.45	20.97
80–100	High Risk	486.18	17.15
100–120	Very High Risk	158.82	5.60

Table 9. Human health risk index (HHRI) distribution from exposure to F^- and NO_3^-.

Class	Risk Index	Index	Area (km^2)	Area (%)
0–3	I	Very Low Health Risk	2168.43	76.49
3–6	II	Low Health Risk	523.10	18.45
6–9	III	Moderate Health Risk	74.65	2.63
9–12	IV	High Heath Risk	44.32	1.56
12–15	V	Very High Health Risk	24.51	0.86

The findings of this study have significant implications for regions facing similar groundwater contamination challenges. Across India, especially in other parts of the Indo-Gangetic Plain and peninsular regions, geogenic fluoride contamination and nitrate pollution from agricultural runoff are pervasive (Kumar and Singh, 2022). Globally, rural areas in arid and semi-arid zones, such as in Sub-Saharan Africa, parts of China, and Latin America, also encounter groundwater contamination due to analogous geogenic and anthropogenic factors [14,30]. This study's integrated approach, employing geospatial, statistical, and health risk assessment techniques, provides a replicable framework for identifying contamination hotspots and prioritizing areas for intervention. Policymakers and water resource managers can leverage this methodology to design targeted mitigation strategies, fostering sustainable groundwater use and public health safeguards worldwide.

4. Limitations and Constraints

The Mahoba district, representing the present study area, is characterized by gently sloping terrain predominantly composed of granite, granodiorite, granitic gneisses, and dolerite, due to its location within the Bundelkhand Massif Terrain. Consequently, the study area primarily comprises impermeable hard rock formations, where groundwater occurs mainly through fracture filling (secondary porosity), making it widely distributed across the region [70]. Given the hard rock-dominated nature of the region, the groundwater sampling strategy for this study was carefully designed, based on (a) the accessibility of the region, (b) groundwater usage by the locals, and (c) avoiding oversampling and ensuring maximum geological heterogeneity.

One of the primary objectives of this study was to identify the controlling factors influencing groundwater chemistry and its spatial variability, using multivariate statistical approaches and geostatistical modeling. Therefore, the samples were collected in a stratified random sampling manner to minimize the uncertainty in the space. Thus, the total number of samples collected (n = 60) through the random sample collection method not only covered the maximum area (geology), as outlined in Figure 1 of this manuscript, but was also uniformly distributed, making it an excellent choice for any interpolation techniques to be adopted. Also, since the area has uniform aquifer conditions (as represented by rocky aquifers in a granitic complex), the groundwater geochemistry in the study area is the same throughout the subsurface region at different levels. This strengthens the idea of applying the interpolation technique for use at a larger spatial extent.

For any interpolation method, the number of samples to be collected and their distribution in space are critical, as this controls the accuracy of the measurement. In the present approach, the number of samples (n = 60) collected was subject to the availability of India Mark-II hand pumps. However, we made sure that all the samples were collected in a stratified random sampling method and so should have a uniform distribution (i.e., they were collected from all parts of the study area), which was a prerequisite of any interpolation method to be performed.

Each sample was analyzed for fluoride (F^-) and nitrate (NO_3^-), which were later considered random variables. To transform these random variables into a continuous raster surface (spatial distribution), a technique called "Kriging", which is a statistically based "optimal" estimator of spatial variables, was adopted. Kriging uses regionalized variable theory to explain the autocorrelation between random variables through a semivariogram function. The best interpolation method depends on the character and assumptions of data behavior. Therefore, in the present approach, several methods have been explored and selected, based on semivariogram parameters and cross-validation estimates for different kriging methods, showing the relationship between the measured and predicted indicator values of F^- and NO_3^- concentrations. The optimal interpolator (indicator kriging (IK); a co-kriging approach is found most applicable in the present case) was chosen, based on a semivariogram parameter called "nugget" that indicates an initial semivariance when autocorrelation is highest (intercept), or just no uncertainty where the distance is close to 0 (ε"). The best fit of experimental semivariogram model for F^- and NO_3^-) was exponential and indicated a nugget variance of 0 (Figure 11a,b; Table 5). In addition, IK was used to measure the uncertainty or random noise representing stochastic variation independent of sample location. IK allowed the probability-based prediction of F^- and NO_3^-) by transforming them into binary indicator codes, based on thresholds.

5. Conclusions and Ways Forward

The area of the study region, characterized by a semi-arid climate and a gneissic-granitic geological complex, faces critical groundwater quality and quantity challenges.

Groundwater depletion exacerbates strong rock–water interactions, leading to higher concentrations of solutes such as fluoride and contributing to the deterioration of water quality. Detailed hydrogeochemical analysis showed that the primary water types are Na-HCO_3, Ca-HCO_3, Ca-Mg-HCO_3, and Ca-Na-HCO_3. These chemical signatures suggest that processes such as silicate weathering and reverse ion exchange mainly drive geogenic contamination. Agriculture dominates the regional economy, but the thin alluvium cover limits productivity, prompting the frequent usage of chemical fertilizers. This practice directly contributes to elevated nitrate contamination in the groundwater. Statistical analysis through factor analysis extracted five principal components of contamination sources: (1) geogenic mineralization (32.3% of the total variance), (2) anthropogenic sources (12.3%), and (3–5) mixed sources. Hierarchical cluster analysis (HCA) corroborated these findings, linking the high nitrate levels in cluster 1 to agricultural activities, while cluster 2, located away from farmland, exhibited lower contamination. Notably, 30% of the samples had high fluoride levels beyond the WHO-permitted limits. A human health risk assessment was conducted to evaluate non-carcinogenic risks, using the hazard quotient of fluoride and nitrate exposure through drinking water. The urban populations exhibited the highest risk probability, due to higher groundwater reliance and population density. However, the current risk assessment lacked a comprehensive analysis of demographic variables such as age, sex, and rural–urban differences, which could provide a more detailed understanding of the health impacts. Based on the above findings, the following mitigation strategies have been proposed:

1. Implementing rooftop rainwater harvesting and recharge wells can dilute fluoride concentrations, lowering them to permissible levels.
2. Affordable government-subsidized methods can help to mitigate fluoride and nitrate contamination effectively.
3. Promoting hybrid crops, precision irrigation (e.g., sprinklers), and optimal groundwater use can reduce nitrate runoff and preserve groundwater quality.
4. Raising awareness about water quality and conservation can encourage better practices and informed decision-making.

By integrating geospatial and statistical approaches with sustainable practices, these measures aim to address groundwater contamination and ensure long-term water security in the studied region and can be implemented on a national scale.

Supplementary Materials: The following supporting information can be downloaded at: https://www.mdpi.com/article/10.3390/w16243683/s1, Table S1: Detailed geochemical analysis of the groundwater samples (n = 60).

Author Contributions: Conceptualization, D.S.G. and A.R.; methodology, D.S.G., A.P. and A.R.; software, D.S.G., A.P. and A.R.; validation, A.K. (Amit Kumar) and S.K.C.; formal analysis, D.S.G. and A.R.; investigation, A.K. (Amit Kumar), V.S. and A.K. (Ankit Kumar); data curation, D.S.G., A.P. and A.R.; writing—original draft, D.S.G., A.P. and A.R.; writing—review and editing, D.S.G., A.P., A.R., V.S., S.K.C., A.K. (Ankit Kumar), R.K. and S.S.R.; funding acquisition, A.K. (Amit Kumar). All authors have read and agreed to the published version of the manuscript.

Funding: This research received no external funding.

Data Availability Statement: Data used in this study are available on reasonable request from the corresponding author.

Acknowledgments: The authors acknowledge the Institute of Eminence (IoE), Banaras Hindu University (BHU), Defence Geoinformatics Research Establishment (DGRE), and Central Instrument Facility, Indian Institute of Technology—BHU for extending their support to conduct this work.

Conflicts of Interest: The authors declare no conflicts of interest.

References

1. Sen Gupta, D.; Ghosh, P.; Rai, S.P.; Tripathi, S. Hydrogeochemical variability and appraisal of water quality of groundwater in Mahoba district, Uttar Pradesh, India. *Environ. Earth Sci.* **2022**, *81*, 149. [CrossRef]
2. Kumar, S.; Singh, A. Geospatial analysis of groundwater contamination in semi-arid regions: A case study from India. *Environ. Monit. Assess.* **2022**, *194*, 567.
3. Shyam, R.; Krishan, G.; Kheraj Kumar, A. Evaluation of groundwater quality for life-supporting activities: A case study of Haryana, India. *Int. J. River Basin Manag.* **2024**, *22*, 297–308. [CrossRef]
4. Singh, A.; Raju, A.; Kumar, S.; Chandniha, S.K.; Singh, L.; Tyagi, I.; Karri, R.R.; Kumar, A. Hydrogeochemical characterization of groundwater and their associated potential health risks. *Environ. Sci. Pollut. Res.* **2022**, *30*, 14993–15008. [CrossRef]
5. Kimambo, V.; Bhattacharya, P.; Mtalo, F.; Ahmad, A. Fluoride occurrence in groundwater systems at global scale and status of defluoridation-State of the art. *Groundw. Sustain. Dev.* **2019**, *9*, 100223. [CrossRef]
6. Raju, A.; Singh, R.P.; Kannojiya, P.K.; Patel, A.; Singh, S.; Sinha, M. Declining groundwater and its impacts along Ganga riverfronts using combined Sentinel-1, GRACE, water levels, and rainfall data. *Sci. Total Environ.* **2024**, *920*, 170932. [CrossRef]
7. Shaji, E.; Sarath, K.V.; Santosh, M.; Krishnaprasad, P.K.; Arya, B.K.; Babu, S.B. Fluoride contamination in groundwater: A global review of the status, processes, challenges, and remedial measures. *Geosci. Front.* **2024**, *15*, 101734. [CrossRef]
8. Singh, G.; Kumari, B.; Sinam, G.; Kriti Kumar, N.; Mallick, S. Fluoride distribution and contamination in the water, soil and plants continuum and its remedial technologies, an Indian perspective– a review. *Environ. Pollut.* **2018**, *239*, 95–108. [CrossRef] [PubMed]
9. Patel, A.; Rai, S.P.; Akpataku, K.V.; Puthiyottil, N.; Singh, A.K.; Pant, N.; Noble, J. Hydrogeochemical characterization of groundwater in the shallow aquifer system of Middle Ganga Basin, India. *Groundw. Sustain. Dev.* **2023**, *21*, 100934. [CrossRef]
10. Patel, A.; Rai, S.P.; Puthiyottil, N.; Singh, A.K.; Noble, J.; Singh, R.; Hagare, D.; Kumar, U.S.; Rai, N.; Akpataku, K.V. Refining aquifer heterogeneity and understanding groundwater recharge sources in an intensively exploited agrarian dominated region of the Ganga Plain. *Geosci. Front.* **2024**, *15*, 101808. [CrossRef]
11. Patel, A.; Singh, A.K.; Singh, R.; Puthiyottil, N.; Rai, S.P. Fluoride Mobilization and Provenance Identification in Semi-arid Conditions: A Hydrochemical and Isotopic Approach. In *Surface and Groundwater Resources Development and Management in Semi-arid Region: Strategies and Solutions for Sustainable Water Management*; Springer International Publishing: Cham, Switzerland, 2023; pp. 97–116.
12. Ali, S.A.; Ali, U. Hydrochemical characteristics and spatial analysis of groundwater quality in parts of Bundelkhand massif. India. *Appl. Water Sci.* **2018**, *8*, 39. [CrossRef]
13. Arslan, H. Spatial and Temporal Mapping of Groundwater Salinity Using Ordinary Kriging and Indicator Kriging: The Case of Bafra Plain, Turkey. *Agric. Water Manag.* **2012**, *113*, 57–63. [CrossRef]
14. Zhang, W.; Wang, S.; Huang, H. Assessment of groundwater quality and health risks in nitrate-affected regions: Insights from a case study in northern China. *Sci. Total Environ.* **2020**, *740*, 140087.
15. Ahmad, S.; Singh, R.; Arfin, T.; Neeti, K. Fluoride contamination, consequences and removal techniques in water: A review. *Environ. Sci. Adv.* **2022**, *1*, 620–661. [CrossRef]
16. Khan, M.Y.A.; El Kashouty, M.; Gusti, W.; Kumar, A.; Subyani, A.M.; Alshehri, A. Geo-Temporal Signatures of Physicochemical and Heavy Metals Pollution in Groundwater of Khulais Region—Makkah Province, Saudi Arabia. *Front. Environ. Sci.* **2022**, *9*, 800517. [CrossRef]
17. Krishan, G.; Rao, M.S.; Vashisht, R.; Chaudhary, A.; Singh, J.; Kumar, A. Isotopic Assessment of Groundwater Salinity: A Case Study of the Southwest (SW) Region of Punjab, India. *Water* **2022**, *14*, 133. [CrossRef]
18. Laxman Kumar, D. Health risk assessment of nitrate and fluoride toxicity in groundwater contamination in the semi-arid area of Medchal, South India. *Appl. Water Sci.* **2022**, *12*, 11.
19. Li, C.; Gao, X.; Wang, Y. Hydrogeochemistry of high-fluoride groundwater at Yuncheng Basin, northern China. *Sci. Total Environ.* **2015**, *508*, 155–165. [CrossRef] [PubMed]
20. Mankikar, T.Y. Geological controls on fluoride contamination of unconfined aquifer in Mahoba district. *Arab. J. Geosci.* **2022**, *15*, 978. [CrossRef]
21. Raju, A.; Singh, A. Assessment of groundwater quality and mapping human health risk in central Ganga alluvial plain, Northern India. *Environ. Process.* **2017**, *4*, 375–397. [CrossRef]
22. Raju, A.; Singh, A.; Srivastava, N.; Singh, S.; Jigyasu, D.K.; Singh, M. Mapping Human Health Risk by Geostatistical Method: A Case Study of Mercury in Drinking Groundwater Resource of the Central Ganga Alluvial Plain, Northern India. *Environ. Monit. Assess.* **2019**, *191* (Suppl. S2), 298. [CrossRef]
23. Raju, N.J.; Ram, P.; Dey, S. Groundwater quality in the lower Varuna River basin, Varanasi district, Uttar Pradesh, India. *J. Geol. Soc. India* **2009**, *73*, 178–192. [CrossRef]
24. Raju, A.; Singh, A.; Chandniha, S.K. A synergetic approach for quantification and analysis of coal fires in Jharia Coalfield, India. *Phys. Chem. Earth* **2023**, *131*, 103441. [CrossRef]
25. Rawat, U.; Awasthi, A.; Gupta, D.S.; Paul, R.S.; Tripathi, S. Morphometric analysis using remote sensing and GIS techniques in the Bagain river basin, Bundelkhand region, India. *Indian J. Sci. Technol.* **2017**. [CrossRef]
26. Skold, A.; Cosco, D.L.; Klein, R. (2011) Methemoglobinemia: Pathogenesis, Diagnosis, and Management. *South. Med. J.* **2011**, *104*, 757–761. [CrossRef] [PubMed]

27. Toolabi, A.; Bonyadi, Z.; Paydar, M.; Najafpoor, A.A.; Ramavandi, B. Spatial distribution, occurrence, and health risk assessment of nitrate, Fluoride, and arsenic in Bam groundwater resource, Iran. *Groundw. Sustain. Dev.* **2021**, *12*, 100543. [CrossRef]
28. Vikas, D.; Sharma, S. Fluoride contamination in drinking water and associated health risk assessment in the Malwa Belt of Punjab, India. *Environ. Adv.* **2022**, *8*, 100242.
29. Abascal, E.; Gómez-Coma, L.; Ortiz, I.; Ortiz, A. Global diagnosis of nitrate pollution in groundwater and review of removal technologies. *Sci. Total Environ.* **2022**, *810*, 152233. [CrossRef]
30. USEPA. *Human Health Risk Assessment*; United States Environmental Protection Agency: Washington, DC, USA, 2019.
31. Dwivedi, L.; Gupta, D.S.; Tripathi, S. Groundwater potential mapping of Ukmeh river watershed area of Upper Vindhyan region using remote sensing and GIS. *Indian J. Sci. Technol.* **2016**, *9*, 1–7. [CrossRef]
32. Handa, B.K. Geochemistry and genesis of Fluoride containing ground waters in India. *Groundwater* **1975**, *13*, 275–281. [CrossRef]
33. Mukherjee, A.; Fryar, A.E. Deeper groundwater chemistry and geochemical modeling of the arsenic affected western Bengal basin, West Bengal, India. *Appl. Geochem.* **2008**, *23*, 863–894. [CrossRef]
34. Narsimha, A.; Li, P.; Qian, H. Evaluation of groundwater contamination for Fluoride and nitrate in semi-arid region of Nirmal Province, South India: A special emphasis on human health risk assessment (HHRA). *Hum. Ecol. Risk Assess. Int. J.* **2018**, *25*, 1107–1124.
35. Pandey, H.K.; Singh, V.K.; Singh, S.K. Multi-criteria decision making and Dempster-Shafer model–based delineation of groundwater prospect zones from a semi-arid environment. *Environ. Sci. Pollut. Res.* **2022**, *29*, 47740–47758. [CrossRef] [PubMed]
36. Paul, R.S.; Rawat, U.; Gupta, D.S.; Biswas, A.; Tripathi, S.; Ghosh, P. Assessment of groundwater potential zones using multi-criteria evaluation technique of Paisuni river Basin from the combined state of Uttar Pradesh and Madhya Pradesh, India. *Environ. Earth Sci.* **2020**, *7*, 340. [CrossRef]
37. Rose, S. Comparative major ion geochemistry of piedmont streams in the Atlanta, Georgia region: Possible efects of urbanization. *Environ. Geol.* **2002**, *42*, 102–113. [CrossRef]
38. Bajpai, R.; Shukla, V.; Raju, A.; Singh, C.P.; Upreti, D.K. A geostatistical approach to compare metal accumulation pattern by lichens in plain and mountainous regions of northern and central India. *Environ. Earth Sci.* **2022**, *81*, 203. [CrossRef]
39. Bhowmik, A.K.; Alamdar, A.; Katsoyiannis, I.; Shen, H.; Ali, N.; Ali, S.M.; Bokhari, H.; Schafer, R.B.; Eqani, S.A.M.A.S. Mapping human health risks from exposure to trace metal contamination of drinking water sources in Pakistan. *Sci. Total Environ.* **2015**, *538*, 306–316. [CrossRef]
40. Abhay, K.S.; Raj, B.; Tiwari, A.K.; Mahato, M.K. Evaluation of hydrogeochemical processes and groundwater quality in the Jhansi district of Bundelkhand region, India. *Environ. Earth Sci.* **2012**, *70*, 1225–1247. [CrossRef]
41. Adhikary, P.P.; Dash, C.; Bej, R.; Chandrasekharan, H. Indicator and Probability Kriging Methods for Delineating Cu, Fe, and Mn Contamination in Groundwater of Najafgarh Block, Delhi, India. *Environ. Monit. Assess.* **2011**, *176*, 663–676. [CrossRef]
42. Adimalla, N.; Rajitha, S. Spatial distribution and seasonal variation in fluoride enrichment in groundwater and its associated human health risk assessment in Telangana State, South India. *Hum. Ecol. Risk Assess. Int. J.* **2018**, *24*, 2119–2132.
43. Gupta, D.S.; Tripathi, S.; Ghosh, P. A Quantitative Morphometric Analysis of Barhar River Watershed of Mahoba district, U.P.; India using Remote Sensing and GIS. *Indian J. Sci. Technol.* **2017**, *10*, 1–5. [CrossRef]
44. Hounslow, A.W. *Water Quality Data: Analysis and Interpretation*; CRC Lewis Publishers: New York, NY, USA, 1995.
45. Jabrinder, S.; Bhardwaj, P.; Abhishek, A. Health implications among school children due to Fluoride in underground aquifers of Haryana state, India. *Environ. Qual. Manag.* **2022**, *31*, 233–240.
46. Sajil Kumar, P.J. Hydrogeochemical and multivariate statistical appraisal of pollution sources in the groundwater of the lower Bhavani River basin in Tamil Nadu. *Geol. Ecol. Landsc.* **2020**, *4*, 40–51. [CrossRef]
47. Sarin, M.M.; Krishnaswami, S.; Dilli, S.B.L.K.; Moore, W.S. Major ion chemistry the Ganga-Brahmaputra River system: Weathering processes and fuxes of the Bengal. *Geochem. Cosmochim Acta* **1989**, *53*, 997–1009. [CrossRef]
48. Karunanidhi, D.; Aravinthasamy, P.; Subramani, T.; Roy, P.D.; Srinivasamoorthy, K. Risk of fluoride-rich groundwater on human health: Remediation through managed aquifer recharge in a hard rock terrain South India. *Nat. Resour. Res.* **2020**, *29*, 2369–2395. [CrossRef]
49. Devi, V.; Atique, M.M.; Raju, A.; Upreti, G.; Jigyasu, D.K.; Yadav, J.K.; Singh, S.; Kar, R.; Singh, M. Mercury transportation dynamics in the Ganga Alluvial Plain, India: Rainwater-groundwater-river water interaction study from hotspot region. *Int. J. Environ. Sci. Technol.* **2022**, *19*, 4891–4900. [CrossRef]
50. Misra, A.K.; Mishra, A. Study of quaternary aquifers in Ganga Plain, India: Focus on groundwater salinity, Fluoride and fluorosis. *J. Hazard. Mater.* **2007**, *144*, 438–448. [CrossRef]
51. Muhammad, S.; Shah, M.T.; Khan, S. Health risk assessment of heavy metals and their source apportionment in drinking water of Kohistan region, northern Pakistan. *Microchem. J.* **2011**, *98*, 334–343. [CrossRef]
52. Ground Water Department, Uttar Pradesh; Central Ground Water Board, Northern Region. *Dynamic Ground Water Resources of Uttar Pradesh*; Ground Water Department, Uttar Pradesh, and Central Ground Water Board: Lucknow, India, 2021.
53. Zhao, Z.; Kumar, A.; Wang, H. Predicting Arsenic Contamination in Groundwater: A Comparative Analysis of Machine Learning Models in Coastal Floodplains and Inland Basins. *Water* **2024**, *16*, 2291. [CrossRef]
54. Gupta, D.S.; Biswas, A.; Ghosh, P.; Rawat, U.; Tripathi, S. Delineation of groundwater potential zones, groundwater estimation and recharge potentials from Mahoba district of Uttar Pradesh, India. *Int. J. Environ. Sci. Technol.* **2021**, *19*, 12145–12168. [CrossRef]

55. Rahman, A. "Petrology and Geochemistry of Bundelkhand Granites Around Mahoba, Dist Hamirpur UP". Available online: https://core.ac.uk/download/pdf/144517481.pdf (accessed on 18 November 2024).
56. WHO. *Guidelines for Drinking-Water Quality: First Addendum to the Fourth Edition*; World Health Organization: Geneva, Switzerland, 2017.
57. *IS 10500*; Bureau of Indian Standards-Indian Standard Specification for Drinking Water. BIS: New Delhi, India, 2012.
58. Gibbs, R.J. Mechanisms controlling world water chemistry. *Science* **1970**, *170*, 1088–1090. [CrossRef] [PubMed]
59. Chadha, D.K. A proposed new diagram for geochemical classification of natural water and interpretation of chemical data. *Hydrogeol. J.* **1999**, *7*, 431–439. [CrossRef]
60. Cloutier, V.; Lefebvre, R.; Therrien, R.; Savard, M.M. Multivariate statistical analysis of geochemical data as indicative of the hydrogeochemical evolution of groundwater in a sedimentary rock aquifer system. *J. Hydrol.* **2008**, *353*, 294–313. [CrossRef]
61. Kumar, P.; Gupta, D.S.; Rao, K.; Biswas, A.; Ghosh, P. Delineation of groundwater potential zones and its extent of contamination from the hard rock aquifers in west-Bengal, India. *Environ. Res.* **2024**, *249*, 12145–12168. [CrossRef]
62. Kannojiya, P.K.; Raju, A.; Singh, A.; Srivastava, N.; Singh, S.; Singh, M. Health risk assessment from exposure to dissolved trace element concentration in drinking groundwater resources of Central Ganga Alluvial Plain: A case study of Lucknow region. *Urban Water J.* **2023**, *19*, 846–858. [CrossRef]
63. Ward, J.H. Ward's method. *J. Am. Stat. Assoc.* **1963**, *58*, 236–246. [CrossRef]
64. CIESIN. Documentation for the Gridded Population of the World, Version 4 (GPWv4), Revision 11 Data Sets. NASA Socioeconomic Data and Applications Center (SEDAC), Center for International Earth Science Information Network, Columbia University. 2018. Available online: https://data.nasa.gov/w/8nsk-ng4h/default?cur=q9bEB2fiVLf (accessed on 18 November 2024).
65. Piper, A.M. A graphical procedure in the geochemical interpretation of water analysis. *Am. Geophys. Union Trans.* **2007**, *25*, 119–129.
66. Datta, P.S.; Tyagi, S.K. Major ion chemistry of groundwater in Delhi area: Chemical weathering processes and groundwater flow regime. *Geol. Soc. India* **1996**, *47*, 179–188. [CrossRef]
67. Gaillardet, J.; Dupre, B.; Louvat, P.; Allegre, C.J. Global silicate weathering and CO_2 consumption rates deduced from the chemistry of large rivers. *Chem. Geol.* **1999**, *159*, 3–30. [CrossRef]
68. UNEPA. *Guidance for Performing Aggregate Exposure and Risk Assessments*; Office of Pesticide Programs: Washington, DC, USA, 1999.
69. Chen, J.; Wu, H.; Qian, H.; Gao, Y. Assessing nitrate and fluoride contaminants in drinking water and their health risk of rural residents living in a semiarid region of Northwest China. *Expo Health* **2017**, *9*, 183–195. [CrossRef]
70. Basu, A.K. Geology of parts of the Bundelkhand granite massif, Central India. *J. Geol. Soc. India* **1986**, *117*, 61e124.

water

MDPI

Article

Assessment of Groundwater Quality and Vulnerability in the Nakivale Sub-Catchment of the Transboundary Lake Victoria Basin, Uganda

Emmanuel Nabala Hyeroba [1,2,*] and Robert M. Kalin [1,*]

1 Department of Civil and Environmental Engineering, University of Strathclyde, Glasgow G1 1XJ, UK
2 Directorate of Water Resources Management, Ministry of Water and Environment, Kampala 20026, Uganda
* Correspondence: nabala.emmanuel91@gmail.com or nabala.emmanuel@mwe.go.ug (E.N.H.); robert.kalin@strath.ac.uk (R.M.K.); Tel.: +256-783553292 (E.N.H.); +44-7772025478 (R.M.K.)

Abstract: This study evaluates the quality and vulnerability of groundwater within the Nakivale Sub-catchment of the transboundary Lake Victoria Basin in Southwestern Uganda. Groundwater quality assessment focuses on its suitability for both drinking and agricultural uses. Hydrochemical analysis of 19 groundwater samples revealed that 90% comply with World Health Organization drinking water standards, although localized contamination was noted, particularly in terms of total iron, nitrate, potassium, magnesium, and sulfates. The drinking groundwater quality index shows that over 90% of the samples fall within the good-to-excellent quality categories. Elevated nitrate levels and chloride–bromide ratios indicate human impacts, likely due to agricultural runoff and wastewater disposal. For irrigation, Sodium Adsorption Ratio analysis revealed medium-to-high salinity hazards in the region, while Sodium Percentage and other parameters indicated low-to-moderate risks of soil degradation. DRASTIC vulnerability assessments identified low contamination risks due to impermeable geological layers, steep terrain, slow groundwater recharge, deep aquifer depth, and clayey soil cover. These findings emphasize the need for conjunctive water resource management, including improved groundwater quality monitoring, public education on sustainable practices, and protective measures for recharge zones and areas highly susceptible to contamination. By addressing these issues, this study aims to preserve groundwater resources for domestic and agricultural use, ensuring long-term sustainability in the region.

Keywords: groundwater quality; groundwater vulnerability; Nakivale; Victoria basin

1. Introduction

Access to clean and safe water is vital for human survival, economic growth, and sustainable development [1]. In many developing regions in Africa, water scarcity and poor water quality pose significant challenges, hindering development and trapping communities in cycles of poverty [2]. Recognizing the importance of clean water, international frameworks such as the United Nations Sustainable Development Goal (SDG) 6, which strives to ensure the availability and sustainable management of water and sanitation for all, have placed water security at the forefront of global and local agendas [3,4]. SDG 6 emphasizes equitable access to clean water, improved water management, and the protection of water-related ecosystems [1].

One region where clean water supply is critically a challenge is the Nakivale Sub-catchment in Southwestern Uganda [5,6]. This area is located within the Isingiro District and lies within Uganda's cattle corridor [5]. The cattle corridor is known for its water stress, particularly during dry spells, which are marked by declining groundwater levels from boreholes and the drying up of seasonal water bodies [7,8].

This situation greatly affects the socio-economic welfare of the 616,700 people in the area, a third of whom are refugees living in Nakivale refugee camp, one of the oldest in Uganda, hosting refugees from Burundi, the Democratic Republic of Congo, Rwanda, Somalia, Sudan, and South Sudan [9]. With an annual population growth rate projected at 3.6% [10], the challenge of accessing clean water is becoming increasingly urgent [9].

For the riparian communities in Nakivale, access to clean water is not only essential for survival but also a crucial factor in breaking the cycle of poverty [11]. Water is a key resource that supports livelihoods in agriculture, livestock farming, and small-scale industries within the project area [7,9,12]. However, scaled groundwater quality and its vulnerability to contamination are not well understood [13]. Evaluating the quality and susceptibility of groundwater is vital to ensure its sustainable use and to safeguard it for future generations [14].

Previous studies by [15,16] focused on the hydrogeochemical evolution of groundwater resources within the region, offering valuable insights into the factors controlling groundwater hydrochemistry. The latter also provided key information on groundwater recharge, flow, and occurrence within the study area. The authors of one study [6] conducted a national groundwater resources availability and demand assessment, which involved an evaluation of groundwater quality in various catchments, including the Rwizi catchment, where the project area is located. This study highlighted emerging groundwater quality issues, such as nitrate contamination in urban areas of Mbarara district, situated upstream of the project area. The authors of [5,17] identified growing water quality concerns in the region, driven by a rapidly increasing population and urbanization. Other studies [5,18,19] also provided important insights into contamination challenges within regional hydrological systems, with a major focus on surface water.

Despite the valuable contributions from these studies, none have specifically focused on scaled assessment of groundwater quality and vulnerability within the Nakivale catchment. Localized assessments are crucial due to the inherent heterogeneity of the underlying hydrological systems [20,21]. Groundwater chemistry can vary significantly over small spatial scales, influenced by local geological, biological, and chemical factors [22]. These assessments account for these intrinsic variations, which are essential for accurately understanding groundwater dynamics that broader assessments might overlook [23]. The groundwater quality and vulnerability assessment is of paramount importance in determining the current status of groundwater resources and the risks they face [24].

Groundwater quality, particularly in regions with limited alternative water sources, is often assessed for its intended purpose, such as drinking and irrigation suitability [25–29]. This study employed a comprehensive approach using the Groundwater Quality Index (GWQI), which consolidates various chemical and physical water parameters into a single value to facilitate a clear assessment of water quality suitability for human consumption [25–33]. The GWQI method was selected due to its ability to integrate multiple water quality parameters [29,30,34,35] and enable targeted comparisons across regions [26,30]. For irrigation, hydrochemical indices such as Sodium Adsorption Ratio (SAR), Kelly's Ratio (KR), Sodium Percentage (SP), Magnesium Adsorption Ratio, Residual Sodium Carbonate (RSC), among others, provided insight into possible impacts on soil health, crop yield, and the risk of soil degradation [28,31,33,36], all of which affect agricultural productivity over time [28].

Complementing quality assessments, groundwater vulnerability assessments determine how susceptible aquifers are to contamination from surface sources [37]. This is especially relevant in Nakivale, where agricultural runoff and land use practices may threaten groundwater resources [7]. For this purpose, the DRASTIC model was chosen for its established capability to assess aquifer vulnerability by incorporating essential hydrogeological parameters that directly influence contaminant percolation, including depth to water, net recharge, aquifer media, soil media, topography, impact of the vadose zone, and hydraulic conductivity [23,34,38,39]. By analyzing these specific factors, the DRASTIC model effectively estimates the likelihood of contaminant migration through soils and aquifers [38,39], providing a robust measure of groundwater susceptibility to surface-based contamination sources [14,24,34,35,37–42].

This study serves as an initial investigation, aiming to establish a baseline understanding of groundwater quality, potential contamination threats, and vulnerability of this vital resource within the Nakivale Sub-catchment. The findings are intended to guide policymakers, local authorities, and stakeholders in implementing targeted interventions for the sustainable management of groundwater resources in both the study area and the broader Victoria basin. Additionally, this research provides a foundational reference for future studies focused on assessing groundwater quality and vulnerability across the region.

The specific objectives of this study are to assess the current groundwater quality for both drinking and agricultural usage in the Nakivale Sub-catchment by analyzing key physio-chemical parameters such as major ions, trace elements, minor ions, nutrients concentrations, and hydrochemical field parameters; to identify key sources of groundwater contamination; to evaluate the vulnerability of groundwater resources to contaminants emerging from the surface; and to provide recommendations for the sustainable management of groundwater resources, with a focus on enhancing groundwater resilience to climate variability and population growth in the Nakivale riparian communities.

2. Materials and Methods

2.1. Study Area

The study area is located between latitudes (9,900,000 N; 9,930,000 N) and longitudes (230,000 E; 280,000 E) within the Isingiro District in Southwestern Uganda (Figure 1). It spans 760 km^2 and is part of the Rwizi catchment, which falls within the greater Victoria Basin [5,6,13]. The landscape is predominantly composed of remnants of lowland surfaces, which account for over 60% of the sub-catchment. These lowland surfaces are significant for groundwater discharge within the region [16].

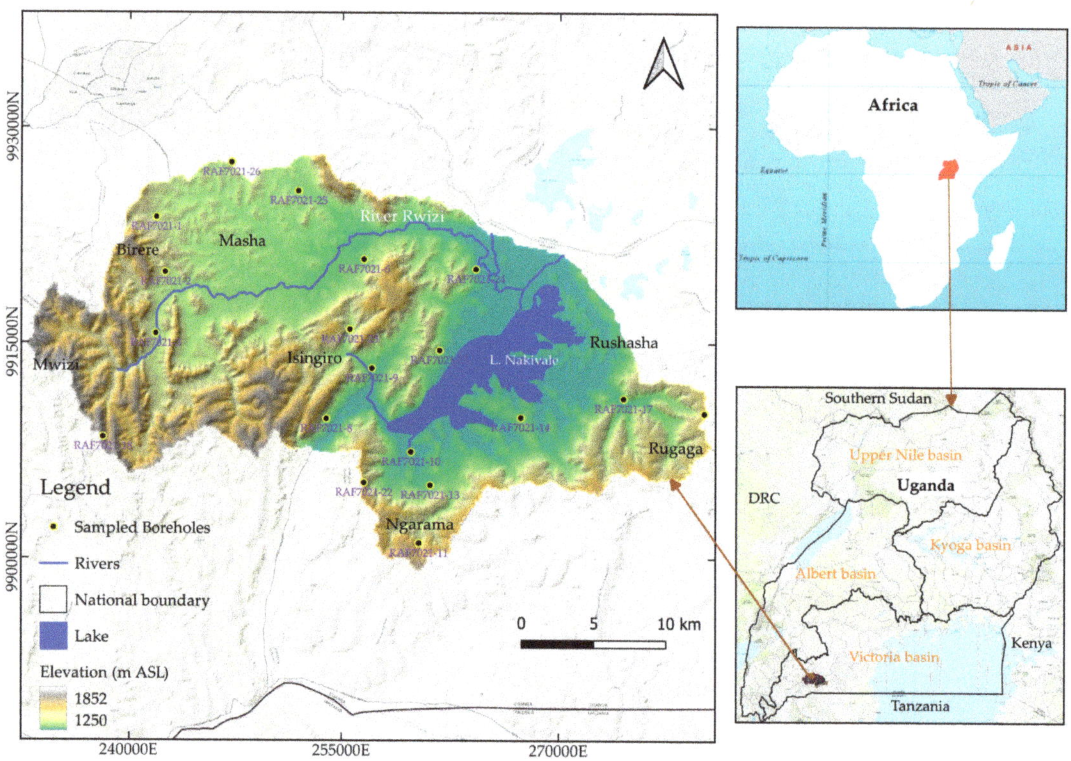

Figure 1. Location map of Nakivale Sub-catchment within Southwestern Uganda.

Infill areas are found in the western and eastern parts of the sub-catchment and are characterized by poorly sorted outwash fans that impact groundwater infiltration and quality due to the heterogeneity of the sediments. High-relief areas are located along the extreme margins of the catchment and are covered by remnants of upland surfaces, which are prone to erosion. According to [16], these elevated areas are zones of groundwater recharge and are associated with high hydraulic head gradients relative to the lowlands.

The riparian communities primarily engage in livestock farming and banana plantations [11]. These social-economic activities are heavily reliant on groundwater resources [5,13]. The local communities tend to settle on the lower flanks of the area in nucleated settlement patterns [7]. The lower lands are rich in thick, loamy soils, which support agricultural productivity [7]. Given the area's reliance on groundwater for both agriculture and domestic purposes [11], understanding its vulnerability to degradation from both geogenic and anthropogenic effects is of paramount importance [24,37].

This understanding is critical for ensuring the protection and sustainable use of groundwater resources in the Nakivale Sub-catchment. Groundwater of good quality is not only essential for human survival [5], but it is also a cornerstone of economic development within the study area [26]. Safeguarding these water resources is, therefore, crucial for maintaining agricultural productivity, supporting livestock, and ensuring the well-being of the communities that depend on them [11,17,24].

2.2. Geological and Hydrogeological Setting

The study area is overlain by the Proterozoic Karagwe-Ankolean system of metasedimentary rocks that exhibit varying degrees of metamorphism (Figure 2). These rocks include shales, slates, and phyllites. These host rocks were intruded by younger granitic rocks of variable mineral compositions [43]. The area's hydrogeology is heavily influenced by these geological formations, which play a crucial role in determining groundwater recharge, flow, storage, and quality [5,6,13]. The predominant aquifer in the region is the weathered and fractured aquifer [5,6], typically found at depths greater than 60 m above sea level [44]. Additionally, other aquifer types include paleo-channels and sedimentary alluvial aquifers [5,6], both of which also contribute to the local groundwater system [44,45]. According to [6], the hydraulic conductivity (K) and specific storage (Ss) values for the fracture domain are 1.7×10^{-2} m/day and 3.76×10^{-2} m^{-1}, respectively, while for the matrix domain, these values are 1.4×10^{-7} m/day and 4.37×10^{-4} m^{-1}. The same study reports that annual recharge in the area ranges between 50 and 100 mm.

Over 80% of the area is covered by fine-grained clayey soils formed from the weathering of in situ argillaceous rocks [16]. These soils are associated with poor hydraulic properties [46], which significantly hinder groundwater flow and storage processes [47]. Active groundwater recharge in the Nakivale Sub-catchment area predominantly occurs in elevated areas that are characterized by a network of lineaments [16], which enhance water infiltration. In contrast, groundwater discharge zones are primarily located in the lowlands, particularly in areas occupied by Lake Nakivale and the Rwizi River (Figure 2). This distribution of recharge and discharge areas underscores the importance of the region's geological features in shaping the flow and storage processes of the underlying local aquifer systems.

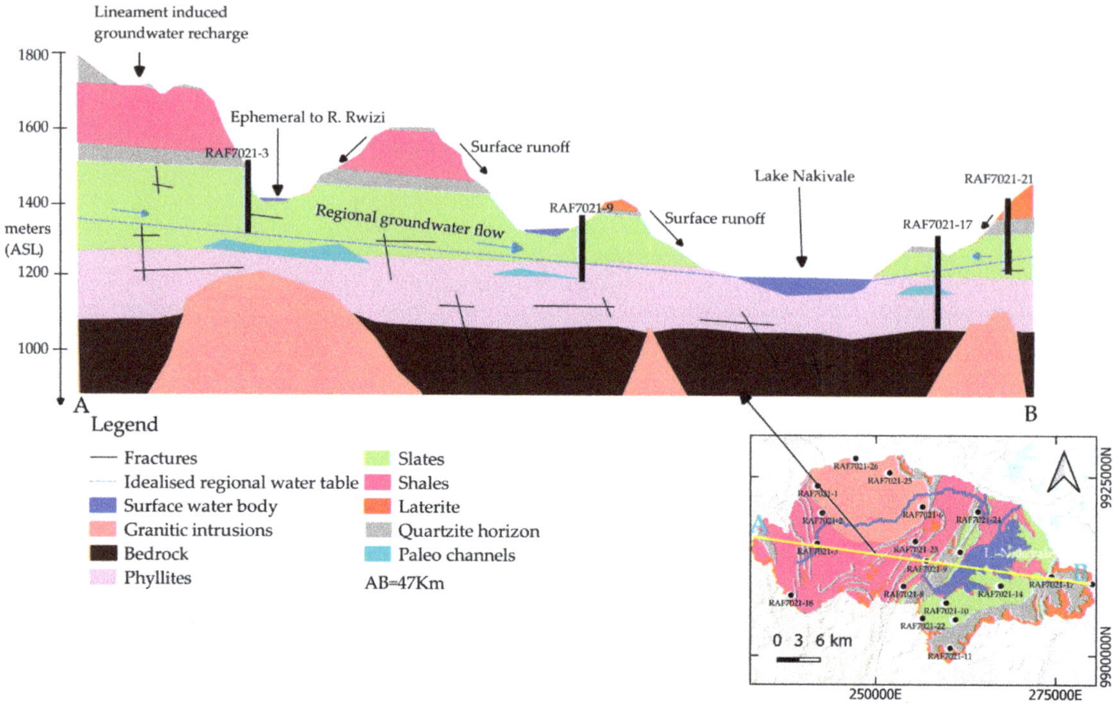

Figure 2. Hydrogeological cross-section between points A and B for Nakivale Sub-catchment, Southwestern Uganda.

2.3. Climate

This area experiences a tropical climate with two distinct rainy seasons: March–April–May and September–October–November, with annual rainfall ranging from 966 mm to 1380 mm [5,13]. These seasonal rainfall patterns are primarily influenced by the movement of the inter-tropical convergence zone (ITCZ), a low-pressure belt [6,16]. The position of the ITCZ in relation to the study area has a significant effect on local seasonal variations [48]. The region also experiences a mean monthly temperature of 27 °C and a mean monthly dew point of 19 °C [5]. These meteorological conditions play a pivotal role in shaping the local hydrological system, as they influence the amount of hydrological input [49], which directly affects both surface and groundwater quantifiable attributes [13]. Rainfall largely determines water availability, while temperatures and dew points affect evapotranspiration rates [50], influencing how much water is lost to the atmosphere [51]. These factors together shape the overall water balance in the area, directly impacting the quantity of water stored in surface reservoirs and subsurface aquifers [52].

2.4. Groundwater Sample Collection and Analysis

This study is based on a total of 19 groundwater samples collected from strategically selected borehole locations to capture variations in geology, land use, geomorphology, and hydrology, ensuring that they represent the intrinsic heterogeneity of the area, as emphasized by [16]. According to the same study, sampling occurred during the dry season, from 15 February to 20 February 2024. The geographic coordinates of each borehole were recorded, and groundwater levels were measured using a dip meter to accurately assess the water table. To guarantee representative samples, each borehole was purged for five minutes prior to sample collection. Water samples were collected in sterile containers to maintain their chemical integrity. Field measurements of key parameters, including dissolved oxygen, pH, temperature, and electrical conductivity, were carried out using a calibrated HANNATM multi-parameter meter (HI 9829_S/N 07100011101) (Hanna Instruments, Woonsocket, RI, USA). Total alkalinity was determined through acid-base titration using 0.02 M hydrochloric acid.

For laboratory analysis, samples for cation and anion measurements were collected in tightly sealed 500 mL HDPE bottles. The cation samples were filtered using GF/C filter papers and acidified with concentrated nitric acid (analytical grade) to reduce the pH to less than 2, ensuring their stability for transport. These samples were then shipped to the Centre National de l'Energie des Sciences et des Techniques Nucléaires (CNESTEN) chemical laboratory in Morocco under the IAEA RAF7021 project for hydrochemical analysis.

The major ions in the water samples were analyzed in mg/L using the ion chromatography method. These ions included Sodium (Na^+), Potassium (K^+), Calcium (Ca^{2+}), Magnesium (Mg^{2+}), Chloride (Cl^-), Nitrate (NO_3^-), and Sulfate (SO_4^{2-}). The bicarbonate concentration was determined through titration against a standardized acid, with the endpoint of the reaction identified using a pH indicator. Additionally, the minor ions, Aluminium (Al), Iron (Fe), Silicon (Si), and Boron (B), were analyzed in their elemental forms using Inductively Coupled Plasma Mass Spectrometry (ICP-MS). This method allows for highly sensitive and precise measurements of elements at very low concentrations [53].

A robust quality control (QC) and quality assurance (QA) system was implemented to ensure the integrity and reliability of hydrochemical analyses. Both the sampling and laboratory hydrochemical analysis process strictly followed the IAEA Groundwater Sampling Procedures, ensuring that each sample was properly collected, labeled, and stored to prevent any contamination or alteration of its hydrochemical composition. Charge balance errors (CBE) were calculated for each sample as a critical QC measure. This was performed using (Equation (1)), with cations and anions expressed in meq/L. This process ensured that the ionic concentrations were correctly measured and balanced within acceptable limits, typically less than ±5%, considered acceptable or 5–15%, and treated with caution [54].

$$\text{CBE} = \left(\frac{\sum \text{Cations} - \sum \text{Anions}}{\sum \text{Cations} + \sum \text{Anions}}\right) \times 100 \quad (1)$$

2.5. *Groundwater Quality Assessment*

The assessment employed a twofold methodological approach, integrating spatial analysis using Quantum Geographic Information System (QGIS) Version 3.36.3 with an index-based assessment of groundwater quality.

2.5.1. Spatial Analysis

The spatial analysis involved generating groundwater quality maps for both major and minor ions. Additionally, it included spatial mapping of hydrochemical parameters, groundwater quality indices, and the chloride–bromide ratio. The maps were created using the Inverse Distance Weighting (IDW) interpolation method, a deterministic technique that assigns greater influence to nearby points based on distance, making it particularly effective for estimating values in areas with irregularly spaced data [55]. IDW is well-suited for groundwater quality mapping of sparse data as it does not rely on spatial autocorrelation, which requires evenly distributed data [29,56,57]. However, due to the complex and unpredictable nature of many underlying groundwater systems, it is essential to interpret the results cautiously, recognizing the basic principles inherent to the IDW method.

2.5.2. Drinking Groundwater Quality Index Assessment

Drinking groundwater quality was assessed through the computation of the groundwater quality index for each groundwater sample. This method has been utilized by various researchers to assess groundwater quality based on its hydrochemical composition [25,26,30]. The approach involved assigning weights w_i on a scale from 1 to 5 to each analyzed chemical species deemed significant for groundwater potability needs. After assigning weights (w_i), the relative weight (W_i) for each parameter was computed using the formula below (Equation (2)).

$$W_i = \frac{w_i}{\sum_i^n w_i} \quad (2)$$

The rating scale for each parameter, denoted by (q_i), was subsequently calculated by dividing the concentration (C_i) in mg/L of each parameter by the World Health (WHO) standard value (S_i) in mg/L for portable water (Equation (3)).

$$q_i = \frac{C_i}{S_i} \times 100 \quad (3)$$

The relative weights (W_i) assigned to each parameter, along with their corresponding rating scales (q_i), were then multiplied to derive the sub-index (SI_i) for each parameter.

$$SI_i = W_i \times q_i \quad (4)$$

Finally, the groundwater quality index (GWQI) for each sample was calculated by summing the sub-indices of each parameter for a given water sample, as indicated by Equation (5).

$$GWQI = \sum_i^n SI_i \quad (5)$$

The Table 1 shows the water quality index range and the corresponding water quality classification for each sample.

Table 1. Groundwater quality index range and class [25,30,58].

Groundwater Quality Index Range	Groundwater Quality Class
<50	Excellent Groundwater
50–100	Good Groundwater
100–200	Poor Groundwater
200–300	Very Poor Groundwater
>300	Unsuitable Groundwater

2.5.3. Irrigation Groundwater Quality Assessment

This assessment centered on analyzing key hydrochemical indices that determine the suitability of water for agricultural use. The parameters evaluated included Electrical Conductivity (EC), Sodium Adsorption Ratio (SAR), Kelly's Ratio (KR), Magnesium Adsorption Ratio (MAR), Sodium Percentage (SP), and Residual Sodium Carbonate (RSC). These indices were examined to assess risks related to soil permeability reduction, nutrient imbalance, and the potential buildup of salinity and alkalinity. Each parameter was classified according to established thresholds (Table 2), facilitating a comprehensive evaluation of groundwater quality for irrigation while highlighting both the individual and combined effects of the indices on soil health and crop productivity. This irrigation water quality indexing assessment approach has been widely recognized and utilized globally to assess the suitability of water resources for irrigation [28,31,33,36,58–61]. The following mathematical models were used to calculate the respective groundwater quality indices for irrigation, with each hydrochemical species expressed in meq/L:

$$SAR = \frac{Na^+}{\sqrt{\frac{Ca^{2+} + Mg^{2+}}{2}}} \quad (6)$$

$$MAR = \frac{Mg^{2+}}{Ca^{2+} Mg^{2+}} \times 100 \quad (7)$$

$$KR = \frac{Na^+}{Ca^{2+} + Mg^{2+}} \quad (8)$$

$$SP = \frac{Na^+ + K^+}{Ca^{2+} + Mg^{2+} + Na^+ + K^+} \times 100 \quad (9)$$

$$RSC = (HCO_3^- + CO_3^{2-}) - (Ca^{2+} Mg^{2+}) \quad (10)$$

Table 2. Classification ranges for the assessed irrigation groundwater quality indices.

Index	Classification Range	Class
KR	<1 >1	Safe Unsuitable
MAR (%)	<50 >50	Suitable Unsuitable
SAR	<10 10–18 18–26 >26	Excellent Good Marginal Poor
SP (%)	<20 20–40 40–60 60–80 >80	Excellent Good Permissible Doubtful Unsuitable
RSC (meq/L)	<0 0–1 1–2.5 >2.5	Low Medium High Very high

2.6. Groundwater Vulnerability Assessment

This study employed the DRASTIC assessment approach to evaluate the vulnerability of groundwater resources to contaminants from surface sources. The DRASTIC model is based on seven critical hydrogeological parameters that influence the movement of contaminants into aquifer systems [23]. These parameters include Depth to groundwater (D), Net recharge (R), Aquifer media (A), Soil media (S), Topography (T), Impact of the vadose zone (I), and Hydraulic conductivity (C). The decision to use DRASTIC over other groundwater vulnerability models, such as GOD and GALDIT, stems from its capability to assess the intrinsic vulnerability of groundwater at each specific location [34,38,39]. While the GOD model is simple, it tends to overgeneralize and is unable to capture localized variations [41,62], and the GALDIT model is designed primarily for coastal aquifers [35,40].

In the DRASTIC model, weights (w) of 5, 4, 3, 2, 1, 5, and 3 were assigned to the respective parameters. Ratings were then applied on a scale of 1 to 10 based on the range of each parameter at each groundwater well, following the guidelines set by [23] for confined aquifers. This approach involved careful yet subjective judgment informed by field experience and available data to ensure that the ratings accurately reflect the specific conditions of each site. The DRASTIC index was subsequently calculated using Equation (11) in Microsoft Excel, where subscripted parameters represent the ratings corresponding to each DRASTIC model input parameter.

$$\text{DRASTIC Index} = (D_r \times 5) + (R_r \times 4) + (A_r \times 3) + (S_r \times 2) + (T_r \times 1) + (I_r \times 5) + (C_r \times 3) \quad (11)$$

3. Results and Discussion

3.1. Hydrochemical Parameter Analysis

All 19 analyzed groundwater samples passed the charge balance error (CBE) test (Table S4). A total of 84% of the samples had a CBE of less than 5%, while 16% of the samples had a CBE between 5 and 10%. The results from the latter were treated with caution. The table below presents a descriptive summary of the hydrochemical test results for major parameters. See Table 3.

Table 3. Summary of descriptive statistical analysis of major water quality parameters (adapted from [16], with additional parameters).

Parameter	Mean	Standard Deviation (SD)	Max.	Min.	WHO Guideline Value (GV)	% of Samples > WHO GV
pH	6.7	1.0	9.3	4.5	6.5–8.5	31
HCO_3^- (mg/L)	141.5	134.0	439.3	10.0	-	-
Cl^- (mg/L)	48.6	25.6	96.1	16.0	250.0	-
NO_3^- (mg/L)	16.4	14.1	46.1	0.0	50.0	-
SO_4^{2-} (mg/L)	204.7	148.9	515.6	29.9	500.0	5
Na^+ (mg/L)	59.9	27.4	134.6	8.7	200.0	-
K^+ (mg/L)	6.6	3.8	15.1	2.8	12.0	5
Mg^{2+} (mg/L)	23.5	15.6	51.5	0.0	50.0	10
Ca^{2+} (mg/L)	59.0	42.3	144.0	0.7	300.0	-
Total Fe (mg/L)	0.8	1.5	5.1	0.02	0.3	47
Alkalinity (mg/L $CaCO_3$)	145.2	136.1	445.3	11.4	-	-
EC (µS/cm)	746.8	366.3	1538.0	297.0	1857	5

3.1.1. Analysis of Major Cations

Sodium and calcium concentrations from all 19 analyzed samples fall within acceptable WHO guideline ranges. Potassium (K^+) concentrations vary from 2.8 to 15.1 mg/L, with a mean of 6.6 mg/L and a standard deviation (SD) of 3.8. The WHO guideline value for potassium is 12 mg/L, with only 5% of the samples exceeding this limit. This suggests minor potassium contamination, which may lead to taste issues in drinking water but generally poses low health risks at these levels [63].

Magnesium (Mg^{2+}) concentrations range from 0.0 to 51.6 mg/L, with an average of 23.5 mg/L and a standard deviation of 15.6. Only 10% of the samples exceed the WHO guideline of 50 mg/L, indicating potential localized magnesium issues. Elevated Mg^{2+} levels can contribute to water hardness, affecting taste and usability for domestic purposes. Additionally, high magnesium concentrations have been linked to an increased risk of ischemic heart disease [64].

Total iron (Fe) levels range from 0.01 to 5.1 mg/L, with a mean of 0.8 mg/L and a standard deviation of 1.5. The WHO guideline value for iron is 0.3 mg/L, and 47% of the samples slightly exceed this limit. Elevated iron levels can cause several issues, including staining of laundry and plumbing fixtures, a metallic taste in drinking water, and potential health effects, such as gastrointestinal irritation [65].

3.1.2. Analysis of Major Anions

Chloride (Cl^-), bicarbonate (HCO_3^-), and nitrate (NO_3^-) concentrations fall within acceptable WHO guideline values, with the exception of sulfate (SO_4^{2-}). Sulfate levels range from 29.9 to 515.6 mg/L, with a mean of 204.7 mg/L and a standard deviation (SD) of 148.9. The WHO guideline for sulfate is 500 mg/L, with 5% of the samples exceeding this threshold. Elevated sulfate levels may lead to a bitter taste in water, laxative effects, and potential corrosion of plumbing systems [66].

The hydrochemical maps indicate both localized point contamination sources and diffuse sources (Figure 3), attributed to a combination of anthropogenic and geogenic factors. According to [15,16], groundwater chemistry in the region is primarily governed by water–rock interaction processes. Groundwater composition is significantly influenced by the mineralization of the underlying geological formations, predominantly composed of mafic and felsic feldspars (such as muscovite, orthoclase, anorthite, and biotite) [16] and sulfide minerals like pyrite and chalcopyrite, particularly evident in the Igayaza area of

Isingiro District [67]. High localized iron concentrations are linked to the potential rusting of galvanized iron commonly used in the installation of handpumps in the Nakivale area.

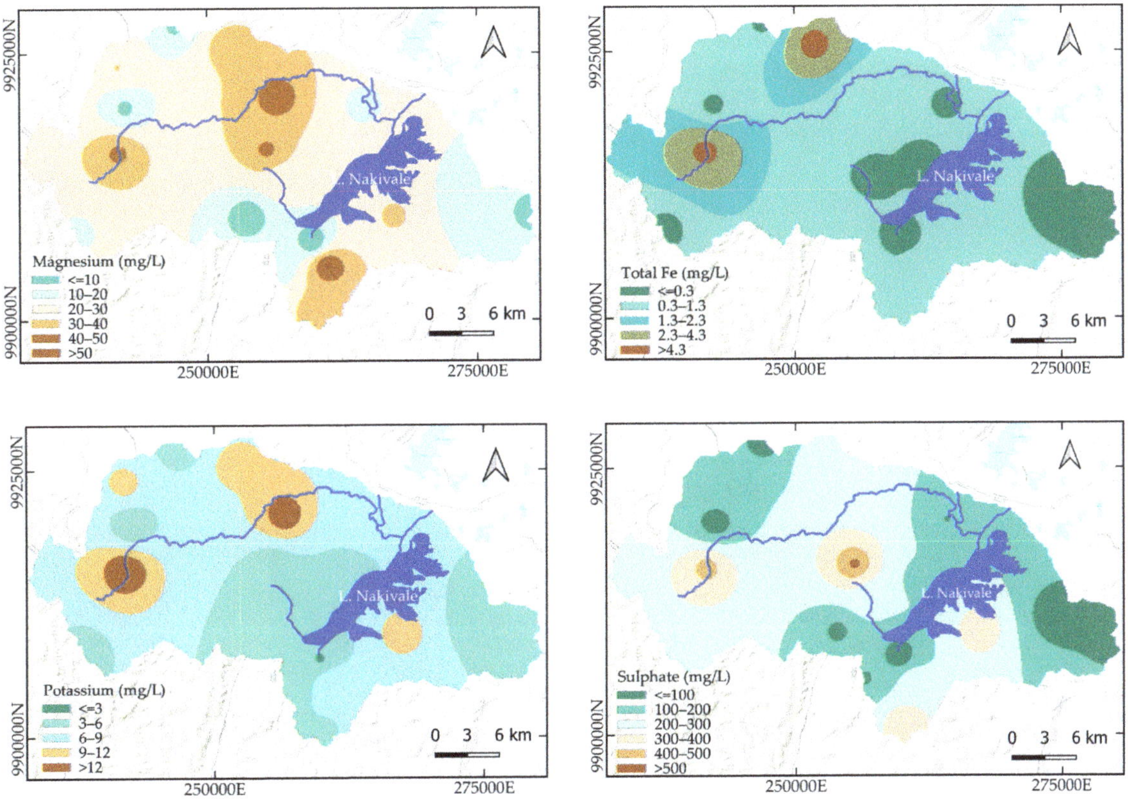

Figure 3. Groundwater quality maps showing spatial distribution of major ions exceeding WHO guideline values for drinking water.

Figure 3 highlights major contamination issues, with levels exceeding WHO guideline values for magnesium (50 mg/L), total iron (0.3 mg/L), potassium (12 mg/L), and sulfate (500 mg/L) in the northwestern and southwestern parts of the study area. Additionally, a central portion exhibits elevated concentrations of magnesium and sulfates, surpassing the WHO guidelines.

3.1.3. Analysis of Other Chemical Parameters

Electrical conductivity (EC) varies from 314 to 1726 μS/cm, with elevated values concentrated in low-lying areas (Figure 4). These high EC levels in the discharge zones suggest chemically enriched, metamorphosed water resulting from rock–water mineralization along groundwater flowlines, as noted by [16]. Additional potential sources of elevated EC include agricultural activities involving fertilizers and pesticides, as well as industrial waste in the region [5].

This analysis also reveals a range of Eh values across the study area, indicating different redox conditions. Lower Eh values (−500 to 0 mV) in the Lake Nakivale region indicate reducing conditions, typically found in older or deeper groundwater with limited oxygen exchange [68]. In contrast, higher Eh values (31 to 178 mV) in elevated areas suggest oxidizing conditions commonly associated with recently recharged or shallow groundwater. A comparable trend is also evident in dissolved oxygen levels. These results align with the findings of [16], which identified groundwater likely older than 20 years characterized by low tritium levels in the lower flanks of the study area.

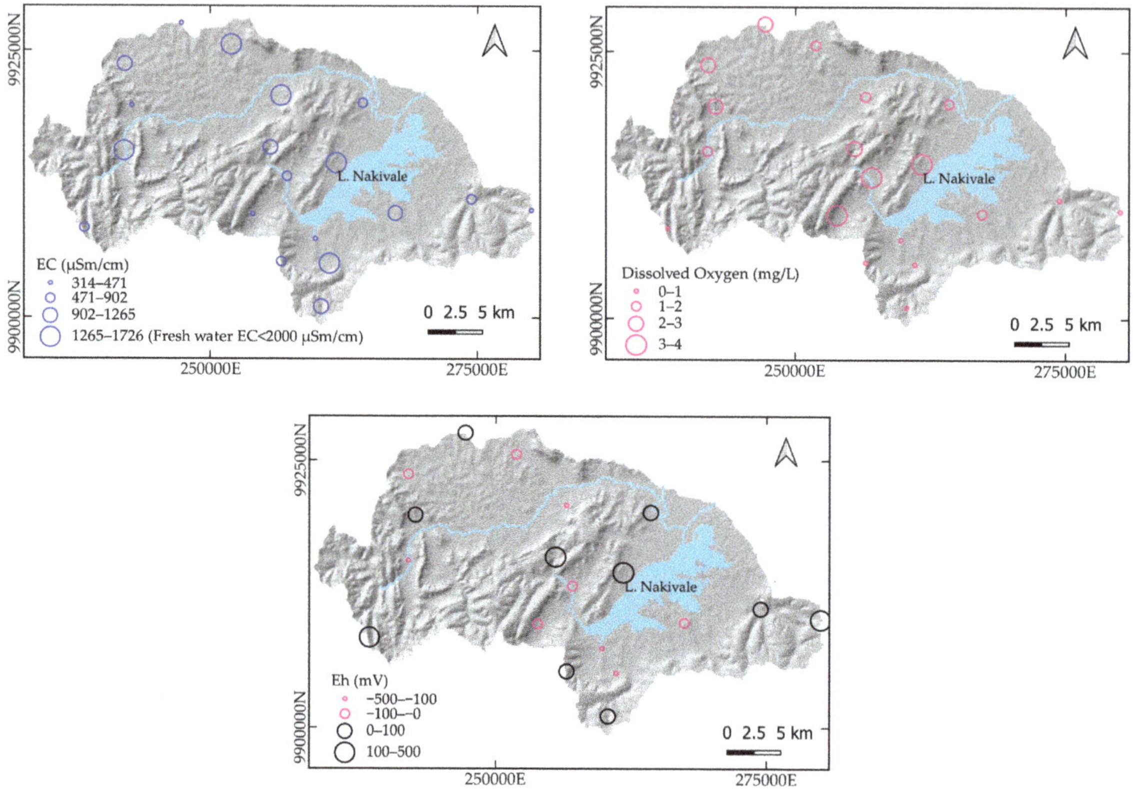

Figure 4. Groundwater quality maps showing Electrical Conductivity, Dissolved Oxygen, and Redox Potential spatial distribution within the project area.

3.2. Drinking Groundwater Quality Index Assessment

The majority of the groundwater samples (90%) have a groundwater quality index of less than 100. Ten groundwater samples (53%) have indices below 50, while seven samples (37%) have indices ranging from 50 to 100 (Figure 5 and Table 4). However, two samples, RAF7021-3 and RAF7021-25, obtained from Masha and Nyamuyanja, respectively, exhibit exceedingly high groundwater quality indices above 200, primarily due to high concentrations of total iron (Figure 5). The elevated concentration of total iron can be attributed to localized anthropogenic sources from the use of galvanized iron pipes commonly installed for handpump boreholes in the study area. However, this assumption needs to be confirmed through fieldwork. Additionally, the high GWQI of these two boreholes is also influenced by relatively elevated levels of magnesium, potassium, and sulfate, which slightly exceed WHO guideline values. According to [15,16], the concentration of these major ions in groundwater within the study area is mainly controlled by rock–water interaction processes. Overall, the study area is underlain by good-to-excellent quality groundwater for drinking, as over 90% of the sampled groundwater points have a groundwater quality index below 100. A total of 90% of all analyzed parameters for all groundwater samples also fall within plausible values stipulated by the WHO for drinking water.

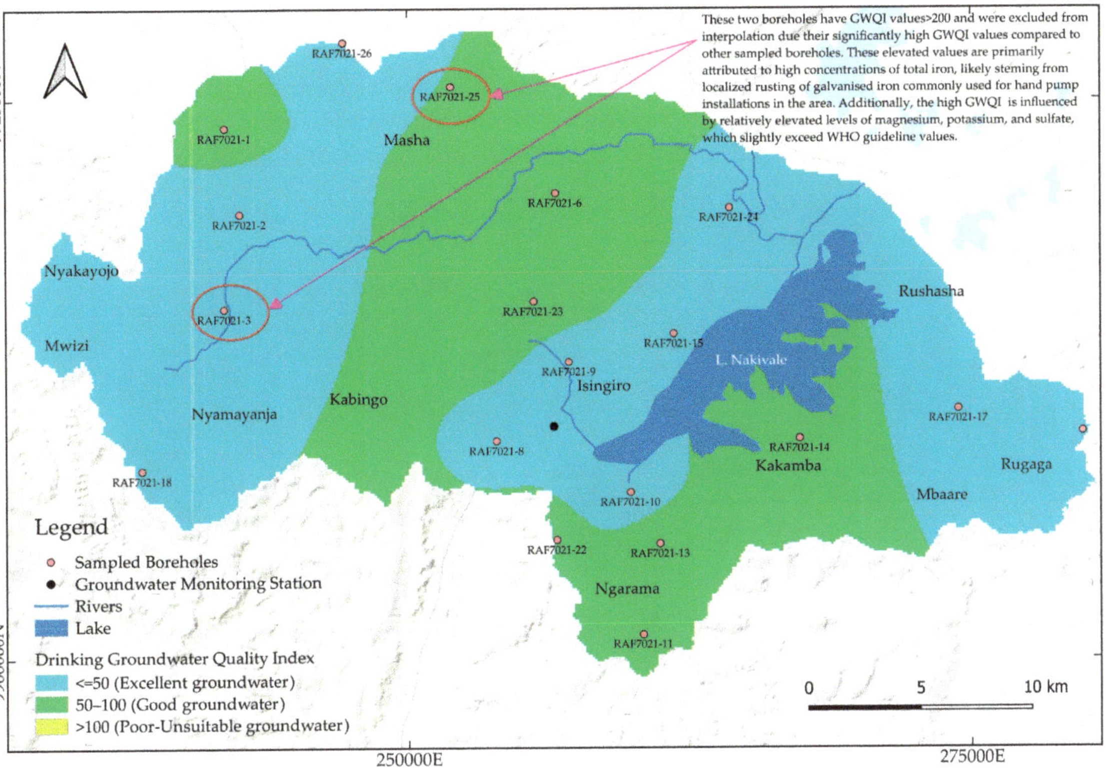

Figure 5. Drinking groundwater quality index map for Nakivale Sub-catchment (excluding samples from RAF7021-3 and RAF7021-25 due to potential localized contamination within these boreholes rather than reflecting the overall groundwater quality of the underlying aquifer system).

Table 4. Groundwater quality index (GWQI)-based classification for groundwater samples collected from 19 hand pump boreholes in Nakivale Sub-catchment, Southwestern Uganda.

Sample Number	GWQI	Groundwater Quality Type	Sample Number	GWQI	Groundwater Quality Type
RAF7021-1	64.2	Good	RAF7021-15	38.2	Excellent
RAF7021-2	25.0	Excellent	RAF7021-17	30.0	Excellent
RAF7021-3	276.0	Poor	RAF7021-18	45.8	Excellent
RAF7021-6	87.9	Good	RAF7021-21	25.2	Excellent
RAF7021-8	47.8	Excellent	RAF7021-22	53.0	Good
RAF7021-9	38.0	Excellent	RAF7021-23	67.5	Good
RAF7021-10	22.1	Excellent	RAF7021-24	42.3	Excellent
RAF7021-11	98.3	Good	RAF7021-25	262.6	Poor
RAF7021-13	76.8	Good	RAF7021-26	34.7	Excellent
RAF7021-14	80.1	Good			

3.3. Analysis of the Chloride–Bromide Ratio

The chloride–bromide (Cl/Br) ratio is a valuable tool for tracing factors that influence groundwater quality due to its conservative nature [61,69,70]. However, its interpretation requires careful consideration of potential overlaps between geogenic and anthropogenic sources [61]. Geogenic processes, such as the dissolution of halite, generally result in high Cl/Br ratios because chloride is more prevalent in these natural sources compared to bromide [69]. Anthropogenic activities, such as industrial waste disposal, agricultural pollution, and landfill leachate, can also introduce significant amounts of chloride with minimal bromide, thereby elevating the Cl/Br ratio [71]. See Figure 6.

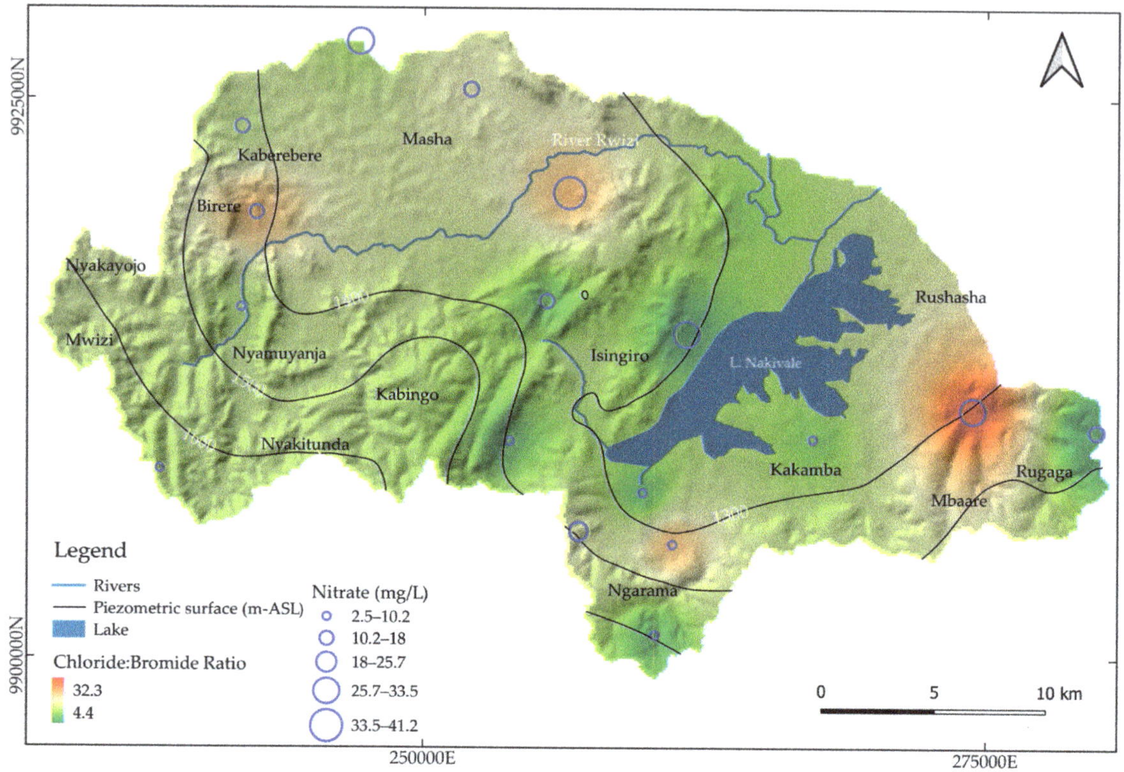

Figure 6. Chloride–bromide ratio map for tracing sources of groundwater contamination.

To accurately differentiate between geogenic and anthropogenic influences, the Cl/Br ratio was analyzed alongside nitrate levels. This combined approach provides a more nuanced understanding of whether high Cl/Br ratios are due to natural processes or human activities [70]. Areas with high nitrate levels and elevated Cl/Br ratios likely indicate anthropogenic impacts on groundwater quality, possibly from nitrate-rich and chloride-rich wastes, respectively [70,72]. Nitrate concentrations vary from 2.5 to 41.2 mg/L, with higher levels linked to existing and emerging anthropogenic effects such as agricultural pollution, septic effluent, and industrial waste discharge [5]. This relationship is also evident in (Figure 7), where elevated nitrate levels correlate with built and farmland land use and land cover (LULC) classes. Furthermore, an increase in the Cl/Br ratio along groundwater flowlines suggests a systematic alteration in groundwater chemistry, potentially due to the preferential adsorption of bromide, which is larger than chloride ions [61].

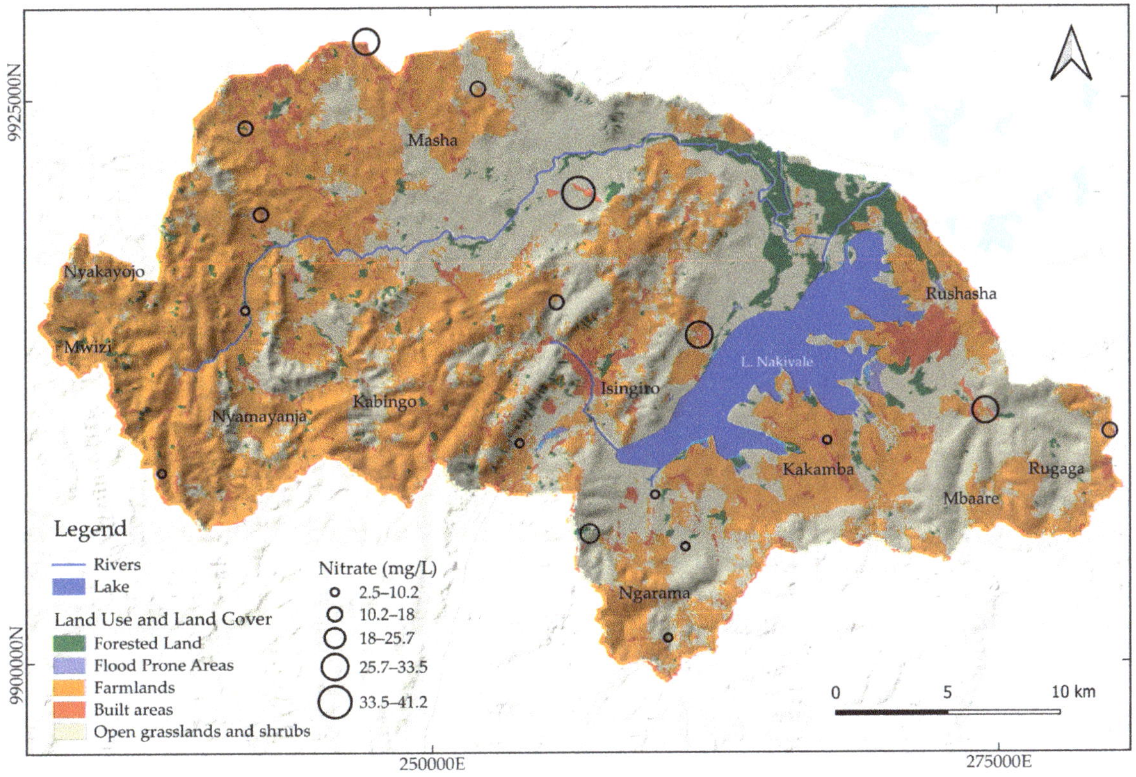

Figure 7. Land Use and Land Cover (LULC) map for Nakivale Sub-catchment (Data Source: FAO, 2024).

3.4. Irrigation Groundwater Quality Assessment

The analysis of the Sodium Adsorption Ratio (SAR) reveals that 94% of the samples fall within the C2-S1 (47%) and C3-S1 (47%) SAR-Electrical Conductivity regions (Figure 8). C2 and C3 indicate medium-to-high salinity hazards, respectively, while S1 reflects a low sodium (alkali) hazard. Despite the low sodium hazard, the results suggest that the region could face emerging issues with reduced crop productivity, driven by potential changes in the osmotic potential of crops if salinity levels increase, which could impede their growth [32,33,73]. One sample falls in the C2-S2 region, indicating medium sodium and salinity hazards, which also highlights the risk of lower crop productivity if sodium and salinity levels rise in the future [19,28].

According to the Sodium Percentage (SP), which measures sodium concentration relative to other cations [73], 74% of samples are within the Good range (20–40% SP); 16% are Permissible (40–60% SP), and 10% are Doubtful (60–80% SP) (Figure 9). While most samples fall in the Good range, indicating safe sodium levels, those in the Doubtful range may require managed irrigation to avoid long-term sodium accumulation in soils [73].

Kelly's Ratio (KR) assesses sodium risk relative to calcium and magnesium [28], with values less than 1 indicating Safe water (79% of samples) (Table S1) and those over 1 classified as Unsuitable (21% of samples) [31,73]. High KR values suggest an elevated sodium content, which could affect soil structure [28], leading to reduced permeability and increased runoff, impacting crop root zones [73].

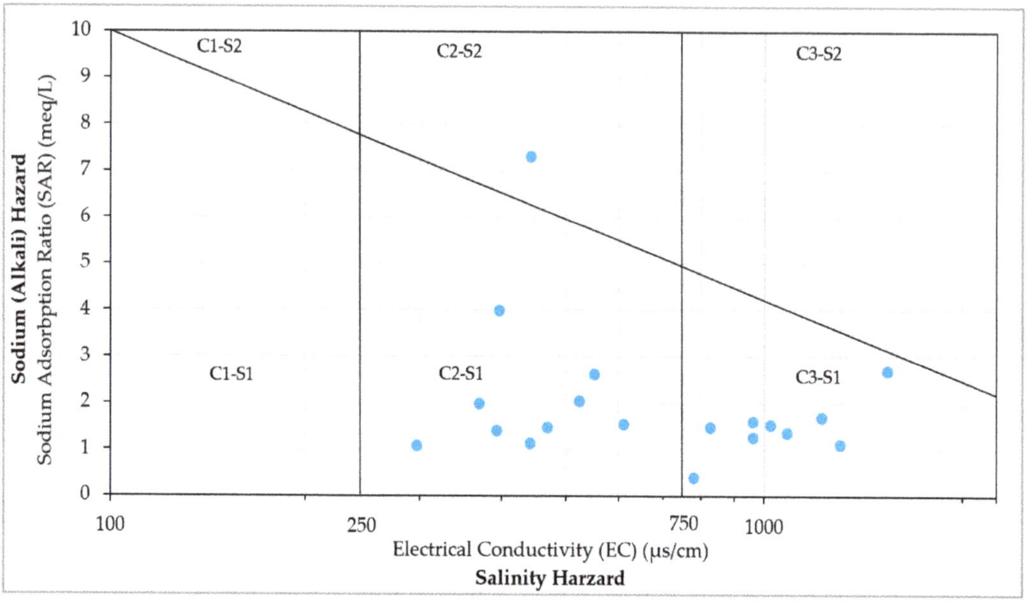

Figure 8. Sodium Adsorption Ratio (SAR) vs. Electrical Conductivity (EC) diagram for Classification of Irrigation water.

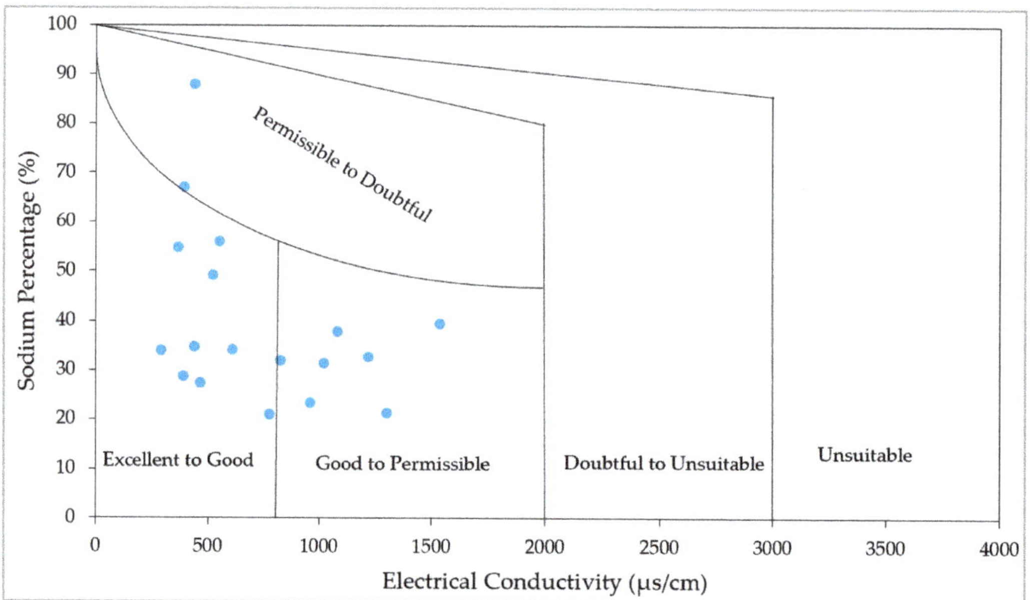

Figure 9. Sodium Percentage (SP) vs. Electrical Conductivity plot for classification of suitability of water for irrigation.

The Magnesium Adsorption Ratio (MAR) evaluates the dominance of magnesium, with values under 50% considered Suitable (89% of samples) (Table S1) and those above deemed Unsuitable (11% of samples) [25,28,73]. Excessive magnesium concentrations in Unsuitable samples may exacerbate soil alkalinity, potentially affecting nutrient availability to crops and, thereby, influencing plant health [28,31,73].

Residual Sodium Carbonate (RSC) measures the relative abundance of carbonate and bicarbonate over calcium and magnesium [28,31], with values of less than zero being Low (89% of samples), 0–1 being Medium (0%), 1–2.5 being High (11%), and above 2.5 being Very High (0%) (Table S1). The 11% of samples with High RSC levels suggest a potential risk for soil alkalinity issues, as excess carbonate may reduce soil permeability and lead to reduced crop productivity [31].

The overall groundwater quality for irrigation in the region shows low-to-moderate risks, with primary concerns centered around potential salinity hazards and isolated cases of high sodium and carbonate levels. This analysis supports the importance of consistent monitoring and management to address salinity and sodium hazards, as well as careful management and control of anthropogenic Na-rich chemical inputs, which may contribute to rising salinity and sodium levels in the region.

3.5. Groundwater Vulnerability Assessment

The seven DRASTIC model parameters influencing groundwater vulnerability, as suggested by [23], were evaluated to assess the susceptibility of groundwater resources within the Nakivale Sub-catchment to contamination from surface sources.

Depth-to-water table (D) varies spatially across the sub-catchment, ranging from 5 to 60 m [6]. Generally, shallower depths to aquifers are situated with lower elevation areas, increasing the potential for contamination. Given the low-to-moderately high depth to the aquifer window and the natural confining conditions, ratings were assigned between 3 and 6, with lower scores assigned to areas where the depth was greater, as deeper water tables offered greater protection from surface contamination [23].

Net recharge (R) within the sub-catchment is low, ranging between 50 and 100 mm/year [6]. This low recharge is attributed to the presence of low-permeability geological formations such as shales, slates, and phyllites, which restrict water infiltration except at structural weaknesses like joints and faults. A rating of 4 was assigned to areas with low recharge, while a slightly higher rating was given to areas near lineament features, where recharge and contamination potential were moderately elevated.

The aquifer media (A) primarily consists of fine-grained formations, such as shales, slates, and phyllites, which limit contaminant transport due to their low permeability (Figure 2). These formations were assigned a rating of 2. However, areas with surface lineament features, fractured and fluvial aquifer types, as identified in historical borehole drill reports, received a higher rating due to their enhanced permeability and vulnerability.

The area is primarily composed of clayey to silty soils, with smaller sandy sections near rivers and wetlands (Figure S2). The clay and silt offer some protection against contamination due to their fine texture, while sandy soils allow for quicker contaminant transport [47]. Consequently, ratings were set between 3 and 6, with higher ratings applied to sandy soils (Table 5).

Slopes range from 0 to 18% across the study area, with flat terrains (0–2%) receiving higher ratings (9–10) due to their increased susceptibility to infiltration (Figure S1). Conversely, steeper slopes were assigned lower ratings, as the risk of groundwater contamination decreased with increased runoff [23].

The vadose zone in Nakivale is primarily composed of impermeable formations like slates and phyllites (Figure 2). These rocks significantly reduce contaminant transport unless fractured. Areas with fractured zones or lineament features received a higher rating of 3 due to their increased permeability and associated vulnerability.

Table 5. The DRASTIC parameters, their corresponding weights, index score intervals, and ratings used in the vulnerability assessment (adapted from [23]).

DRASTIC Factor	Weight	DRASTIC Factor Interval	Rate
1. Depth-to-water table (D)	5	<1.50 m	10
		1.50–4.50 m	9
		4.51–9.00 m	7
		9.01–15.00 m	5
		15.01–22.50 m	4
		22.51–30.00 m	3
		30.01–50.00 m	2
		>50.00 m	1
2. Net Recharge (R)	4	<30.00 mm/year	1
		30.01–50.00 mm/year	2
		50.01–70.00 mm/year	4
		70.01–90.00 mm/year	5
		>90.00 mm/year	6
3. Aquifer Media (A)	3	Shales, slates, phyllites	2
		Fractured bedrock (crystalline rocks)	4
4. Soil Media (S)	2	Clay	3
		Silt	4
		Loam	5
		Sand	6
5. Topography (Slope) (T)	1	<2.00%	10
		2.01–6.00%	9
		6.01–12.00%	5
		12.01–18.00%	3
		>18.00%	1
6. Impact of Vadose Zone (I)	5	Sedimentary and metasedimentary (Shales, slates, phyllites)	3
7. Hydraulic Conductivity (C)	3	<0.050 m/day	1
		0.051–0.100 m/day	2
		0.101–0.500 m/day	3
		0.501–1.000 m/day	4
		1.001–10.000 m/day	7
		>10.000 m/day	8

The hydraulic conductivity varies spatially across the project area and inherent hydrogeological domains. According to [6], the fracture domain possesses a hydraulic conductivity of 1.7×10^{-2} m/day, and the matrix domain possesses a much lower value of 1.4×10^{-7} m/day. Due to this variability, areas within the fracture domain were given higher scores (2), reflecting a higher potential for contaminant transport, while a lower score (1) was assigned to matrix-dominated areas.

This assessment reveals low DRASTIC indices for the Nakivale Sub-catchment (<110) (Table S3). These low values generally indicate a reduced risk of groundwater contamination from surface sources [23,37–39]. This is primarily due to the area's varied steep slopes, moderately thick vadose zone, clayey soils, and the presence of impervious shales, slates, and phyllites, which cover over 90% of the region, consequently impacting groundwater recharge (Figure 1).

Groundwater vulnerability varies across the sub-catchment, with different areas exhibiting distinct levels of susceptibility (Figure 10). Birere, Nyamuyanja, Mwizi, and Kakamba have very low vulnerability, attributed to protective geological and soil characteristics. In contrast, Masha and Rugaga demonstrate moderate vulnerability, driven by factors such as slightly more permeable soils, a thin vadose zone, a relatively flat landscape, and shallow groundwater depths, all of which facilitate contaminant infiltration into underlying aquifers [23]. Kabingo, Ngarama, and parts of Kaberebere exhibit moderately low vulnerability, balancing protective and risk-enhancing characteristics.

The DRASTIC model results were validated by overlaying nitrate concentration data onto the groundwater vulnerability map (Figure 10). This overlay revealed that areas with high nitrate concentrations tended to correspond with regions of greater groundwater susceptibility compared to those with lower contamination risks. It is important to note that while there is a correlation between groundwater vulnerability and nitrate levels, nitrate concentrations remain below the WHO guideline value (50 mg/L) overall, which highlights the limited vulnerability of groundwater resources to surface contaminants. This resilience is due to protective hydrogeological conditions within the area. However, it is also prudent to note that the DRASTIC model may not directly identify areas with groundwater contamination issues but, instead, provides insights into the spatial vulnerability of groundwater resources across the study area based on inherent hydrogeological conditions [23,24,37,38,42].

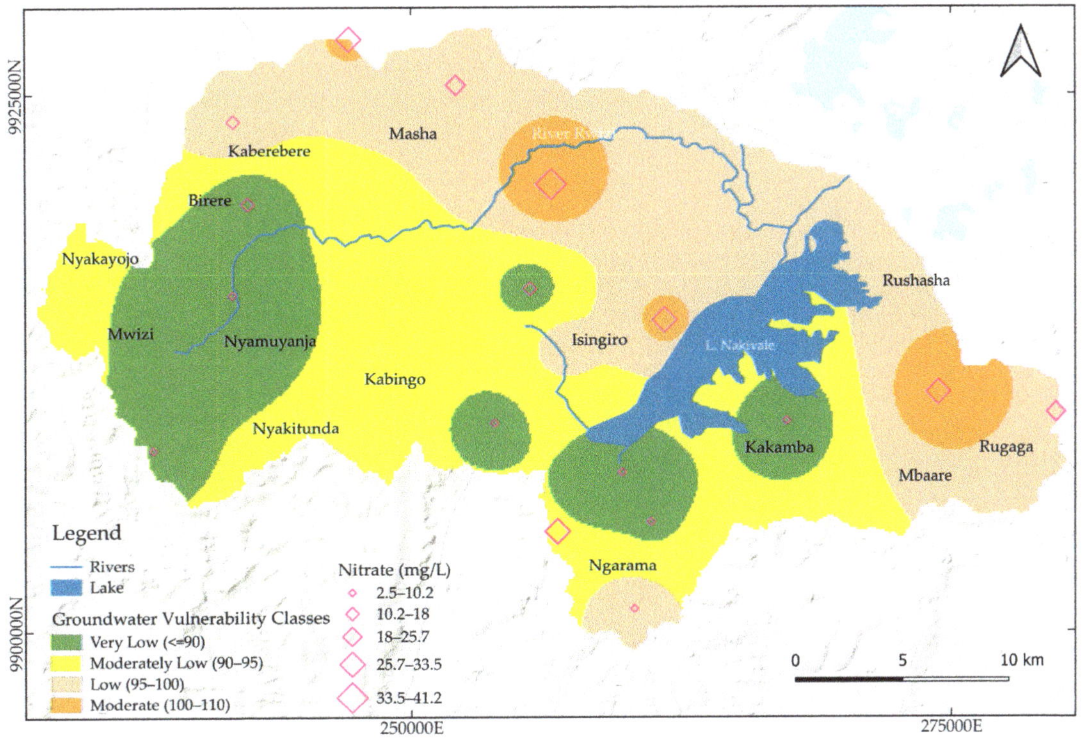

Figure 10. Groundwater Vulnerability DRASTIC Index map for Nakivale Sub-catchment.

4. Conclusions and Recommendations

4.1. Conclusions

This study assessed the groundwater quality and vulnerability in the Nakivale Sub-catchment, Southwestern Uganda, using a combination of hydrochemical analysis, groundwater quality indices, irrigation suitability parameters, and the DRASTIC groundwater vulnerability model. The results indicate that the majority of groundwater samples exhibit good-to-excellent water quality, with more than 90% of the samples falling below the groundwater quality index (GWQI) threshold of 100, suggesting generally safe drinking water. However, two samples showed high GWQI values (above 200), primarily due to elevated concentrations of total iron and other ions (Mg^{2+}, K^+, and SO_4^{2-}). These elevated iron levels are potentially linked to localized anthropogenic contamination, particularly from the use of galvanized iron pipes in handpump boreholes.

The hydrochemical analysis revealed some localized contamination concerns, particularly elevated concentrations of iron, magnesium, potassium, and sulfate in certain areas, which may impact water aesthetics and usability. In addition, the chloride–bromide (Cl/Br) ratio analysis revealed potential anthropogenic influences, especially in regions with agricultural or industrial activities. Nitrate concentrations, while within WHO limits, showed correlations with land use activities, further suggesting anthropogenic contributions to groundwater quality in specific zones.

From an irrigation perspective, the groundwater in the region generally falls within safe ranges for sodium and salinity hazards, though emerging salinity risks and localized sodium issues were identified in a few samples. The Sodium Adsorption Ratio (SAR) and related irrigation quality parameters, including Sodium Percentage (SP) and Kelly's Ratio (KR), suggest that while most of the groundwater is suitable for irrigation, certain areas may require careful management to avoid long-term soil degradation. Additionally, elevated carbonate concentrations in some areas point to the potential for increased soil alkalinity, which could affect agricultural productivity.

The DRASTIC model, used to assess groundwater vulnerability, revealed low vulnerability indices for the Nakivale Sub-catchment, reflecting the protective geological and hydrogeological conditions, such as clayey soils and impermeable formations (shales, slates, and phyllites). However, areas with shallower groundwater tables and more permeable soils, such as Masha and Rugaga, were identified as having moderate vulnerability. The validation of the DRASTIC model with nitrate concentration data highlighted that regions with higher nitrate levels correspond to areas with greater groundwater vulnerability, although nitrate concentrations remained below the WHO guideline value, indicating limited contamination from surface sources.

Overall, the groundwater resources in the Nakivale Sub-catchment provide safe drinking water for the majority of the population, with localized concerns related to specific contaminants, primarily iron and magnesium. While the area demonstrates resilience to surface contamination, there is a need for ongoing monitoring to address potential emerging issues, particularly with respect to salinity, sodium, and carbonate levels. This study underscores the importance of sustainable groundwater management, particularly in areas with anthropogenic influences, to ensure long-term water quality and safeguard agricultural productivity in the region.

4.2. Policy Recommendations

The findings of this research highlight the need for conjunctive management of groundwater resources in the Nakivale Sub-catchment area. Below are policy recommendations to ensure the sustainable management of groundwater resources within the study area:

- The elevated concentrations of total iron in localized areas of the Nakivale Sub-catchment highlight the importance of regulating anthropogenic contaminants, particularly from infrastructure such as galvanized iron pipes commonly used in handpump boreholes. Policies should be introduced to mandate the use of corrosion-resistant and non-reactive materials for water infrastructure. This will reduce the release of metals like iron into groundwater, enhancing drinking water quality. Additionally, there should be a focused effort to assess and minimize the impact of industrial waste, particularly from agricultural and industrial practices, which contribute to the contamination of groundwater;
- To ensure the ongoing safety and sustainability of groundwater resources, a robust monitoring and surveillance system should be established. Regular testing of water quality parameters, such as heavy metals (including iron), major ions, and microbial contaminants, is critical. The data gathered would not only inform public health interventions but also guide future management practices. Areas identified with higher concentrations of contaminants should be prioritized for more frequent testing, and local authorities should be trained to interpret and act on water quality data promptly;

- The findings underscore the role of agricultural activities in groundwater contamination, particularly through nitrate pollution. Policies should focus on encouraging sustainable farming practices, such as the adoption of organic fertilizers, efficient irrigation techniques, and integrated pest management strategies. Additionally, farmers should be educated on the importance of reducing the use of chemical fertilizers and pesticides to prevent nutrient leaching into groundwater. Financial incentives or subsidies for environmentally friendly practices could also support the transition to sustainable agriculture;
- Given the identified risks related to salinity, especially in areas falling within the C2-S1 and C3-S1 regions, it is important to implement comprehensive irrigation management plans. These should include guidelines for optimizing water use and minimizing salinity buildup. Policies can promote the use of salt-tolerant crop varieties, the application of soil conditioners, and techniques like drip irrigation to reduce water wastage and prevent salinization of soil. Training and extension services for farmers in these areas are crucial for ensuring that these practices are adopted effectively;
- Based on the vulnerability assessment using the DRASTIC model, regions with shallow water tables, low permeability, and high recharge rates should be designated as groundwater protection zones. Within these zones, stricter land-use regulations should be enforced to limit activities that could introduce contaminants into the groundwater system, such as industrial waste disposal, large-scale agricultural runoff, and urban expansion. These zones should be monitored closely to ensure compliance with environmental protection standards and prevent any activities that could compromise groundwater quality;
- Lastly, there should be an integrated approach to water quality management that combines local, regional, and national policies. Public awareness campaigns are essential to educating communities about the importance of protecting groundwater resources and the risks posed by poor land-use practices, over-extraction, and contamination. Local stakeholders, including community leaders, farmers, and water managers, should be involved in decision-making processes to ensure policies are context-specific and effectively address local needs.

Supplementary Materials: The following supporting information can be downloaded at https://www.mdpi.com/article/10.3390/w16233386/s1, Figure S1: Spatial variation in topographical slope within Nakivale Sub-catchment; Figure S2: Soil texture within Nakivale Sub-catchment, Southwestern Uganda; Table S1: Calculated Irrigation groundwater quality indices each groundwater sample; Table S2: Computed groundwater quality sub-index for each hydrochemical parameter and overall groundwater quality index (GWQI) for each sample; Table S3: Assigned weights and ratings for each respective parameter within the applied DRASTIC aquifer vulnerability assessment model; Table S4: Estimated Change Balance Error (CBE) for each sample.

Author Contributions: Conceptualization, E.N.H. and R.M.K.; methodology, E.N.H. and R.M.K.; software, E.N.H.; validation, E.N.H. and R.M.K.; formal analysis, E.N.H. and R.M.K.; investigation, E.N.H.; resources, E.N.H. and R.M.K.; data curation, E.N.H. and R.M.K.; writing—original draft preparation, E.N.H.; writing—review and editing, E.N.H. and R.M.K.; visualization, E.N.H.; supervision, R.M.K.; project administration, E.N.H.; funding acquisition, E.N.H. and R.M.K. All authors have read and agreed to the published version of the manuscript.

Funding: This research received funding from the International Atomic Energy Agency (IAEA) through the RAF7021 project in collaboration with Uganda's Ministry of Water and Environment. This study was carried out during a fellowship (Code: EVT2304936-0001-UGA) at the University of Strathclyde in the United Kingdom, spanning from 9 October 2023 to 8 November 2024.

Data Availability Statement: The data utilized in this research are available upon request from the corresponding authors.

Acknowledgments: The authors gratefully acknowledge IAEA TC Project RAF7021 for the invaluable financial support extended throughout all stages of the project's implementation. The authors are also thankful for the financial support of the Scottish Government under the Climate Justice Fund Water Futures Program (research grant HN-CJF-03), awarded to the University of Strathclyde (R.M. Kalin PI).

Conflicts of Interest: The authors declare no conflicts of interest.

References

1. United Nations Environment Programme. Goal 6: Clean Water and Sanitation [Internet]. 2024. Available online: https://www.unep.org/explore-topics/sustainable-development-goals/why-do-sustainable-development-goals-matter/goal-6 (accessed on 2 September 2024).
2. African Union. *The African Union Commission Agenda 2063 Framework Document*; African Union: Addis Ababa, Ethiopia, 2015; pp. 1–172.
3. de Lázaro Torres, M.L.; Borderias Uribeondo, P.; Morales Yago, F.J. Citizen and educational initiatives to support Sustainable Development Goal 6: Clean water and sanitation for all. *Sustainability* **2020**, *12*, 2073. [CrossRef]
4. Cai, J.; Zhao, D.; Varis, O. Match words with deeds: Curbing water risk with the Sustainable Development Goal 6 index. *J. Clean. Prod.* **2021**, *318*, 128509. [CrossRef]
5. Ministry of Water and Environment. *Rwizi Catchment Management Plan*; Ministry of Water and Environment: Kampala, Uganda, 2020.
6. Ministry of Water and Environment. *Baseline Groundwater Resources Assessment Report: National Groundwater Study*; Ministry of Water and Environment: Kampala, Uganda, 2022.
7. Adonia, K.; Kakurungu, B. Assessment of the effects of changing land use from pastoralism to crop farming on Lake Nakivale wetland system in Isingiro District, Uganda. *J. Afr. Stud.* **2014**, *6*, 56–66. [CrossRef]
8. Nanfuka, S.; Mfitumukiza, D.; Egeru, A. Characterisation of ecosystem-based adaptations to drought in the central cattle corridor of Uganda. *Afr. J. Range Forage Sci.* **2020**, *37*, 257–267. [CrossRef]
9. Ronald, M.A. An assessment of economic and environmental impacts of refugees in Nakivale, Uganda. *Migr. Dev.* **2022**, *11*, 433–449. [CrossRef]
10. Uganda Bureau of Statistics. Census Population Counts (2002 and 2014) by Region, District, and Mid-Year Population Projections (2015–2021) [Internet]. 2023. Available online: https://www.ubos.org/explore-statistics/20/ (accessed on 7 September 2024).
11. Aturihaihi, C.; Tumwesigye, W.; Opio, F.; Beyihayo, G.A. Knowledge, Attitude and the Practice of Climate-Smart Agriculture among Smallholder Farmers in Isingiro District, South Western Uganda. *East Afr. J. Agric. Biotechnol.* **2023**, *6*, 82–97. [CrossRef]
12. Bjørkhaug, I. Revisiting the refugee–host relationship in Nakivale Refugee Settlement: A dialogue with the Oxford Refugee Studies Centre. *J. Migr. Hum. Secur.* **2020**, *8*, 266–281. [CrossRef]
13. Ministry of Water and Environment. *National Water Resources Assessment Report*; Ministry of Water and Environment: Kampala, Uganda, 2013.
14. Chenini, I.; Zghibi, A.; Kouzana, L. Hydrogeological investigations and groundwater vulnerability assessment and mapping for groundwater resource protection and management: State of the art and a case study. *J. Afr. Earth Sci.* **2015**, *109*, 11–26. [CrossRef]
15. Owor, M.; Muwanga, A.; Tindimugaya, C.; Taylor, R.G. Hydrogeochemical processes in groundwater in Uganda: A national-scale analysis. *J. Afr. Earth Sci.* **2021**, *175*, 104113. [CrossRef]
16. Nabala, E.H. Hydrochemical and Isotopic Characterisation of Groundwater Resources in Nakivale Sub-Catchment in Isingiro District, Southwestern Uganda. Master's Thesis, University of Strathclyde, Glasgow, UK, 2024.
17. Songa, P.; Rumohr, J.; Musota, R. A shared water risk assessment for a vulnerable river basin: River Rwizi in Uganda. *WIT Trans. Ecol. Environ.* **2015**, *197*, 213–224.
18. Nimusiima, D.; Byamugisha, D.; Omara, T.; Ntambi, E. Physicochemical and microbial quality of water from the Ugandan stretch of the Kagera Transboundary River. *Limnol. Rev.* **2023**, *23*, 157–176. [CrossRef]
19. Walter, O.; Wasswa, J.; Nakiguli, C.K.; Ntambi, E. *Spatial Variation in Physicochemical Surface Water Quality in River Rwizi, Western Uganda*; Mbarara University of Science & Technology: Kampala, Uganda, 2019.
20. Chatterjee, R.; Ray, R.K. A proposed new approach for groundwater resources assessment in India. *J. Geol. Soc. India* **2016**, *88*, 357–365. [CrossRef]
21. Rojas, R.; Commander, P.; McFarlane, D.; Ali, R.; Dawes, W.; Barron, O.; Hodgson, G.; Charles, S. Groundwater resource assessment and conceptualization in the Pilbara Region, Western Australia. *Earth Syst. Environ.* **2018**, *2*, 345–365. [CrossRef]
22. Lions, J.; Devau, N.; Elster, D.; Voutchkova, D.D.; Hansen, B.; Schullehner, J.; Petrović Pantić, T.; Samolov, K.A.; Camps, V.; Arnó, G.; et al. A broad-scale method for estimating natural background levels of dissolved components in groundwater based on lithology and anthropogenic pressure. *Water* **2021**, *13*, 1531. [CrossRef]
23. Aller, L. *Thornhill JDRASTIC: A Standardized System for Evaluating Ground Water Pollution Potential Using Hydrogeologic Settings*; Robert, S., Ed.; Kerr Environmental Research Laboratory, Office of Research and Development, US Environmental Protection Agency: Ada, OK, USA, 1987.
24. Tufano, R.; Allocca, V.; Coda, S.; Cusano, D.; Fusco, F.; Nicodemo, F.; Pizzolante, A.; De Vita, P. Groundwater vulnerability of principal aquifers of the Campania region (southern Italy). *J. Maps* **2020**, *16*, 565–576. [CrossRef]

25. Ibrahim, M.N. Assessing groundwater quality for drinking purpose in Jordan: Application of water quality index. *J. Ecol. Eng.* **2019**, *20*, 101–111. [CrossRef]
26. Balan, I.; Shivakumar, M.; Kumar, P.M. An assessment of groundwater quality using water quality index in Chennai, Tamil Nadu, India. *Chron. Young Sci.* **2012**, *3*, 146. [CrossRef]
27. Alqarawy, A.; El Osta, M.; Masoud, M.; Elsayed, S.; Gad, M. Use of hyperspectral reflectance and water quality indices to assess groundwater quality for drinking in arid regions, Saudi Arabia. *Water* **2022**, *14*, 2311. [CrossRef]
28. Vranešević, M.; Zemunac, R.; Grabić, J.; Salvai, A. Hydrochemical Characteristics and Suitability Assessment of Groundwater Quality for Irrigation. *Appl. Sci.* **2024**, *14*, 615. [CrossRef]
29. Ketata, M.; Gueddari, M.; Bouhlila, R. Use of geographical information system and water quality index to assess groundwater quality in El Khairat deep aquifer (Enfidha, Central East Tunisia). *Arab. J. Geosci.* **2012**, *5*, 1379–1390. [CrossRef]
30. Shuaibu, A.; Kalin, R.M.; Phoenix, V.; Banda, L.C.; Lawal, I.M. Hydrogeochemistry and water quality index for groundwater sustainability in the Komadugu-Yobe Basin, Sahel Region. *Water* **2024**, *16*, 601. [CrossRef]
31. El Osta, M.; Masoud, M.; Alqarawy, A.; Elsayed, S.; Gad, M. Groundwater suitability for drinking and irrigation using water quality indices and multivariate modeling in makkah Al-Mukarramah province, Saudi Arabia. *Water* **2022**, *14*, 483. [CrossRef]
32. Worwornyo, S.K. *Assessment of Groundwater Quality and Its Suitability for Domestic and Agricultural Purposes in the North-Western Part of the Central Region, Ghana*; University of Ghana: Accra, Ghana, 2014.
33. Alogayell, H.M.; EL-Bana, E.M.M.; Abdelfattah, M. Groundwater Quality and Suitability Assessment for Irrigation Using Hydrogeochemical Characteristics and Pollution Indices: A Case Study of North Al-Quwayiyah Governorate, Central Saudi Arabia. *Water* **2023**, *15*, 3321. [CrossRef]
34. Gonçalves, V.; Albuquerque, A.; Carvalho, P.; Almeida, P.; Cavaleiro, V. Groundwater Vulnerability Assessment to Cemeteries Pollution through GIS-Based DRASTIC Index. *Water* **2023**, *2023*, 812. [CrossRef]
35. Recinos, N.; Kallioras, A.; Pliakas, F.; Schuth, C. Application of GALDIT index to assess the intrinsic vulnerability to seawater intrusion of coastal granular aquifers. *Environ. Earth Sci.* **2015**, *73*, 1017–1032. [CrossRef]
36. Markus, U.I.; Ilori, O.P.; Wada, I.M.; Musa, S.T.; Peter, J.E. Hydrogeochemical evaluation and geospatial distribution modeling of the major ion chemistry of groundwater and their suitability for drinking and irrigation in Lagelu, Southwestern Nigeria. *Appl. Water Sci.* **2022**, *12*, 160. [CrossRef]
37. Abu-Bakr, H.A.E.A. Groundwater vulnerability assessment in different types of aquifers. *Agric. Water Manag.* **2020**, *240*, 106275. [CrossRef]
38. Barbulescu, A. Assessing groundwater vulnerability: DRASTIC and DRASTIC-like methods: A review. *Water* **2020**, *12*, 1356. [CrossRef]
39. Malakootian, M.; Nozari, M. GIS-based DRASTIC and composite DRASTIC indices for assessing groundwater vulnerability in the Baghin aquifer, Kerman, Iran. *Nat. Hazards Earth Syst. Sci.* **2020**, *20*, 2351–2363. [CrossRef]
40. Pisciotta, A.; Suozzi, E.; Tiwari, A.K. A modified GALDIT-NUTS index to assess Favignana Island aquifer vulnerability. *Geocarto Int.* **2022**, *37*, 11706–11731. [CrossRef]
41. Kaddour, K.; El Hacen, B.; Hlima, D.; Yasmina, D. Groundwater vulnerability assessment using GOD method in Boulimat coastal District of Bejaia area North east Algeria. *J. Biodivers. Environ. Sci.* **2018**, *13*, 109–116.
42. El Osta, M.; Niyazi, B.; Masoud, M. Groundwater evolution and vulnerability in semi-arid regions using modeling and GIS tools for sustainable development: Case study of Wadi Fatimah, Saudi Arabia. *Environ. Earth Sci.* **2022**, *81*, 248. [CrossRef]
43. Stheeman, H.A.; Stheeman, H.A. Various Phases of Folding-Nature of Contact-Manner of Intrusion. In *The Geology of Southwestern Uganda: With Special Reference to the Stanniferous Deposits*; Springer: Berlin/Heidelberg, Germany, 1932; pp. 53–64.
44. Tindimugaya, C. *Groundwater Flow and Storage in Weathered Crystalline Rock Acquifer Systems of Uganda: Evidence from Environmental Tracers and Acquifer Responses to Hydraulic Stress*; UCL (University College London): London, UK, 2008.
45. Taylor, R.G.; Howard, K.W.F. Post-Palaeozoic evolution of weathered landsurfaces in Uganda by tectonically controlled deep weathering and stripping. *Geomorphology* **1998**, *25*, 173–192. [CrossRef]
46. Kohno, M. Effects of hydraulic gradient and clay type on permeability of clay mineral materials. *Minerals* **2020**, *10*, 1064. [CrossRef]
47. Boynton, S.S.; Daniel, D.E. Hydraulic conductivity tests on compacted clay. *J. Geotech. Eng.* **1985**, *111*, 465–478. [CrossRef]
48. Toma, V.E.; Webster, P.J. Oscillations of the intertropical convergence zone and the genesis of easterly waves. Part I: Diagnostics and theory. *Clim. Dyn.* **2010**, *34*, 587–604. [CrossRef]
49. Riedel, T.; Weber, T.K. The influence of global change on Europe's water cycle and groundwater recharge. *Hydrogeol. J.* **2020**, *28*, 1939–1959. [CrossRef]
50. Bui, A.; Johnson, F.; Wasko, C. The relationship of atmospheric air temperature and dew point temperature to extreme rainfall. *Environ. Res. Lett.* **2019**, *14*, 74025. [CrossRef]
51. Vicente-Serrano, S.M.; McVicar, T.R.; Miralles, D.G.; Yang, Y.; Tomas-Burguera, M. Unraveling the influence of atmospheric evaporative demand on drought and its response to climate change. *Wiley Interdiscip. Rev. Clim. Change* **2020**, *11*, e632. [CrossRef]
52. Conicelli, B.; Hirata, R.; Galvão, P.; Bernardino, M.; Simonato, M.; Abreu, M.C.; Aranda, N.; Terada, R. Determining groundwater availability and aquifer recharge using GIS in a highly urbanized watershed. *J. S. Am. Earth Sci.* **2021**, *106*, 103093. [CrossRef]
53. Montaser, A.; Golightly, D.W. (Eds.) *Inductively Coupled Plasmas in Analytical Atomic Spectrometry*; John Wiley & Sons: Hoboken, NJ, USA, 1996.

54. Fritz, S.J. A survey of charge-balance errors on published analyses of potable ground and surface waters. *Groundwater* **1994**, *32*, 539–546. [CrossRef]
55. Lu, G.Y.; Wong, D.W. An adaptive inverse-distance weighting spatial interpolation technique. *Comput. Geosci.* **2008**, *34*, 1044–1055. [CrossRef]
56. Mahalingam, B.R.M.D.B.; Bhauso, R.; Deepali, M.; Jayashree, P. Assessment of groundwater quality using GIS techniques: A case study of Mysore City. *Int. J. Eng. Innov. Technol.* **2014**, *3*, 117–122.
57. Gorantla, S.S.; Pradeep Kumar, G.N. Spatial distribution of groundwater quality indices using decision support system in Cheyyeru Watershed, YSR District, India. *Spat. Inf. Res.* **2021**, *29*, 297–307. [CrossRef]
58. Maurya, P.K.; Ali, S.A.; Zaidi, S.K.; Wasi, S.; Tabrez, S.; Malav, L.C.; Ditthakit, P.; Son, C.T.; Cabral-Pinto, M.M.; Yadav, K.K. Assessment of groundwater geochemistry for drinking and irrigation suitability in Jaunpur district of Uttar Pradesh using GIS-based statistical inference. *Environ. Sci. Pollut. Res.* **2023**, *30*, 29407–29431. [CrossRef] [PubMed]
59. Hasan, M.; Shang, Y.; Akhter, G.; Jin, W. Evaluation of groundwater suitability for drinking and irrigation purposes in Toba Tek Singh District, Pakistan. *Irrig. Drain. Syst. Eng.* **2017**, *6*, 185.
60. Taherian, P.; Joodavi, A. Hydrogeochemical characteristics and source identification of salinity in groundwater resources in an arid plain, northeast of Iran: Implication for drinking and irrigation purposes. *Acque Sotter. Ital. J. Groundw.* **2021**, *10*, 21–31. [CrossRef]
61. Goldowitz, J. *The Chloride to Bromide Ratio as an Environmental Groundwater Tracer, with a Field Study at the Wellton-Mohawk Irrigation and Drainage District*; The University of Arizona: Tucson, AZ, USA, 1989.
62. Huang, C.C.; Yeh, H.F.; Lin, H.I.; Lee, S.T.; Hsu, K.C.; Lee, C.H. Groundwater recharge and exploitative potential zone mapping using GIS and GOD techniques. *Environ. Earth Sci.* **2013**, *68*, 267–280. [CrossRef]
63. Aizman, R.I.; Gerasev, A.D.; Aizman, O.; Krasheninina, G.I.; Trofimovich, E.M. Physiological and hygienic basis of potassium norm in drinking water: Theoretical and practical implications. *Novosib. State Pedagog. Univ. Bull.* **2018**, *8*, 266–284. [CrossRef]
64. Marx, A.; Neutra, R.R. Magnesium in drinking water and ischemic heart disease. *Epidemiol. Rev.* **1997**, *19*, 258–272. [CrossRef]
65. Liap, Z. *Environmental Chemistry of Trace Elements and Their Biological Effect*; Chinese Environmental Science Press: Beijing, China, 1992; p. 320.
66. Burlingame, G.A.; Dietrich, A.M.; Whelton, A.J. Understanding the basics of tap water taste. *J. Am. Water Work. Assoc.* **2007**, *99*, 100–111. [CrossRef]
67. Nagudi, B. *Status of Geological Resources in Uganda*; For the Embassy of the Republic of Korea in Uganda: Kampala, Uganda, 2011.
68. Glynn, P.D.; Plummer, L.N. Geochemistry and the understanding of ground-water systems. *Hydrogeol. J.* **2005**, *13*, 263–287. [CrossRef]
69. Andreasen, D.C.; Fleck, W.B. Use of bromide: Chloride ratios to differentiate potential sources of chloride in a shallow, unconfined aquifer affected by brackish-water intrusion. *Hydrogeol. J.* **1997**, *5*, 17–26. [CrossRef]
70. Ratri, D.; Putra, D.P.E.; Wilopo, W. Nitrate source detection using nitrate/chloride ratios and chloride/bromide ratios in shallow groundwater, Bantul. In *AIP Conference Proceedings*; AIP Publishing: Yogyakarta, Indonesia, 2023; Volume 2598.
71. Katz, B.G.; Eberts, S.M.; Kauffman, L.J. Using Cl/Br ratios and other indicators to assess potential impacts on groundwater quality from septic systems: A review and examples from principal aquifers in the United States. *J. Hydrol.* **2011**, *397*, 151–166. [CrossRef]
72. Hudak, P.F. Chloride/bromide ratios in leachate derived from farm-animal waste. *Environ. Pollut.* **2003**, *121*, 23–25. [CrossRef]
73. Amwele, H.R.; Kgabi, N.A.; Kandjibi, L.I. Sustainability of groundwater for irrigation purposes in semi-arid parts of Namibia. *Front. Water* **2021**, *3*, 767496. [CrossRef]

water

MDPI

Article

Hydrogeochemical Characterization and Processes Controlling Groundwater Chemistry of Complex Volcanic Rock of Jimma Area, Ethiopia

Wagari Mosisa Kitessa [1,2], Adisu Befekadu Kebede [1,2], Fayera Gudu Tufa [1,2], Beekan Gurmessa Gudeta [1,2], Alemu Yenehun [1], Bekan Chelkeba [1,2], Seifu Kebede Debela [1,2], Fekadu Fufa Feyessa [2] and Kristine Walraevens [1,*]

1 Laboratory for Applied Geology and Hydrogeology, Department of Geology, Ghent University, 9000 Ghent, Belgium; wagarimosisa.kitessa@ugent.be (W.M.K.); adisubefekadu.kebede@ugent.be (A.B.K.); fayeragudu.tufa@ugent.be (F.G.T.); beekan.gudeta@ugent.be (B.G.G.); alemuyenehun.beyene@ugent.be (A.Y.); bekanchelkeba.tumsa@ugent.be (B.C.); seifukebede.debela@ugent.be (S.K.D.)

2 Faculty of Civil and Environmental Engineering, Jimma Institute of Technology, Jimma University, Jimma 378, Ethiopia; fekadu.fufa@ju.edu.et

* Correspondence: kristine.walraevens@ugent.be

Abstract: The sustainable management of groundwater in the Jimma area is complicated by a lack of comprehensive studies on its chemical makeup and the geochemical processes influencing its hydrochemistry. This research aims to fill that gap by examining 51 groundwater samples from various sources, including deep groundwaters, shallow groundwaters, hand-dug well groundwaters, surface waters, and springs within the area primarily consisting of complex volcanic rocks. The goal is to describe the hydrogeochemical characteristics and determine the key processes affecting groundwater composition in this volcanic area. The study identifies clear patterns in cation and anion concentrations. For deep groundwaters, the average cation concentration is ranked as $Na^+ > Ca^{2+} > Mg^{2+} > K^+$, while shallow groundwaters, hand-dug well groundwaters, surface waters, and springs show a ranking of $Ca^{2+} > Na^+ > Mg^{2+} > K^+$. The major anions are typically ordered as $HCO_3^- > NO_3^- > Cl^- > SO_4^{2-}$. The quantitative hydrogeochemical analysis indicates that the freshwater types in the region are primarily Ca-HCO_3 and Ca-Mg-HCO_3, with some highly mineralized Na-HCO_3 waters also detected. The weathering of silicate minerals mainly drives the geochemical processes affecting groundwater chemistry. An increase in mineralization, suggested by saturation indices, points to a longer residence time underground, with deep groundwaters exhibiting the highest saturation levels and springs the lowest. This mineralization is especially significant for Mg-silicates and carbonates. Stability diagrams for feldspar minerals further demonstrate groundwater evolution along flow paths, revealing that shallow systems are in equilibrium with minerals like gibbsite, whereas deeper systems achieve stability with albite, Ca-montmorillonite, and microcline. Higher CO_2 levels ($10^{-1.5}$ to $10^{0.5}$ atm), likely from mantle-magma degassing, add more HCO_3^- to the deeper aquifers. This study offers the first thorough characterization of the groundwater composition in the Jimma area and provides important insights into the Jimma area's hydrogeochemical development, establishing a basis for enhanced groundwater management within this intricate volcanic aquifer system.

Keywords: anthropogenic activities; complex volcanic rock; hydrogeochemical characteristics; groundwater evolution; saturation indices; silicate minerals

1. Introduction

Groundwater serves as a vital freshwater source in numerous developing countries, where it is typically utilized for drinking, irrigation, and other domestic purposes [1,2]. As a dynamic system, the composition and quality of groundwater can be influenced by various geochemical and environmental factors, including atmospheric contributions of solutes [3], gases [4], and aerosols [5]. Furthermore, groundwater experiences considerable chemical changes as it interacts with different geological materials along its flow path [6,7]. These interactions, involving chemical and biological reactions, can alter the composition and properties of the groundwater [7–9]. Some key processes contributing to groundwater quality changes include cation exchange [10,11], redox reactions and decomposition [12], evapo-concentration [13], adsorption and desorption [14], biological reactions [15], weathering and the hydrolysis of silicate and other minerals, precipitation, mixing, CO_2 dissolution and redistribution [3,16,17].

The Jimma area in Ethiopia is located in a geologically intricate region characterized by tertiary volcanic rocks and Quaternary alluvial deposits [18]. These rocks exhibit varying degrees of weathering and are composed of fractured basalt, pyroclastic materials, rhyolite, tuffs, and paleosoils. Quaternary alluvial deposits exist in lowland regions, resulting in a heterogeneous aquifer system with the potential for diverse groundwater composition. Despite the crucial role of groundwater in the water supply of the Jimma region, few thorough studies have been conducted on the area's hydrogeochemistry; existing research has primarily focused on water quality parameters rather than the fundamental geochemical processes involved.

Past studies in the Jimma area have established a foundation for understanding groundwater composition, but significant gaps persist. Beyene et al. [19] assessed the suitability of groundwater for drinking and irrigation but did not explore the hydrogeochemical processes leading to elevated fluoride and sodium levels, which have implications for water safety and usability. Bakure et al. [20] studied Gilgel Gibe catchment streams located in the Jimma zone and focused on the investigation of dissolved oxygen (DO), temperature, pH, electrical conductivity (EC), biochemical oxygen demand (BOD), chemical oxygen demand (COD), ammonium (NH_4-N), soluble reactive phosphorous, and nitrate nitrogen (NO_3-N). Also, these authors did not provide any information regarding the processes governing the geochemistry of the groundwater of the Jimma area and its surroundings. In addition, there is the work of the hydrogeochemical study of the Jimma Zone (which is a huge area) investigated at the map sheet level [21]. Thus, parameters like Mg^{2+}, Ca^{2+}, Na^+, HCO_3^-, SO_4^{2-}, Cl^-, K^+, NO_3^-, F^-, CO_2, SiO_2, EC, and pH were investigated. The report found that the dominant water types in the Jimma map sheet are calcium-bicarbonate ($CaHCO_3$). However, the authors did not discuss a comprehensive hydrogeochemical evolution of groundwater, even within the map sheet.

This study seeks to address the gaps identified in previous research by conducting a thorough analysis of the hydrogeochemical evolution and the processes that shape groundwater composition in the Jimma area. Specifically, it will examine the sources and mechanisms of groundwater mineralization and chemical variability, particularly within a volcanically influenced aquifer system. This research is innovative in its integration of extensive chemical data with geochemical modeling to enhance the understanding of factors affecting groundwater quality. By considering the distinctive volcanic and other influences and geochemical reactions, the study offers essential insights for the effective management of groundwater resources in Jimma and similar hydrogeological contexts.

2. Description of the Study Area

2.1. Location and Geographical Settings

The Jimma area is geologically located in a complex of volcanic rocks and alluvial deposits, covering approximately 231 km^2 in the southwestern Oromia Region, Ethiopia. It lies within UTM zone 37, between 36°40′40″ to 36°54′22″ longitude and 7°37′06″ to 7°46′51″ latitude (Figure 1). Jimma city, the largest city in the area, has an estimated population of 239,022 as reported by the population projection for towns as of July 2021. Jimma city is situated within the Oromiya regional state.

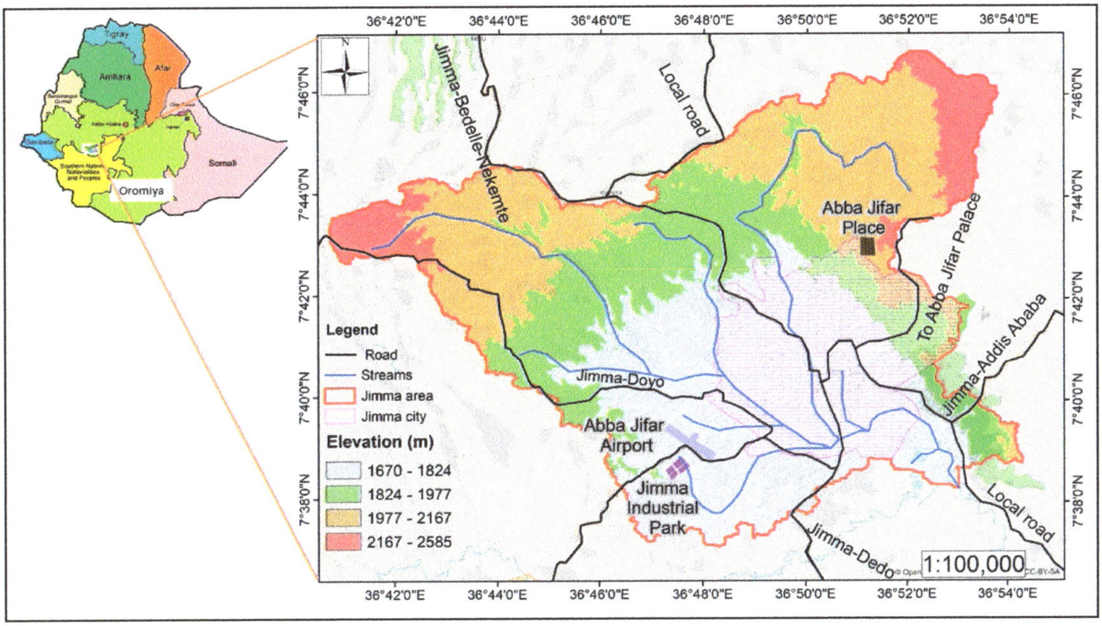

Figure 1. Location map of the Jimma area.

The climate of the Jimma area can be classified as a tropical and sub-tropical climate, distinguished due to the elevation difference [22]. There are four seasons in the study area: from December to February, Winter ('Bona'); March to May, Spring ('Arfaasaa'); June to August, Summer ('Ganna'); and September to November, Autumn ('Birraa') [23]. The average monthly rainfall at the two stations Jimma and Yebu is 152.6 mm and 198.4 mm, respectively, while the average monthly temperature typically ranges from 18 to 20 °C (Figure 2) as described by Tufa et al. [24]. This variation is also observed seasonally, with maximum temperatures from December to February, which are the dry months.

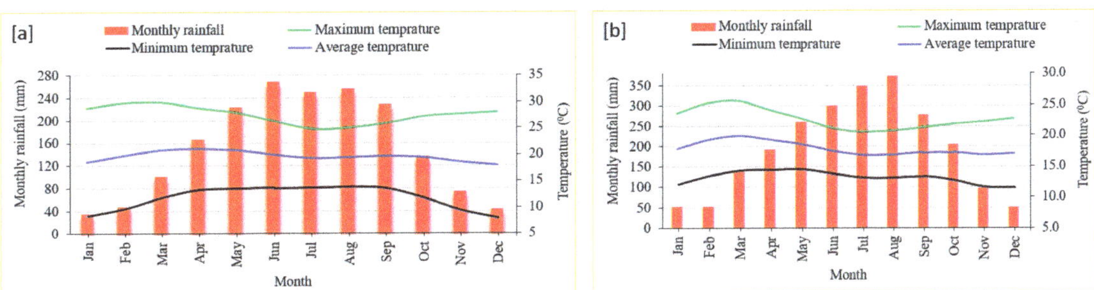

Figure 2. Average monthly rainfall and temperature for the Jimma station (**a**) and for the Yebu station (**b**).

2.2. Geological Settings

The Jimma area complex volcanic rocks include different types, such as basalt, ignimbrite, and tuff, while Quaternary deposits include alluvial and residual covers in a small part of the study area. The volcanic rocks are mainly represented by basalts and are traditionally referred to as the trap succession [25]. A study by Haro et al. [26] clearly identified regionally the geology of the area as a part of the southwestern Ethiopian highlands and associated lowlands. In addition, the classification of the geologic setting of the Jimma map sheet is described by Sima and Agezew [21], Haro et al. [26] and Regassa et al. [27]. Based on their description, the geologic setting of the Jimma area is a combination of Sentema basalt, middle basalt, middle trachyte, and Quaternary alluvial deposits (Figure 3).

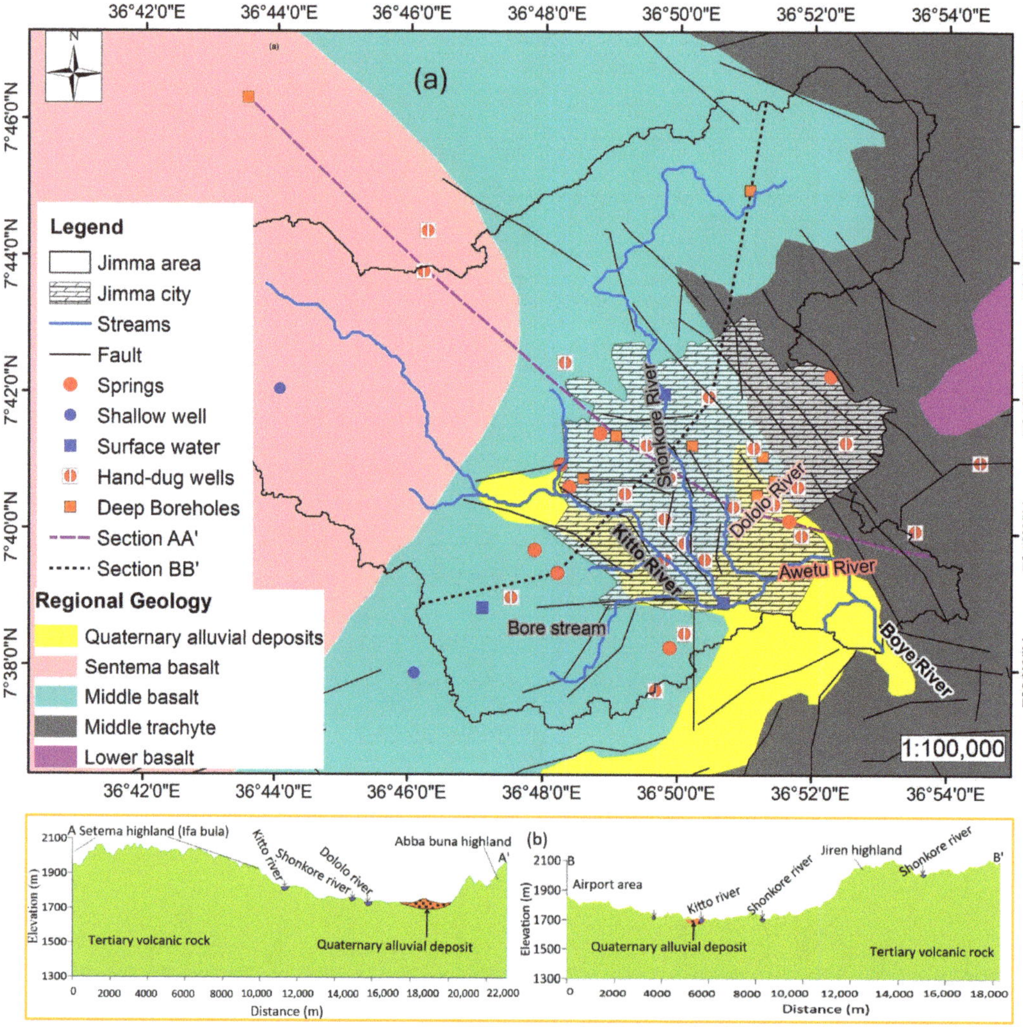

Figure 3. (**a**) Simplified geological map of the Jimma area with sampling points; (**b**) cross-section along AA′ and BB′ after [26].

The weathering of basalt frequently releases ions like calcium, magnesium, and bicarbonate into groundwater, leading to the prevalent formation of calcium-bicarbonate and calcium-magnesium-bicarbonate water types in volcanic areas [28]. Studies in hydrogeochemistry indicate that the dissolution of basalt minerals is a major source of these ions, affecting water chemistry by enhancing alkalinity and hardness through the processes of mineral dissolution and ion exchange. This effect is particularly significant in regions with volcanic geology, where silicate minerals in basalts interact with groundwater, resulting in unique ionic compositions that vary according to environmental and geological factors such as temperature, rainfall, and residence time [28]. Furthermore, the weathering of feldspar minerals found in basalts and trachytes can release sodium and potassium ions, which help form Na-HCO_3 water types typically seen in deeper groundwater systems [29]. Therefore, the mineralogical composition of volcanic rocks is essential for geochemical evolution, as interactions with silicate minerals significantly influence the ionic composition of groundwater.

Physiographically, the Jimma area is located on the plateau of moderate relief [27,30]. It is characterized by a decreasing topographic altitude from the North, culminating in a large plain in the South. The Northern highland, an extension of the Sentema highland, transitions into the Ifa Bula highland, which has a maximum elevation of about 2100 m above mean sea level (a.m.s.l.). Moving eastward from the plain area, the East Kochi highland precedes the Jiren highland, with maximum elevations of about 2080 m above sea level.

As seen in Figure 3a,b, the Kitto Valley, situated to the right of the Aweytu Valley, is drained by the perennial Kitto Surface water. Originating northwest of Jimma city, it flows through an alluvial plain in a southeast direction, with the valley floor lying at a lower elevation of approximately 1708 m a.m.s.l. All streams converge to form the Boye Surface water in the lower part of the Jimma area. The Boye Surface water then turns eastward and, after covering a certain distance, exits the Jimma area to join the Gilgel Gibe Surface water.

2.3. *Hydrogeology of Jimma Area*

The hydrogeological units in the Jimma area comprise three main types of aquifers, each significantly affecting groundwater chemistry. These aquifers include the fractured rock aquifer, the Quaternary alluvial deposit aquifer, and the weathered volcanic rock aquifer (Figure 4). Each type imparts unique chemical properties to the groundwater, reflecting the mineralogical and hydrological dynamics at play within these systems.

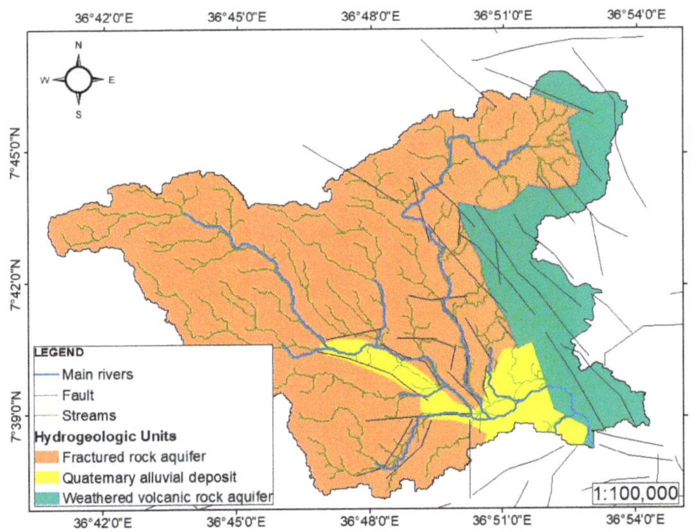

Figure 4. Simplified hydrogeological map of the Jimma area after [26].

The fractured rock aquifer is the largest and most extensive system in the area, primarily formed from basalt and trachyte. Groundwater flows through fissures and open faults, allowing for substantial water movement over considerable distances [31]. While basalt is the dominant composition, the presence of trachyte and rhyolites adds geochemical complexity. The weathering of these rocks introduces ions such as Ca^{2+}, Mg^{2+}, and HCO_3^- into the groundwater, with well-fractured areas promoting groundwater movement and increased mineral dissolution.

The Quaternary alluvial deposit aquifer consists of clayey, sandy, and gravelly materials, forming a thin yet highly porous layer from loose deposits of the Quaternary period. This aquifer effectively transmits water, serving as a productive groundwater source. Stratified tuffs within this layer enhance permeability, facilitating water movement and impacting groundwater chemistry through the dissolution of silicate minerals. The prevalence of sandy and gravelly materials leads to the release of ions like Na^+, K^+, and HCO_3^-, while clay layers can influence cation exchange processes, thereby altering water chemistry [32].

The weathered volcanic rock aquifer (regolith) consists of weathered trachyte lava flows, ignimbrites, rhyolites, and porous tuffs. Other materials deposited by surface waters and landslides create a mixture of loose sediments and fractured rocks, enhancing both water storage and flow. The aquifer's permeability arises from intergranular spaces and rock fractures, facilitating interactions between water and mineral surfaces. This interaction promotes the release of ions such as Ca^{2+}, Mg^{2+}, Na^+, and HCO_3^-. Intercalated sediment layers, although not separate aquifers, contribute to the overall water yield and geochemical input [33].

3. Materials and Methods

3.1. Sample Collection

A total of 51 groundwater samples were gathered from various sources in the Jimma area, which included 10 deep groundwaters, 2 shallow groundwaters, 27 hand-dug well groundwaters, 8 springs, 3 surface waters, and 1 wetland, over the period from March 2023 to January 2024. The sampling locations were carefully chosen to reflect the wider hydrological and geological features of the Jimma area, encompassing a variety of water sources and depths. This method aimed to capture the differences in water chemistry linked to various aquifer types, recharge zones, and mineralization levels in both urban and rural environments.

To ensure that the samples accurately represent in situ groundwater quality, a pumping duration of 10–15 min was employed using a submersible pump for the shallow, hand-dug well, and deep groundwaters prior to sample collection. This procedure was crucial for flushing out stagnant water and obtaining fresh samples. Each water sample was collected in a high-density polyethylene bottle (50 mL capacity), which was rinsed thoroughly with water from the same sampling source before final collection to reduce contamination.

For cation analysis, a subset of samples was filtered through a 0.45 μm membrane filter immediately after collection to eliminate suspended particles. After filtration, these samples were acidified to pH 2 using nitric acid (HNO_3) to stabilize dissolved metal ions and prevent precipitation. Additionally, in situ measurements of pH, electrical conductivity (EC), and temperature were recorded at each sampling site to capture baseline water quality parameters directly at the source.

Samples from deep and shallow groundwaters were analyzed at the Jimma Agriculture Research Center laboratory in Ethiopia, while samples from hand-dug well groundwaters, springs, surface waters, and the wetland were analyzed at the Laboratory for Applied Geology and Hydrogeology of Ghent University in Belgium. To verify the accuracy and reliability of analytical data, the ionic balance (IB %) was assessed using Equation (1), where $\sum c$ and $\sum a$ are the summation of cation and summation of anion concentrations (in meq/L) [34].

$$\mathrm{IB}(\%) = \left(\frac{\sum \mathrm{c} - \sum \mathrm{a}}{\sum \mathrm{c} + \sum \mathrm{a}}\right) \times 100 \quad (1)$$

Further, the analytical data were interpreted using various conventional techniques, including Microsoft Excel 2019 and XLStat, Python 3, ArcGIS 10.8, Surfer 5, Piper [35], and AquaChem 2014.2.

3.2. Hydrogeochemical Correlation Analysis

Davis [36] stated that correlation analysis is used to understand parameters that share a greater resemblance, resulting in their clustering together and forming primary pairs of interest. The correlation coefficient is represented by 'r' which ranges from −1 to +1. A value of −1 denotes a strong negative correlation, suggesting that as one parameter increases, the other decreases consistently. Conversely, a value of +1 signifies a strong positive correlation, indicating that as one parameter increases, so does the other. So this coefficient was conducted to understand the association between physicochemical parameters of the Jimma area.

The correlation results emphasize important geochemical processes and contamination risks that are significant for groundwater management. For example, a strong link between calcium and bicarbonate ions suggests the weathering of calcium aluminosilicate minerals, which affects water hardness and its suitability for use. Likewise, correlations between nitrate sodium and chloride may point to contamination from domestic sewage anthropogenic pollution. Understanding these correlations is crucial for effective groundwater management, as they reveal how both natural processes and human activities can influence groundwater composition. In particular, strong correlations between nitrate and electrical conductivity may indicate agricultural impacts, highlighting the necessity for nitrate monitoring to ensure water safety.

Principal component analysis (PCA) was used to examine variability in groundwater ion concentrations and identify key processes affecting groundwater chemistry as demonstrated by Schiavo et al. [37]. The PCA was conducted with XLSTAT using varimax rotation for better interpretability. The dataset was normalized to ensure comparability among variables with different units. PCA extracted principal components based on eigenvalues over 1, and loading and score plots were analyzed to assess variable contributions and groundwater sample distribution.

3.3. Saturation Indices

The saturation index (SI) is used to determine the equilibrium state (the tendency of the precipitation or dissolution) of minerals in aqueous environments, such as groundwater and surface water [38]. The PHREEQC (version 3) software was used to calculate the SI values of minerals in water [39]. The results of SI were used to understand the trend of water and mineral chemical equilibrium and water–rock interaction. If undersaturated (SI < 0), the mineral will be continuously weathered (dissolution) by the groundwater; if supersaturated (SI > 0), the mineral will precipitate; and if SI is close to 0, the mineral phase will remain in an equilibrium state [40]. Equation (2) is applied to calculate a mineral's SI where IAP represents the ion activity product of chemical species in solution and Ks is the equilibrium solubility product.

$$SI = \log\left(\frac{IAP}{Kmineral}\right) \quad (2)$$

4. Results and Discussion

4.1. Statistical Analysis of Physicochemical Parameters

In this study, groundwater samples show significant variability in ionic balance errors (0.24% to 5.74%). Surface and spring waters exhibit narrower ranges in ionic balance errors (surface water: 1.48% to 4.72%; spring water: 2.33% to 4.89%). Overall, all water samples are suitable for analysis based on these IB ranges [41].

4.1.1. Major Ion Distribution

In Table 1, hand-dug, deep, and shallow wells are categorized as groundwaters for clarity. The table shows pH levels ranging from 4.8 to 8.6. Notably, deep groundwaters in Jimma exhibit high pH values (>8.0), while shallow groundwater shows low pH and HCO_3^- values. TDS values from various sources in Jimma vary significantly. Groundwater samples, including deep and hand-dug wells, range from 36 mg/L to 1184 mg/L, averaging 507 mg/L with a standard deviation of 311 mg/L, indicating variability due to surface and subsurface processes. In contrast, surface water sources have lower TDS values, ranging from 79 mg/L to 136 mg/L, averaging 115 mg/L with a standard deviation of 28 mg/L. These lower readings are expected as surface water is more influenced by rainfall and by shallow local groundwater flow systems, providing less base flow from greater depths. Spring water samples show intermediate TDS values, from 49 mg/L to 176 mg/L, averaging 75 mg/L with a standard deviation of 51 mg/L, indicating that spring water reflects shallow groundwater flow systems and is affected by rainfall but also has subsurface input. TH in the Jimma area ranges from 13.2 to 365.3 mg/L as $CaCO_3$. Groundwater generally has higher TH than surface and spring waters. Deep wells, like DW9, often show increased TH due to prolonged water–rock interactions with calcium and magnesium minerals in a regional flow system, while surface waters have lower TH (36.4–61.4 mg/L) due to limited mineral interaction. Springs have intermediate TH values (16.6–82.1 mg/L). Most surface and spring waters are soft to moderately hard, while groundwater is usually hard to very hard. Elevated TH can affect domestic use by causing scaling and impacting soil quality during irrigation [42,43]. These differences highlight the geological and human factors influencing water quality and the need for targeted management strategies.

Table 1. Descriptive statistics of hydrochemical parameters for water samples' major ions.

Parameters	Groundwaters (Hand-Dug, Shallow, and Deep Well) (n = 39)				Surface Water (River) (n = 3)				Springs (n = 8)			
	Min	Max	Mean	Std	Min	Max	Mean	Std	Min	Max	Mean	Std
pH	4.8	8.6	6.4	0.9	6.2	6.4	6.3	0.1	4.9	6.6	5.5	0.5
EC (µS/cm)	40.1	1601.0	569.0	398.6	82.1	176.3	132.4	38.7	45.3	236.0	100.0	57.6
TDS (mg/L)	36.0	1184.0	340.8	247.5	79.0	136.0	114.7	25.4	40.0	156.0	74.9	36.4
Na^+ (mg/L)	4.0	282.0	39.3	53.9	4.0	10.0	7.0	2.4	2.0	12.0	5.5	2.8
K^+ (mg/L)	1.0	41.0	6.9	7.6	1.0	4.0	2.0	1.4	1.0	3.0	1.6	0.9
Ca^{2+} (mg/L)	2.0	115.0	34.8	27.8	8.0	18.0	12.0	4.3	2.0	23.0	9.8	6.2
Mg^{2+} (mg/L)	1.0	29.0	8.4	6.7	3.0	4.0	3.7	0.5	1.0	6.0	2.9	1.9
Cl^- (mg/L)	0.0	144.0	32.7	38.1	2.0	8.0	4.7	2.5	1.0	18.0	7.9	5.9
$SO_4{}^{2-}$ (mg/L)	0.0	86.0	12.2	22.5	0.5	5.2	2.3	2.1	0.0	4.5	1.3	1.6
$NO_3{}^-$ (mg/L)	0.0	194.0	34.0	53.7	0.1	5.0	1.9	2.2	0.0	17.0	7.0	6.2
$HCO_3{}^-$ (mg/L)	8.0	874.0	170.7	191.1	47.0	89.0	74.7	19.6	18.0	81.0	37.9	21.2
TH (mg/L as $CaCO_3$)	13.2	365.3	121.3	91.6	36.4	61.4	45.1	11.6	16.6	82.1	36.2	22.2

Min—Minimum, Max—Maximum, Std—Standard Deviation, EC—Electrical Conductivity, TDS—Total Dissolved Solids, TH—Total Hardness.

Electrical conductivity (EC) ranges from 40 µS/cm to 1601 µS/cm @ 25 °C as documented in Table 1. Most deep groundwaters exhibit high EC values. This could be attributed to the expanded residence time of groundwater within the geological strata [41]. This enables prolonged interactions with subsurface rocks and minerals, subsequently influencing the ion concentrations in the sampled water in the case of deep groundwaters. But for hand-dug well groundwater Bishishe (HD13), the high EC could be from both evapo-concentration and anthropogenic activities and sewage effects, as the well is found in the flood plain of the Jimma area (Table 1). Total dissolved solids (TDS) obtained from laboratory results are a comprehensive indicator of mineralization in groundwater composition representing the sum of all solutes in the solution [44]. It represents interactions between the water and the geological strata it passes through as well as contributions from the atmosphere and surface water sources. TDS levels vary from 36 mg/L (HD27) to 1184 mg/L (DW9), as reported in Table 1. This variation could be from complex hydrogeological processes (e.g., aluminosilicate dissolution), agricultural runoff, urbanization, and evapo-concentration).

The average dominance of major cations (mg/L) is $Na^+ > Ca^{2+} > Mg^{2+} > K^+$ for deep groundwaters, while for shallow and hand-dug groundwaters, surface waters, and springs, it is $Ca^{2+} > Na^+ > Mg^{2+} > K^+$ (Table 1). In some wells, this prevalence has shifted, illustrating localized differences within the Jimma area despite short distances between sampling, indicating hydrogeochemical heterogeneity. Na+ ranges from 2 mg/L at SP7 to 282 mg/L at DW9, while Ca^{2+} ranges from 2 mg/L at HD16 to 115 mg/L at HD13. Similarly, Mg^{2+} ranges from 1 mg/L at HD26 and HD27 to 29 mg/L at SW2 (Table 1). This concentration is mainly influenced by Mg-bearing aluminosilicates like augite and amphiboles. K+ ranges from 1 to 41 mg/L due to the weathering of aluminosilicate minerals and local pollution from fertilizers [45].

From Table 1, groundwater in Jimma exhibits a varied ion distribution influenced by geological conditions and human activities. The pH levels range significantly from acidic (4.89) to alkaline (8.59); deep groundwaters (e.g., DW1 and DW3) typically have higher pH due to extended water–rock interactions, whereas lower pH in shallow groundwaters (e.g., HD24) indicates surface contamination. EC and TDS peak in deep groundwaters (e.g., DW4 and DW9), suggesting prolonged mineral dissolution. Hand-dug groundwaters located near urban regions show increased NO_3^- and Cl^- levels, likely due to contamination, while higher SO_4^{2-} concentrations in lowland areas suggest agricultural runoff. By linking these observations to the specific hydrological and geological contexts of the Jimma area, these findings can inform targeted groundwater management strategies, such as identifying critical areas for pollution prevention and promoting sustainable groundwater utilization.

In the Jimma area, HCO_3^- is the dominant anion making up about 69% of all anions whereas NO_3^- and Cl^- cover about 13.1% and 12.9%, respectively, making $HCO_3^- > NO_3^- > Cl^- > SO_4^{2-}$. The sources of HCO_3^- are CO_2 dissolution and aluminosilicate weathering [45,46] and biogenic activity [47]. But, NO_3^- exceeds HCO_3^- in hand-dug well groundwaters HD7, HD12, HD18, and HD25 [(150 mg/L vs. 12 mg/L), (194 mg/L vs. 12 mg/L), (31 mg/L vs. 16 mg/L), and (177 mg/L vs. 8 mg/L) NO_3^- and HCO_3^-, respectively]. These wells are located in Jimma city. Therefore, the cause for NO_3^- surpassing HCO_3^- in these wells is probably groundwater contamination by sewage from human and animal wastes [48]. In addition, Cl^- significantly surpasses HCO_3^- in the same hand-dug groundwaters HD7 (48 mg/L vs. 12 mg/L), HD12 (67 mg/L vs. 12 mg/L), HD18 (75 mg/L vs. 16 mg/L), and HD25 (76 mg/L vs. 8 mg/L Cl^- and HCO_3^-, respectively). This can be described as sewage [49]. Sewage could enter the groundwater through spills, leaks, or improper disposal. Moreover, higher SO_4^{2-} is observed in the lowland area (e.g., HD13 SO_4^{2-} = 86 mg/L). This may be due to pollution from various sources, such as agricultural runoff, waste disposal, and sewage [50,51]. Poor waste disposal around the upland areas of Jiren and Ifa bula may contain H_2S which is further oxidized and transported to this flood plain causing the elevated SO_4^{2-} concentrations [52].

4.1.2. Correlation Analysis of Physicochemical Parameters

From Table 2, TDS exhibits strong correlations with Na^+ (r = 0.84) and HCO_3^- (r = 0.89), has a moderate correlation with Ca^{2+} (r = 0.58), and has a positive correlation with Mg^{2+} (r = 0.48). K^+ shares low correlation with Na^+ (r = 0.15) indicating both ions could have different origins [53]. Ca^{2+} shows a very poor correlation with Na^+ (r = 0.06). This suggests that different geochemical processes govern the concentrations of these ions in the Jimma area. Anorthite and albite aluminosilicate minerals weathering could be their sources [54]. On the other hand, Ca^{2+} exhibits a strong association with Mg^{2+} (r = 0.76). This indicates similar sources and/or both are affected by the geochemical processes in the aquifer system. Ca^{2+} and Mg^{2+} may originate from aluminosilicate minerals such as pyroxene, amphibole, and augite. EC has a negative correlation with Cl^- and NO_3^-. Since Cl^- and NO_3^- concentrations in deep groundwaters are very low, they are unlikely to originate from deep sources. Instead, the weathering of minerals contributes to increased EC and water alkalinity. However, Cl^- and NO_3^- concentrations in groundwater could result from anthropogenic activities. The positive correlation between NO_3^- and Cl^- (r = 0.52) suggests common sources of anthropogenic pollution.

Table 2. Correlation matrix of physicochemical parameters.

	pH	EC	TDS	Na^+	K^+	Ca^{2+}	Mg^{2+}	Cl^-	SO_4^{2-}	NO_3^-	HCO_3^-
pH	1										
EC	0.48	1									
TDS	0.46	0.83	1								
Na^+	0.35	0.62	0.84	1							
K^+	0.17	0.38	0.37	0.15	1						
Ca^{2+}	0.37	0.58	0.58	0.06	0.38	1					
Mg^{2+}	0.04	0.61	0.48	0.04	0.30	0.76	1				
Cl^-	−0.13	0.34	0.34	0.04	0.36	0.60	0.71	1			
SO_4^{2-}	0.20	0.32	0.37	0.05	0.51	0.60	0.51	0.61	1		
NO_3^-	−0.42	0.15	0.09	−0.02	0.06	0.19	0.39	0.52	−0.03	1	
HCO_3^-	0.59	0.70	0.89	0.86	0.21	0.34	0.17	−0.08	0.13	−0.30	1

4.2. Graphical Presentation of Water Types

The Piper diagram developed by Piper [35] is used to analyze and characterize groundwater types and the governing geochemical processes (Figure 5). Figure 5b highlights several water types based on ion composition, including CaCl, $CaNO_3$, $CaSO_4$, $CaHCO_3$, $NaHCO_3$, and mixed types (e.g., Ca-Mg-HCO_3, Ca-NaHCO, Mg-$NaHCO_3$, $CaHCO_3$-Cl-NO_3, and $CaHCO_3$-SO_4-Cl). Dominant water samples are of the $CaHCO_3$ type. There is a clustering of the NaHCO3 type in the bottom corner of Figure 5b, suggesting older, highly mineralized water types in the Jimma area. Additionally, water samples with both $NaHCO_3$ and $CaHCO_3$ types result from aluminosilicate weathering. These waters have TDS between 370 and 1184 mg/L. For example, groundwaters such as DW1 (TDS = 587 mg/L), DW6 (TDS = 472 mg/L), DW8 (TDS = 992 mg/L), DW9 (TDS = 1184 mg/L), DW10 (TDS = 370 mg/L), HD3 (TDS = 693 mg/L), and HD11 (TDS = 385 mg/L) show high concentrations of Na^+ (27–230 mg/L), Ca^{2+} (17–81 mg/L), K^+ (1.8–10 mg/L), Mg^{2+} (2.2–11 mg/L), and HCO_3^- (153–875 mg/L), while NO_3^- and SO_4^{2-} concentrations are very low. These values confirm the importance of the dissolution of aluminosilicate minerals, possibly supported by the evapo-concentration process.

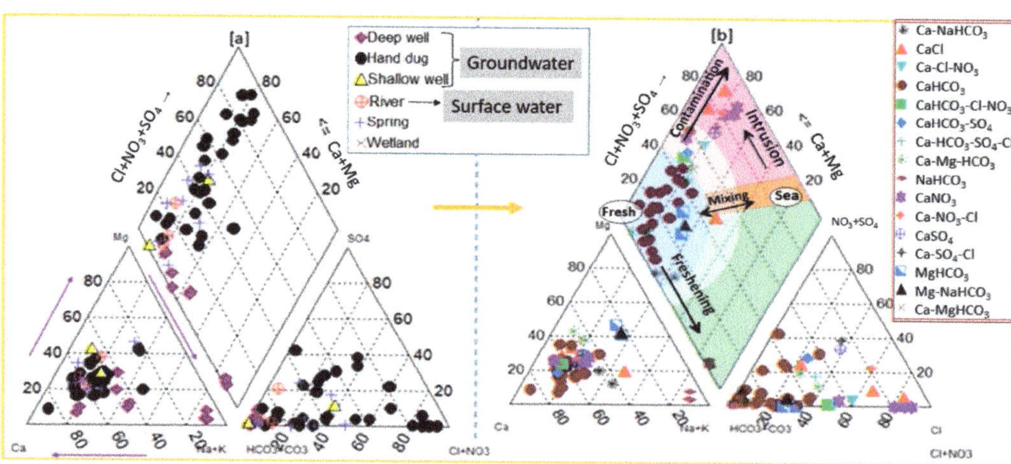

Figure 5. Piper diagram showing (**a**) location of water samples and (**b**) groundwater types.

Water types $CaHCO_3$, $MgHCO_3$, and Ca-Mg-HCO_3 with low mineralization, characterized by low levels of Na^+, Ca^{2+}, HCO_3^-, Mg^{2+}, K^+, NO_3^-, and SO_4^{2-}, are observed mostly in springs, surface waters, and some groundwaters (Table 1). These waters are immature, originating from rainwater infiltration and short residence times in the aquifer [41]. They may evolve from the NaCl end member to $CaHCO_3$ and $MgHCO_3$ due to CO_2 dissolution and limited aluminosilicate weathering.

Moreover, this Piper diagram shows some hand-dug well groundwaters and springs classified as chloride, sulfate, and nitrate-related water types (e.g., CaCl, $CaNO_3$, Ca-SO_4-Cl, Ca-HCO_3-SO_4-Cl, and $CaHCO_3$-Cl-NO_3), associated with anthropogenic pollution from agricultural fertilizers and sewage, contaminating immature infiltrating waters. In these waters, NO_3^- and SO_4^{2-} are not reduced by anaerobic bacteria due to short, shallow flow paths in the oxic zone [41]. Generally, the dissolution of aluminosilicates, evaporative concentration, and anthropogenic activities control the groundwater types in the Jimma area.

The Piper diagram in Figure 5 shows distinct clustering patterns that highlight important hydrogeochemical processes and water types in Jimma's groundwater. The grouping of $NaHCO_3$ type water in the lower corner indicates older, highly mineralized groundwater with extensive water–rock interaction, suggesting long residence times and aluminosilicate weathering in deeper aquifers. Conversely, $CaHCO_3$ and Ca-Mg-HCO_3 waters, which are less mineralized, are typical of younger groundwater from recent precipitation and shorter aquifer flow paths, commonly found in surface waters, springs, and shallow groundwaters. Water types like CaCl, $CaNO_3$, and Ca-SO_4-Cl, found with high concentrations of Cl-, NO_3^-, and SO_4^{2-}, suggest anthropogenic contamination from agricultural runoff and wastewater, particularly in shallow hand-dug well groundwaters with limited filtration. Overall, the clustering reflects a combination of natural geochemical evolution and anthropogenic influences on Jimma's groundwater composition.

Finally, Figure 5b enhances the classification of groundwater by differentiating between fresh, mixing, and high-mineralization water types. This differentiation assists in recognizing and utilizing various water qualities in the Jimma area. High-mineralization water, generally found in deeper aquifers, indicates extended interactions between water and rock, as well as longer residence times, such as $NaHCO_3$. Mixing water serves as an intermediate type, displaying features of both fresh and mineralized waters due to moderate residence times and processes like aluminosilicate weathering. Fresh water encompasses low-mineralization types, such as $CaHCO_3$ and Ca-Mg-HCO_3, which are indicative of recent recharge from precipitation and short flow paths in surface waters, springs, and some hand-dug wells. The chemical composition trends shown in the Piper diagram imply that the increased salinity in certain groundwater samples is a result of inland geogenic processes, such as mineral dissolution, rather than any association with coastal seawater or seawater-like end members.

4.3. Principal Component Analysis

The role of principal component analysis (PCA) in understanding the variability of ion concentrations in groundwater is demonstrated [37]. The authors employed geostatistical analysis through PCA to identify the components responsible for variability in arsenic concentrations. PCA enhances the analysis and interpretation of groundwater chemistry data in the Jimma area. Figure 6 summarizes the PCA analysis and groundwater chemistry findings. Figure 6a, the PCA scree plot, illustrates the eigenvalues for each principal component, aiding in component selection based on explained variance. Figure 6b, the PCA loading plot for F1 (43.77%) and F2 (24.08%), identifies variable contributions, showing that Na^+ and HCO_3^- significantly influence F1 (associated with silicate weathering and with the action of CO_2), while Cl^- and SO_4^{2-} relate to F2 (linked to evaporative concentration and pollution). Figure 6c, the PCA score plot for F1 and F2, displays the distribution of groundwater samples, distinguishing between deep groundwaters, shallow groundwaters, and springs, highlighting different geochemical processes and residence times.

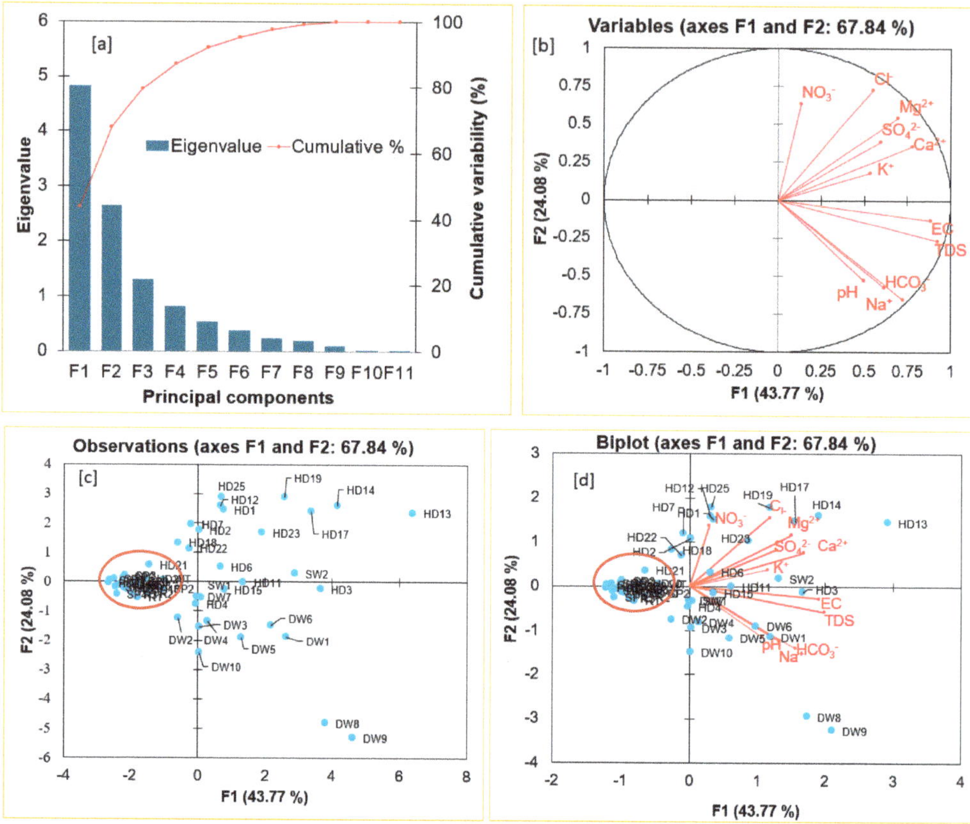

Figure 6. PCA analysis and hydrogeochemical characteristics (**a**–**d**).

Figure 6d, the biplot, presents samples and variables defined by F1 and F2, emphasizing contributions and distributions. Clusters indicate similar chemical compositions, while the direction and length of the vectors reflect variable influence, revealing patterns in the groundwater chemistry of the Jimma area. In interpreting the PCA results, F1 (aluminosilicate weathering and CO_2 influence) shows high loadings of Na^+ and HCO_3^-, indicating that silicate weathering dominates, with CO_2 from mantle-magma degassing contributing to bicarbonate levels in deeper aquifers. The influence of Ca^{2+} and Mg^{2+} suggests other processes play a role, though less than silicate weathering. F2 (collective processes like evapo-concentration and anthropogenic pollution) displays loadings of Cl^- and SO_4^{2-}. The high loading of NO_3^- on F2 highlights the impact of anthropogenic activities, such as agricultural runoff. In Figure 6c,d, samples on the left exhibit lower F1 and F2 scores (indicated by dark red), reflecting reduced values for variables with high positive loadings. This grouping suggests chemical similarity and shared geochemical influences, such as diminished impacts from silicate weathering, the dissolution of CO_2, evapo-concentration, or anthropogenic activities, or possibly other processes. To sum up, the identification of 11 components (F1–F11) suggests a complex system with various factors affecting the variance in groundwater data (Figure 6a). This complexity relates to the diverse volcanic aquifer system in the Jimma area, along with factors such as geochemical processes including silicate weathering, CO_2 effects, and anthropogenic pollution, which are examined in the subsequent sections of this work.

4.4. Hydrogeochemical Processes

The changes in groundwater chemistry in the Jimma area are heavily affected by the weathering of aluminosilicate minerals found in volcanic rocks, a process intensified by the dissolution of CO_2 in water. When CO_2 dissolves in water, it creates carbonic acid (H_2CO_3), which then breaks down into H^+ and HCO_3^- ions, resulting in lower pH levels and increased acidity in the water, as illustrated in Equations (3) and (4) [3,55]. This acidity facilitates the disintegration of aluminosilicate minerals by promoting ion release through hydrolysis, wherein H^+ ions target the silicon–oxygen bonds in silica tetrahedra, leading to the breakdown of mineral structures.

$$CO_2 + H_2O \rightleftharpoons H_2CO_3 \quad (3)$$

$$H_2CO_3 \rightleftharpoons H^+ + HCO_3^- \rightleftharpoons 2H^+ + CO_3^{2-} \quad (4)$$

The dissolution processes of significant aluminosilicate minerals in the Jimma area, such as anorthite, albite, amphibole, biotite, and pyroxene, are detailed in Equations (5)–(12) for anorthite, amphibole, augite, and pyroxene weathering. This reaction releases calcium and other cations into the groundwater, which contributes to the high concentrations of Ca^{2+} and HCO_3^- observed in the area. Albite and other aluminosilicate feldspar minerals undergo similar dissolution processes, adding Na+ and HCO_3^- to the water as well [56].

Moreover, in the highly mineralized waters from deep wells in the Jimma area, elevated total dissolved solids (TDS) and ion levels (Na^+, Ca^{2+}, and HCO_3^-) signify extended interaction with aluminosilicate rocks, a characteristic of older, well-mixed waters. These findings align with research conducted in other volcanic regions [45,56], reinforcing the notion that prolonged residence times and mineral weathering are crucial to the area's hydrochemical evolution. These relationships between CO_2-driven mineral weathering and anthropogenic pollution shape the hydrochemical characteristics observed in the Jimma area.

$$2CaAl_2Si_2O_8 + 4CO_2 + 6H_2O \rightleftharpoons 2Al_2Si_2O_5(OH)_4 + 2Ca^{2+} + 4HCO_3^- + H_4SiO_4 \quad (5)$$

Anorthite to Kaolinite

$$Ca_2Mg_5Si_8O_{22}(OH)_8 + 14CO_2 + 22H_2O \rightleftharpoons 2Ca^{2+} + 5Mg^{2+} + 14HCO_3^- + 8H_4SiO_4 \quad (6)$$

Amphibole

$$2NaAlSi_3O_8 + 9H_2O + 2H_2CO_3 \rightleftharpoons Al_2Si_2O_5(OH)_4 + 2Na^+ + 2HCO_3^- + 4H_4SiO_4 \quad (7)$$

Albite to Kaolinite

$$2K(Mg_3Fe)(AlSi_3)O_{10}(OH)_2 + 7H_2O + 5H_2CO_3 \rightleftharpoons Al_2Si_2O_5(OH)_4 + 2K^+ + 4Mg^{2+}$$

$$\text{Biotite} + 2Fe(OH)_3 + 5HCO_3^- + 4H_4SiO_4 \quad (8)$$

$$Na_2KAl_4Si_{12}O_{16} + 4H_2CO_3 + 2H_2O \rightleftharpoons 2Al_2Si_{12}O_5(OH)_4 + 3Na^+ + K^+ + 4HCO_3^- + H_4SiO_4 \quad (9)$$

Nepheline

$$(CaMg_{0.7}Al_{0.6}Si_{1.7})O_6 + 3.4CO_2 + 4H_2O \rightleftharpoons 0.3Al_2Si_2O_5(OH)_4 + Ca^{2+} + 0.7Mg^{2+} + 3.4HCO_3^- + 1.1H_4SiO_4 \quad (10)$$

Augite

$$CaMg(Si_2O_6) + 4CO_2 + 6H_2O \rightleftharpoons Ca^{2+} + Mg^{2+} + 4HCO_3^- + 2H_4SiO_4 \quad (11)$$

Pyroxene

$$2KAlSi_3O_8 + 9H_2O + 2H_2CO_3 \rightleftharpoons Al_2Si_2O_5(OH)_4 + 2K^+ + 2HCO_3^- + 4H_4SiO_4 \quad (12)$$

Orthoclase feldspar (Sanidine) to Kaolinite

4.5. Bivariate Plots and Their Interpretation

The chemical composition of groundwater samples reveals the geochemical processes shaping groundwater composition. Bivariate plots in Figures 7 and 8 illustrate the relationships between electrical conductivity (EC, μS/cm at 25 °C) and major ions (Na^+, Ca^{2+}, Mg^{2+}, K^+, and HCO_3^-). High EC values generally indicate mature waters with extensive mineral dissolution due to longer residence times [45,56]. Thus, EC in the Jimma area reflects aluminosilicate weathering and is further increased by evapo-concentration and anthropogenic contamination. Aluminosilicate weathering releases Na^+, Ca^{2+}, Mg^{2+}, K^+, and HCO_3^- into groundwater, as shown in Equations (5)–(12) [41,46]. Data in Figure 7a–d show a strong correlation between EC and Na^+ ($r^2 = 0.99$ and 0.62), K^+ ($r^2 = 0.71$), Mg^{2+} ($r^2 = 0.69$), Ca^{2+} ($r^2 = 0.61$), and HCO_3^- ($r^2 = 0.94$, Figure 8a), indicating that EC and ion concentrations increase together, aligning with TDS patterns. Aluminosilicate minerals release these ions through interactions with water, where HCO_3^- results from H_2CO_3 dissociation, facilitating mineral breakdown (Equations (3) and (4)). Figure 7c,d indicate that Mg^{2+} and Ca^{2+} increase with EC, though some encircled samples show weaker relationships, likely due to rock composition.

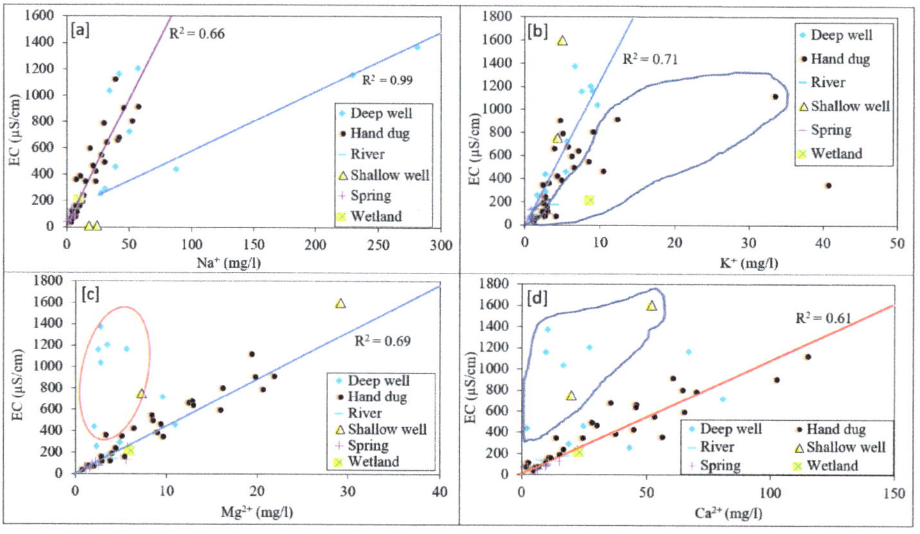

Figure 7. Bivariate plots of EC vs. major ion hydrogeochemical parameters in mg/L (**a**–**d**).

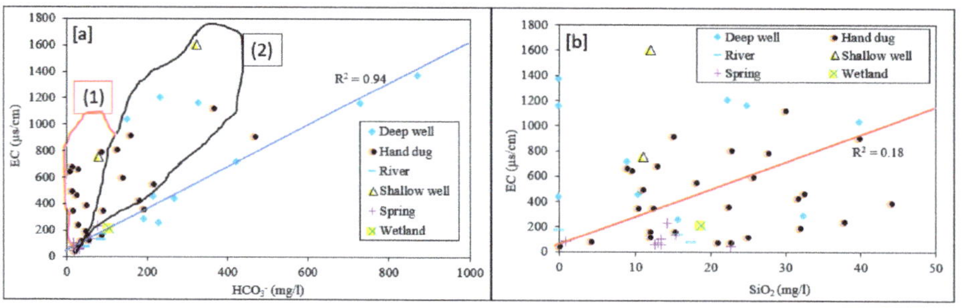

Figure 8. Bivariate plots of (**a**) EC vs. HCO_3^-, and (**b**) EC vs. SiO_2.

Encircled samples in Figure 7c,d also show differences in Mg^{2+}/Ca^{2+} with rising EC, indicating groundwater evolution. Increasing mineralization from the $CaHCO_3$ type stems from the weathering of anorthite, a calcium-rich feldspar. With deeper weathering, albite (sodium-rich feldspar) dominates, producing $NaHCO_3$ type water. At deeper levels (DW7, DW8, DW9, and DW10), cation exchange processes may occur, where Na^+ replaces Ca^{2+} [56]. This cation exchange, shown in Equations (13) and (14), may alter ion concentrations as they adsorb onto clay minerals, such as montmorillonite.

$$Ca^{2+} + 2Na - M \Rightarrow 2Na^{+} + Ca - M \quad (13)$$

$$Mg^{2+} + 2Na - M \Rightarrow 2Na^{+} + Mg - M \quad (14)$$

Although cation exchange is often cited as the mechanism for the shift from a $CaHCO_3$ to $NaHCO_3$ water type, it requires an external source of Na^+ to initiate exchange. In marine aquifers, seawater supplies Na^+, but this is irrelevant in Jimma. Instead, in Jimma's volcanic setting, albite mineral could supply Na^+.

Figure 7b reveals a weak correlation between SiO_2 and EC (r^2 = 0.18), indicating that high silica concentrations do not necessarily correspond with increased mineralization. This results from silica's low solubility and slow silicate dissolution rates, causing much of it to remain in secondary minerals rather than dissolve. Shallow water samples (surface waters, springs, wetlands, and some hand-dug well groundwaters) show low EC, with SiO_2 at shallow depths typically ranging from 0 to 20 mg/L [41]. SiO_2 variation at low mineralization may be due to processes like amorphous silica dissolution or kaolinite to gibbsite transformation.

Figures 9 and 10 display bivariate plots of HCO_3^- versus major cations (Na^+, Mg^{2+}, K^+, and Ca^{2+}), indicating these ions primarily result from the aluminosilicate weathering of feldspars, amphiboles, augite, biotite, and nepheline. In Figure 9a, the Na^+–HCO_3^- relationship reflects albite (1:1) and nepheline (3:4) weathering. Some hand-dug groundwaters and springs plot near the 1:1 line, indicating albite weathering as a key source of Na^+ and HCO_3^-. Most samples plot above the 1:1 and 3:4 lines, showing HCO_3^- enrichment from other processes. A correlation coefficient (r^2 = 0.58, Table 2) suggests mixed sources for these ions.

Figure 9b explores the HCO_3^-–Mg^{2+} relationships from biotite (8:5), augite (7:17), and pyroxene (1:2) weathering. Deep groundwaters mostly plot above the 7:17 line, indicating more HCO_3^- than these minerals release. Some samples align or are near these lines, showing that biotite, pyroxene, amphiboles, and augite contribute Mg^{2+} and some HCO_3^-. However, feldspar weathering (Figures 7 and 8) and mixed sources likely explain the excess HCO_3^-.

Figure 10a illustrates the HCO_3^- and K^+ relationship as predicted by the weathering of sanidine (1:1 line) and nepheline (1:4 line), based on Equations (9) and (12). When zoomed out, the pattern resembles that in Figure 9a. Most samples plot above the lines for the weathering of nepheline and sanidine, indicating a substantial excess of HCO_3^- from processes other than the weathering of these minerals, likely from other aluminosilicate minerals, as shown in Figures 7 and 8 [57,58].

Figure 9b shows a scatter plot of HCO_3^- versus Ca^{2+}, with Equations (5), (6), (10), and (11) representing the weathering processes that release Ca^{2+} and HCO_3^- to groundwater. Some samples above the 1:1 line (anorthite) indicate an excess of HCO_3^- compared to what is predicted by the weathering of anorthite. A few samples, such as those from shallow groundwaters, wetlands, surface waters, springs, and hand-dug well groundwaters, align with the 1:1 line. This suggests that the weathering of anorthite is contributing Ca^{2+} to the groundwater, along with other processes contributing more HCO_3^-. The 10:17 line, representing the weathering of augite, indicates that augite may also contribute Ca^{2+} and HCO_3^- to the groundwater, particularly for deep groundwaters (DW2, DW4, DW5, and DW7). Samples from DW8, DW9, and DW10, plotting above the lines for anorthite, augite, pyroxene, and amphibole, suggest higher HCO_3^- inputs than those predicted by the weathering of these minerals.

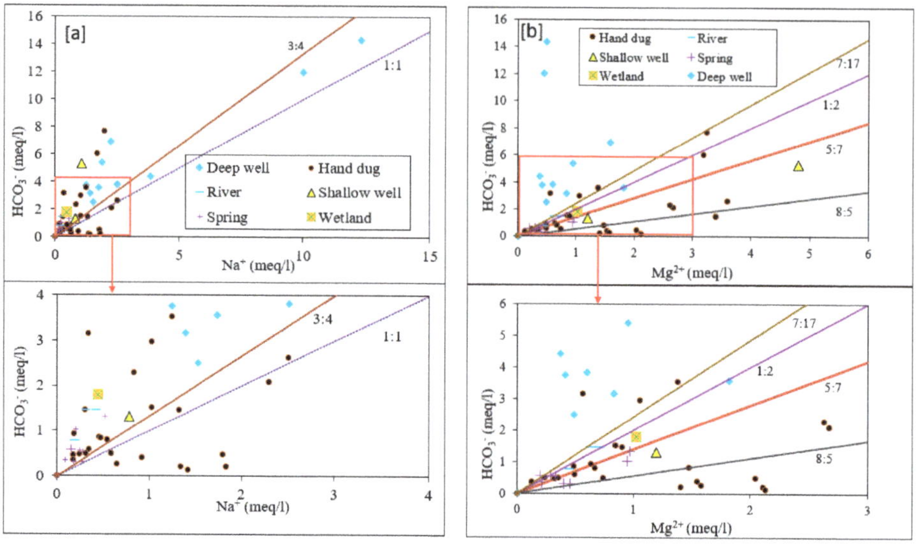

Figure 9. Bivariate plots of (**a**) HCO_3^- vs. Na^+ and (**b**) HCO_3^- vs. Mg^{2+}.

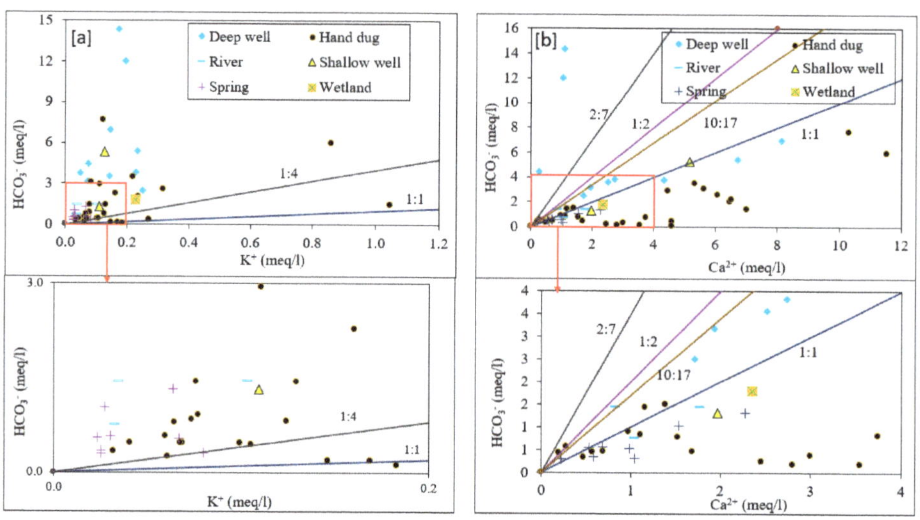

Figure 10. Bivariate plots of (**a**) HCO_3^- vs. K^+ (meq/l) and (**b**) HCO_3^- vs. Ca^{2+} (meq/l).

4.6. Saturation Index

Saturation indices (SIs) can be used in geochemical modeling to predict the precipitation or dissolution of minerals in various environments, such as groundwater, surface water, soil, and rock systems [38]. The PHREEQC software can be used to calculate the SI values of minerals in groundwater [39]. The results of SI suggest the trend of water and mineral chemical equilibrium and water–rock interaction. If undersaturated (SI < 0), the mineral will be continuously weathered (dissolution) by the groundwater; if supersaturated

(SI > 0), the mineral will precipitate; and if SI is close to zero, the mineral phase will remain in an equilibrium state [54,59].

The SI calculation shows that all water samples are undersaturated with respect to anhydrite ($CaSO_4$) except Ifa Bula (HD8) (which has no SI for anhydrite as the value of SO_4^{2-} concentration is zero), Abba Jifar (HD17 SI = 0.31), S/Semero (HD23 SI = 0.13), and JIT Spring (SP8 SI = 0), which are considered to be at equilibrium (Figure 11). On average, most groundwaters in the Jimma area are typically found to be oversaturated with respect to calcite ($CaCO_3$), dolomite ($CaMg(CO_3)_2$), chalcedony (SiO_2), and quartz. This means that these minerals have a tendency to precipitate from the groundwater. Gypsum ($CaSO_4{:}2H_2O$) and anhydrite ($CaSO_4$) minerals are undersaturated. This is because these minerals are not present in the groundwater reservoir, and therefore cannot dissolve. All water samples are undersaturated with respect to chrysotile ($Mg_3Si_2O_5(OH)_4$) except three deep groundwaters (DW1 SI = 6.62, DW5 SI = 3.25, and DW6 SI = 1.26), which are oversaturated. As observed from Figure 11, all deep groundwaters, except DW7 SI = −2.7 and DW9 SI = −2.99, are oversaturated with respect to sepiolite ($Mg_2Si_3O_{7.5}OH{:}3H_2O$). In addition to these two deep groundwaters, DW7 and DW9, all springs, shallow groundwaters, wetlands, and hand-dug groundwaters are undersaturated, except HD3 SI = 1.06, HD4 SI = 1.29, HD11 SI = 0.39, HD13 SI = 1.03, HD14 SI = 1.11, and HD15 SI = 2.23, which are oversaturated (Figure 11). Moreover, all deep groundwaters are oversaturated with respect to talc ($Mg_3Si_4O_{10}(OH)_2$) except DW7 and DW9.

This SI analysis highlights the environmental and practical implications for mineral stability and groundwater quality in the Jimma area. The oversaturation of minerals such as calcite, dolomite, chalcedony, and quartz indicates a significant potential for precipitation, which could result in scaling within infrastructure, impacting water distribution and storage systems [38]. Conversely, the undersaturation of minerals like gypsum, anhydrite, and chrysotile points to active dissolution, releasing ions into the groundwater that affect its hardness and suitability for domestic and agricultural use. The variation in SI values among water sources reveals that certain deep groundwaters are oversaturated with minerals like sepiolite and talc, while shallow groundwaters and other sources are generally undersaturated, reflecting different water–rock interactions based on aquifer depth and mineral exposure. These results underscore the environmental and practical importance of mineral saturation patterns, reinforcing the necessity for targeted groundwater management strategies to prevent scaling and regulate ion concentrations for sustainable water use in the Jimma area.

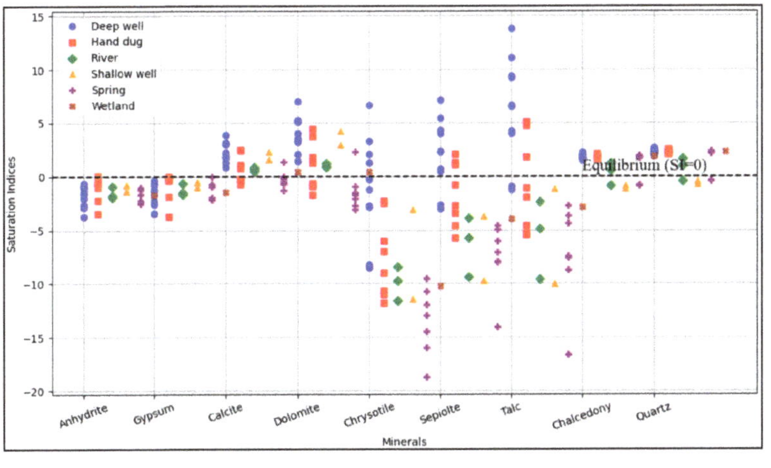

Figure 11. Saturation index of different minerals.

4.7. Hydrochemical Stability Diagram

Stability diagrams were developed for the three feldspar groups to see hydrogeochemical evolution [39,45]. PHREEQC [39] was used to calculate the ion activities of H^+, Ca^{2+}, K^+, Na^+, Mg^{2+}, and H_4SiO_4. In the stability diagrams, the *x*-axis is represented by $\log [a_{H_4SiO_4}]$, and the *y*-axis is represented by $\log [(a_{K^+})/(a_{H^+})]$, $\log [(a_{Na^+})/(a_{H^+})]$ and $\log [(a_{Ca^{2+}})/(a_{H^+})^2]$, respectively (Figure 12).

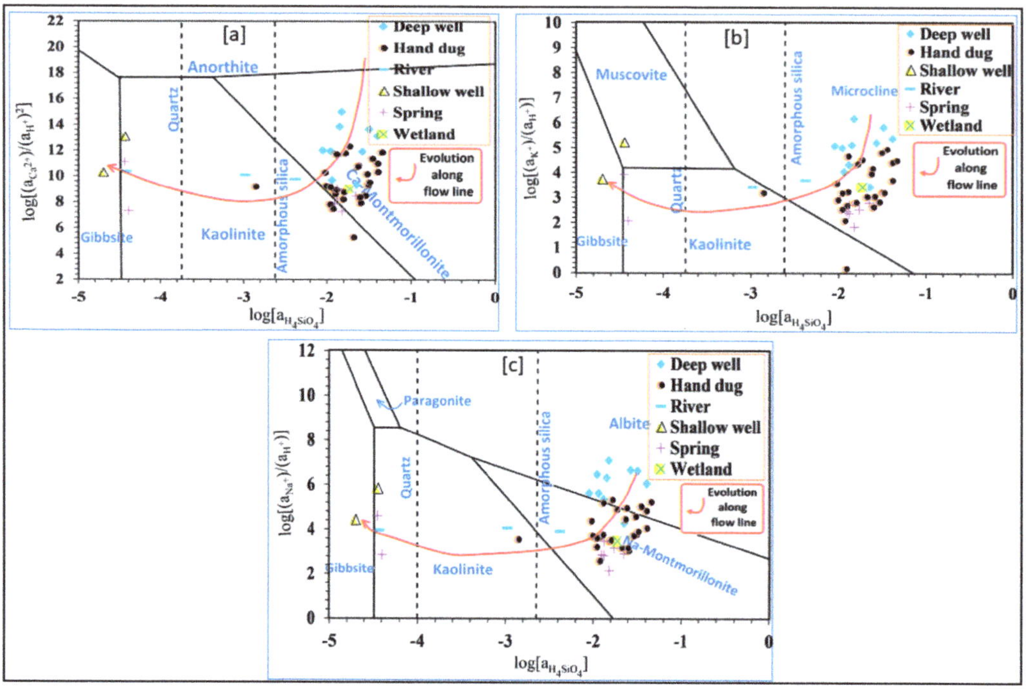

Figure 12. Stability diagrams of aluminosilicate feldspars and their weathering products in the Jimma area after [41,54]. (**a**) Most samples are stable with Ca-montmorillonite; a few align with kaolinite, while anorthite is unstable and water is oversaturated with quartz (**b**) Samples largely fall within the microcline stability field; muscovite alters to kaolinite (**c**) About 65% of samples are stable with Na-montmorillonite, while deep groundwater aligns with albite.

Figure 12 again shows how groundwater chemistry evolves along the flow path, more or less similar to Figure 10. The most strongly leached parts of the flow lines are in equilibrium with gibbsite: this is found in some springs, some surface water samples and the shallow groundwaters. The deep groundwaters belong to the least weathered parts of the flow lines, in which albite, Ca-montmorillonite and microcline are stable. Intermediate weathering is characterizing the hand-dug groundwaters, where both Na- and Ca-montmorillonite are stable, as well as the K-feldspar microcline.

Figure 12a illustrates that few samples of the groundwater are in equilibrium with kaolinite (R1, R2, HD9, and HD10) and most of the groundwater samples are stable with Ca-montmorillonite. Anorthite is unstable for all water samples and will be weathered to Ca-montmorillonite or further advanced to kaolinite. All samples are at the right side of the quartz saturation line, indicating the groundwater is oversaturated with respect to quartz except for five samples at the left side of the quartz saturation line. These samples are R3, SW1, SW2, SP1, and SP2 as clearly justified by the negative saturation indices of quartz for these water samples in Figure 11. Moreover, most samples are oversaturated with respect to amorphous silica as they fall to the right side of the amorphous silica solubility line except for the nine samples on the left side (Figure 11).

Figure 12b indicates that most samples fall within the microcline stability field. Water samples R2, HD2, and HD10 fall within the kaolinite stability field, indicating equilibrium with kaolinite; kaolinite is stable for these samples. Additionally, one sample (SP1) is in equilibrium with both gibbsite and kaolinite, while SW1 is in equilibrium with both gibbsite and muscovite. Generally, the less mineralized groundwater samples lie within the stability field of kaolinite, while the more mineralized ones lie within the stability field of microcline (almost all of the samples). In addition, as depicted in Figure 12b, muscovite would undergo dissolution according to Equation (15) in the presence of water and dissolve under any condition in contact with groundwater, resulting in the alteration of muscovite into kaolinite [60].

Figure 12c shows that six of the groundwater samples are in equilibrium with kaolinite (SW1, SP1, SP2, R2, R3, and HD10). However, about 65% of groundwater samples are in equilibrium with Na-montmorillonite. In addition, the less mineralized waters (six samples) lie within the kaolinite stability field while the more mineralized samples lie within the albite and Na-montmorillonite stability fields. Albite is stable for nine deep groundwaters and four hand-dug groundwaters, but for the rest of the samples, it will be weathered to Na-montmorillonite. Only one shallow groundwater lies within the stability field of gibbsite, reflecting poor drainage conditions, and therefore stronger leaching [41]. Moreover, in Figure 12c, paragonite would undergo dissolution in the presence of water and dissolve under any condition in contact with groundwater, leading to the alteration of paragonite into kaolinite (Equation 16)). Finally, as observed in Figure 12a–c, the ion activity of H_4SiO_4 in shallow groundwaters is lower than in deep groundwaters. This is because shallow groundwaters have lower SI values for SiO_2 (SW1 SI of $SiO_2 = -1.72$ and SW2 SI of $SiO_2 = -1.97$), whereas in deep groundwaters, the SI values of SiO_2 range from 0.82 to 1.34, in line with the maturation of water along the flow line, as a result of longer rock–water interaction. This results in a lower ion activity of H_4SiO_4 in shallow groundwaters.

Generally, the stability fields show that anorthite, muscovite, and paragonite minerals are unstable within the Jimma area aquifer system and will be weathered. This implies that these three minerals are not in equilibrium with groundwater and would decompose under any condition upon contact with the groundwater. Finally, the increase of $\log\left[(a_{K^+})/(a_{H^+})\right]$, $\log\left[(a_{Na^+})/(a_{H^+})\right]$ and $\log\left[(a_{Ca^{2+}})/(a_{H^+})^2\right]$ and $\log\left[a_{H_4SiO_4}\right]$ shows a progressive dissolution of aluminosilicate minerals along the flow directions.

$$2KAl_3Si_3O_{10}(OH)_2 + 9H_2O + 2H_2CO_3 \rightleftharpoons Al_2Si_2O_5(OH)_4 + 2K^+ + 2HCO_3^- + 4H_4SiO_4 \quad (15)$$

Muscovite to kaolinite

$$4NaAl_2(AlSi_3O_{10})(OH)_2 + 4H_2CO_3 \rightleftharpoons 8Al_2Si_2O_5(OH)_4 + 2Na^+ + 8HCO_3^- + 16H_4SiO_4 \quad (16)$$

Paragonite to kaolinite

4.8. Sources of Carbon Dioxide

In this study, aluminosilicate minerals dissolution by dissolution CO_2 is the primary source of HCO_3^- in the groundwater. The pCO_2 (Figure 13b) and measured HCO_3^- in this study were determined using PHREEQC [39]. In Figure 13a, due to pH's logarithmic scale, a scatter plot of pH and HCO_3^- is not expected to display a linear relationship; instead, an increase in pH is typically accompanied by an increase in HCO_3^- which looks like a parabolic relationship. Very high HCO_3^- values are typically found in deep groundwater wells (DW1, DW2, DW3, DW4, DW5, DW6, and DW7) (Figure 13a). From the figure, it is observed that, from shallow groundwater to deep groundwater, pH and HCO_3^- generally tend to increase due to long residence time and increased rock–water interaction [61]. In Figure 13b, two areas are displayed in a scatter plot of HCO_3^- vs. $\log(pCO_2)$ saturation to help visualize the sources of carbon dioxide. Accordingly, the plot is defined as the range between $10^{-3.5}$ and $10^{1.5}$. Typically, the groundwater system receives CO_2 from both atmospheric and biogenic sources. In addition, the water may receive an extra surplus of CO_2 from deep sources of thermal metamorphism of rock minerals or mantle-magma degassing [62].

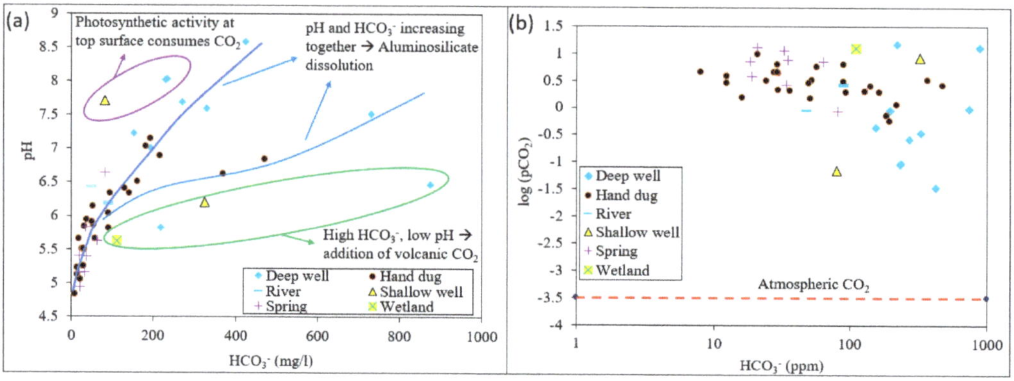

Figure 13. Scatter plot (**a**) of HCO_3^- vs. pH and cross plot (**b**) of HCO_3^- vs. $logpCO_2$.

Deep groundwaters generally have relatively high pH values. For instance, in deep groundwaters like DW1 and DW3, with pH values of 8.59 and 8.02, respectively, more CO_2 is converted into HCO_3^- and CO_3^{2-} due to carbonate equilibria, leading to a lower $\log(pCO_2)$. Thus, higher pH (indicating lower acidity) suggests that CO_2 may have already degassed or been neutralized, resulting in a lower $\log(pCO_2)$. Additionally, CO_2 may escape from deep groundwaters through cracks in overlying confining layers due to deep high pressure, further contributing to the low $\log(pCO_2)$ [41]. Because most of the deep groundwaters in the Jimma area are confined, this may contribute to a low $\log(pCO_2)$ due to their isolation from the surface, longer residence time, and limited biological activity. These factors help maintain lower levels of CO_2 in confined aquifers [54].

4.9. Anthropogenic Pollution

In Figure 14a, the concentrations of Cl^- and Na^+ show poor correlation, indicating that these ions originate from different sources. Na^+ primarily comes from aluminosilicate weathering, while Cl- is from anthropogenic pollution. High concentrations of NO_3^-, Cl^-, and SO_4^{2-} are mainly observed in hand-dug groundwaters rather than in deep groundwaters, shallow groundwater, springs, and wetlands (Figure 14a–c). This is likely due to (1) shorter residence times in hand-dug groundwaters compared to deeper aquifers; (2) smaller volumes in hand-dug groundwaters, resulting in less dilution of contaminants than in springs, surface waters, and wetlands; (3) construction that allows surface pollutant input, often near pit latrines or septic tanks; and (4) being open, leading to direct evapo-concentration. Some groundwater samples are tilted toward the Cl^- axis, while others tilt toward the NO_3^- and SO_4^{2-} axes.

NO_3^- concentration levels above 50mg/L are not recommended for human consumption [63]. This concentration was observed in groundwaters like HD1 (166 mg/L), HD2 (116 mg/L), HD7 (150 mg/L), HD12 (194 mg/L), HD17 (52 mg/L), HD19 (60 mg/L), HD22 (95 mg/L), HD23 (50 mg/L), and HD25 (177 mg/L). As the NO_3^- levels in these wells are high, so is the Cl^- level for most wells. This suggests that there is some local human-caused contamination [40,64]. For instance, the highest Cl^- concentration is observed in HD13, HD14, and HD19. HD13 is found at the lowest part of the Jimma area and market center (Bishishe community market), so this concentration could probably be from sewage, waste pollution, and evapo-concentration. In addition, the Cl^- concentration of HD14 and HD19 is also indicating the effect of anthropogenic activities. Moreover, HD13, HD14, HD15, and

HD17 are four wells having maximum SO_4^{2-} concentrations in the Jimma area. These wells are found in areas where sewage accumulates from different directions. Sewage may contains a significant amount of organic matter, which can decompose and release SO_4^{2-} into the groundwater [65].

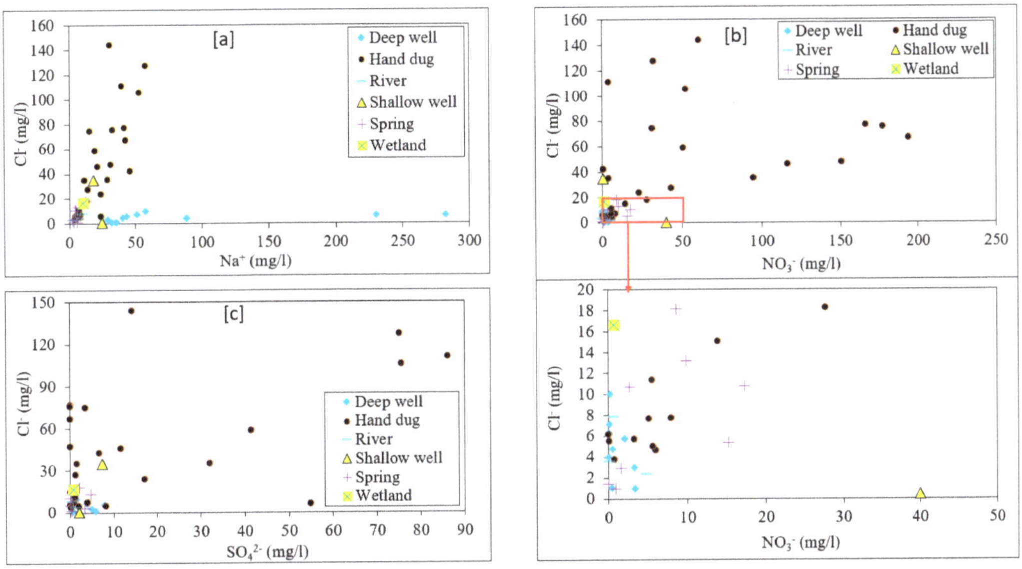

Figure 14. Scatter plot of (**a**) Na^+ vs. Cl^-, (**b**) NO_3^- vs. Cl^-, and (**c**) SO_4^{2-} vs. Cl^- ions concentrations.

The levels of Cl^-, NO_3^-, and SO_4^{2-} are significantly elevated in hand-dug well groundwaters, indicating human-caused pollution from multiple sources, especially in urban and agricultural regions. These pollutants could change the chemistry of the groundwater and may cause issues such as methemoglobinemia (blue baby syndrome) from high NO_3^- levels, while increased Cl^- can influence taste and cause the corrosion of infrastructure, and high SO_4^{2-} can make the water unpalatable. To address these problems, remediation strategies should focus on enhancing wastewater treatment systems, encouraging sustainable agricultural practices to minimize fertilizer runoff, and improving wellhead protection. Additionally, boosting groundwater recharge through rainwater harvesting and implementing stricter monitoring and regulation of groundwater quality can aid in reducing contamination. These actions, along with raising community awareness, will be crucial for improving groundwater quality and alleviating the negative impacts of human pollution on public health.

5. Conclusions

An assessment of the hydrogeochemical processes in the tertiary volcanic rock aquifer of the Jimma area is conducted using data from fifty-one water samples: ten from deep groundwaters, twenty-seven from hand-dug groundwaters, two from shallow groundwaters, eight from springs, three from surface waters, and one from a wetland. The spatial distribution of various physicochemical and hydrogeochemical parameters provides insights into the processes in the aquifer systems in the Jimma area.

The Piper diagram analyzes groundwater types and geochemical processes. Key water types include CaCl, $CaNO_3$, $CaSO_4$, $CaHCO_3$, $NaHCO_3$, and mixed types like Ca-Mg-HCO_3, with $CaHCO_3$ as the dominant type. $NaHCO_3$ types cluster at the bottom corner of the diamond, indicating older, highly mineralized water. Samples with $NaHCO_3$ and $CaHCO_3$ show high TDS from 370 to 1184 mg/L, suggesting aluminosilicate weathering.

Low mineralization types, including $CaHCO_3$, $MgHCO_3$, and Ca-Mg-HCO_3, are found in springs, surface waters, and some wells, indicating immature waters from rainwater infiltration. Certain types, such as CaCl and $CaNO_3$, are linked to pollution from fertilizers and sewage. Overall, groundwater types in the Jimma area are influenced by aluminosilicate dissolution, evaporative concentration, and human activities.

Saturation indices reveal that groundwater is undersaturated with anhydrite across all samples except in specific locations where equilibrium is indicated. Calcite, dolomite, chalcedony, and quartz are typically oversaturated, indicating potential mineral precipitation. Negative SI for carbonate minerals suggests a CO_2-driven dissolution of aluminosilicate minerals, releasing cations and influencing mineral saturation states.

Stability diagrams of feldspars show that anorthite, muscovite, and paragonite minerals are unstable within the Jimma area volcanic aquifer system and will weather. The increase of $\log\left[(a_{K^+})/(a_{H^+})\right]$, $\log\left[(a_{Na^+})/(a_{H^+})\right]$ and $\log\left[(a_{Ca^{2+}})/(a_{H^+})^2\right]$ and $\log\left[a_{H_4SiO_4}\right]$ shows a progressive dissolution of minerals along the flow directions. Aluminosilicate mineral dissolution by dissolved CO_2 contributes to elevated HCO_3^- in groundwater, with deeper wells showing higher levels possibly due to volcanic CO_2 influx. Generally, the pCO_2 suggests atmospheric and biogenic sources in addition to CO_2 from deep thermal metamorphism or mantle-magma degassing. The analysis reveals a poor correlation between Cl^- and Na^+ concentrations, indicating distinct sources, with Na^+ primarily from silicate weathering and Cl^- from anthropogenic pollution. The variation in groundwater quality among different well types suggests localized contamination, with hand-dug groundwaters exhibiting higher concentrations of NO_3^-, Cl^-, and SO_4^{2-} due to anthropogenic input and limited dilution capacity.

The findings of this study are crucial for managing groundwater in the Jimma area. They emphasize the need for targeted approaches to address contamination in shallow groundwaters, which are particularly vulnerable to human activities, and highlight the importance of protecting these vulnerable water sources. Effective management should include improving sanitation infrastructure to reduce sewage infiltration, regulating agricultural runoff to address fertilizer-related pollution, and performing regular assessments of groundwater quality. Furthermore, the study's results are relevant to other areas with similar volcanic aquifer systems, providing valuable insights into the connections between local geology, human actions, and groundwater chemistry. These findings can inform groundwater management strategies in regions facing similar challenges, offering a framework for tackling both natural and human-induced contamination and ensuring the long-term viability of groundwater resources.

Author Contributions: Conceptualization, W.M.K., A.Y., F.F.F. and K.W.; Methodology, W.M.K., A.B.K., B.G.G. and K.W.; Software, W.M.K., A.B.K., S.K.D. and F.F.F.; Validation, W.M.K., F.G.T., F.F.F. and K.W.; Formal analysis, W.M.K., F.F.F. and K.W.; Investigation, W.M.K., A.Y., F.F.F. and K.W.; Resources, W.M.K., B.G.G. and K.W.; Data curation, W.M.K., F.G.T., B.C., S.K.D. and K.W.; Writing—original draft, W.M.K.; Writing—review & editing, W.M.K., F.F.F. and K.W.; Visualization, W.M.K. and K.W.; Supervision, F.F.F. and K.W.; Funding acquisition, K.W. All authors have read and agreed to the published version of the manuscript.

Funding: This study was conducted in the Jimma area, Ethiopia, and was undertaken at Ghent University with support from various sponsorships, including the Global Mind Fund of Ghent University, which facilitates collaboration between Ghent University and partners in the Global South to address sustainable development challenges, and the Network for Advancement of Sustainable Capacity in Education and Research in Ethiopia (NASCERE), funded by the Ministry of Education of Ethiopia, and concluded with Jimma University and Ghent University. In addition, the first author received a BOF-COVID-19 extension grant from Ghent University, which was crucial for this research. This grant was pivotal in completing and advancing the research.

Data Availability Statement: The datasets supporting the findings of this article will be provided at any time upon request.

Acknowledgments: During the study, Wagari Mosisa received assistance from dedicated colleagues, as well as support from local communities, whose contributions were invaluable during the data collection phase. Additionally, the encouragement and advice of Wagari Mosisa's wife, Bikiltu Shifera, will never be forgotten.

Conflicts of Interest: We confirm that this manuscript is original, has not been previously published, and is not under consideration for publication elsewhere. We acknowledge that the corresponding author is the primary point of contact for the editorial process, responsible for communicating with the co-authors regarding the manuscript's progress, submission of revisions, and final approval of proofs. Additionally, we declare that there are no conflicts of interest related to this manuscript.

References

1. Ghani, J.; Ullah, Z.; Nawab, J.; Iqbal, J.; Waqas, M.; Ali, A.; Almutairi, M.H.; Peluso, I.; Mohamed, H.R.H.; Shah, M. Hydrogeochemical Characterization, and Suitability Assessment of Drinking Groundwater: Application of Geostatistical Approach and Geographic Information System. *Front. Environ. Sci.* **2022**, *10*, 874464. [CrossRef]
2. Barman, J.; Biswas, B.; Soren, D.D.L. Groundwater Trend Analysis and Regional Groundwater Drought Assessment of a Semi-Arid Region of Rajasthan, India. *Environ. Dev. Sustain.* **2023**, 1–25. [CrossRef]
3. Hem, J.D. *Study and Interpretation of the Chemical Characteristics of Natural Water*, 3rd ed.; U.S. Geological Survey, Water Supply Paper 2254: Alexandria, VA, USA, 1985; p. 22304.
4. Sharma, S.; Bhattacharya, A. Drinking Water Contamination and Treatment Techniques. *Appl. Water Sci.* **2017**, *7*, 1043–1067. [CrossRef]
5. Elmore, A.J.; Kaste, J.M.; Okin, G.S.; Fantle, M.S. Groundwater Influences on Atmospheric Dust Generation in Deserts. *J. Arid Environ.* **2008**, *72*, 1753–1765. [CrossRef]
6. Carrillo-Rivera, J.J.; Varsányi, I.; Kovács, L.Ó.; Cardona, A. Tracing Groundwater Flow Systems with Hydrogeochemistry in Contrasting Geological Environments. *Water. Air Soil Pollut.* **2007**, *184*, 77–103. [CrossRef]
7. Li, P.; Wu, J.; Qian, H.; Zhang, Y.; Yang, N.; Jing, L.; Yu, P. Hydrogeochemical Characterization of Groundwater in and Around a Wastewater Irrigated Forest in the Southeastern Edge of the Tengger Desert, Northwest China. *Expo. Health* **2016**, *8*, 331–348. [CrossRef]
8. Barry, D.A.; Prommer, H.; Miller, C.T.; Engesgaard, P.; Brun, A.; Zheng, C. Modelling the Fate of Oxidisable Organic Contaminants in Groundwater. *Adv. Water Resour.* **2002**, *25*, 945–983. [CrossRef]
9. Laaksoharju, M.; Gascoyne, M.; Gurban, I. Understanding Groundwater Chemistry Using Mixing Models. *Appl. Geochem.* **2008**, *23*, 1921–1940. [CrossRef]
10. Brunke, M.; Gonser, T.O.M. The Ecological Significance of Exchange Processes between Rivers and Groundwater. *Freshw. Biol.* **1997**, *37*, 1–33. [CrossRef]
11. Lakshmanan, E.; Kannan, R.; Kumar, M.S. Major Ion Chemistry and Identification of Hydrogeochemical Processes of Ground Water in a Part of Kancheepuram District, Tamil Nadu, India. *Environ. Geosci.* **2003**, *10*, 157–166. [CrossRef]
12. Van Huissteden, J.; van den Bos, R.; Alvarez, I.M. Modelling the Effect of Water-Table Management on CO_2 and CH_4 Fluxes from Peat Soils. *Neth. J. Geosci.* **2006**, *85*, 3–18. [CrossRef]
13. Yenehun, A.; Nigate, F.; Belay, A.S.; Desta, M.T.; Van Camp, M.; Walraevens, K. Groundwater Recharge and Water Table Response to Changing Conditions for Aquifers at Different Physiography: The Case of a Semi-Humid River Catchment, Northwestern Highlands of Ethiopia. *Sci. Total Environ.* **2020**, *748*, 142243. [CrossRef] [PubMed]
14. Goldberg, S.; Criscenti, L.J.; Turner, D.R.; Davis, J.A.; Cantrell, K.J. Adsorption–Desorption Processes in Subsurface Reactive Transport Modeling. *Vadose Zone J.* **2007**, *6*, 407–435. [CrossRef]
15. Rivett, M.O.; Buss, S.R.; Morgan, P.; Smith, J.W.N.; Bemment, C.D. Nitrate Attenuation in Groundwater: A Review of Biogeochemical Controlling Processes. *Water Res.* **2008**, *42*, 4215–4232. [CrossRef] [PubMed]
16. Jin, L.; Hamilton, S.K.; Walter, L.M. Mineral Weathering Rates in Glacial Drift Soils (SW Michigan, USA): New Constraints from Seasonal Sampling of Waters and Gases at Soil Monoliths. *Chem. Geol.* **2008**, *249*, 129–154. [CrossRef]
17. Monjerezi, M.; Vogt, R.D.; Aagaard, P.; Saka, J.D.K. Hydro-Geochemical Processes in an Area with Saline Groundwater in Lower Shire River Valley, Malawi: An Integrated Application of Hierarchical Cluster and Principal Component Analyses. *Appl. Geochem.* **2011**, *26*, 1399–1413. [CrossRef]
18. Kebede, S.; Travi, Y.; Asrat, A.; Alemayehu, T.; Ayenew, T.; Tessema, Z. Groundwater Origin and Flow along Selected Transects in Ethiopian Rift Volcanic Aquifers. *Hydrogeol. J.* **2008**, *16*, 55–73. [CrossRef]
19. Beyene, G.; Aberra, D.; Fufa, F. Evaluation of the Suitability of Groundwater for Drinking and Irrigation Purposes in Jimma Zone of Oromia, Ethiopia. *Groundw. Sustain. Dev.* **2019**, *9*, 100216. [CrossRef]
20. Bakure, B.Z.; Fikadu, S.; Malu, A. Analysis of Physicochemical Water Quality Parameters for Streams under Agricultural, Urban and Forest Land-Use Types: In the Case of Gilgel Gibe Catchment, Southwest Ethiopia. *Appl. Water Sci.* **2020**, *10*, 234. [CrossRef]
21. Šíma, J.; Agezew, T. *Hydrogeological and Hydrochemical Maps of Jima NB 37-1 Sheet: Explanatory Notes*; Aquatest, Geological Survey of Ethiopia: Prague, Addis Ababa, 2015.

22. Chalchissa, F.B.; Feyisa, G.L. Frequency and Geospatial Vulnerability Indices of Rainfall and Temperature Extremes in the Jimma Zone, Ethiopia. *Environ. Monit. Assess.* **2022**, *194*, 176. [CrossRef]
23. Fazzini, M.; Bisci, C.; Billi, P. The Climate of Ethiopia. In *Landscapes and landforms of Ethiopia*; Billi, P., Ed.; Springer: Dordrecht, The Netherlands, 2015; pp. 65–87.
24. Tufa, F.G.; Feyissa, F.F.; Kebede, A.B.; Gudeta, B.G.; Kitessa, W.M.; Debela, S.K.; Tumsa, B.C.; Yenehun, A.; Van Camp, M.; Walraevens, K. Estimation of Groundwater Recharge in a Volcanic Aquifer System Using Soil Moisture Balance and Baseflow Separation Methods: The Case of Gilgel Gibe Catchment, Ethiopia. *Hydrology* **2024**, *11*, 109. [CrossRef]
25. Abbate, E.; Bruni, P.; Sagri, M. *Geology of Ethiopia: A Review and Geomorphological Perspectives World Geomorphological Landscapes Book Series (WGLC) Landscapes*; Bill, P., Ed.; Springer Science+Business Media Dordrecht: Florence, Italy, 2015; Volume 53, ISBN 978-94-017-8025-4.
26. Haro, W.; Besufikad, A.; Bulto, D.; Belete, F.; Edris, M. *The Federal Democratic Republic of Ethiopia*; Geological Survey of Ethiopia: Addis Ababa, Ethiopia, 2012.
27. Regassa, A.; Van Daele, K.; De Paepe, P.; Dumona, M.; Deckers, J.; Asrat, A.; Van Ranst, E. Characterizing Weathering Intensity and Trends of Geological Materials in the Gilgel Gibe Catchment, Southwestern Ethiopia. *J. Afr. Earth Sci.* **2014**, *99*, 568–580. [CrossRef]
28. Arroyo-Figueroa, C.; Chalá, D.C.; Gutiérrez-Ribon, G.; Quiñones-Bolaños, E. A Framework to Evaluate Groundwater Quality and the Relationship between Rock Weathering and Groundwater Hydrogeochemistry in the Tropical Zone: A Case Study of Coastal Aquifer Arroyo Grande, in the Caribbean Region of Colombia. *Water* **2024**, *16*, 1650. [CrossRef]
29. Liu, Y.; Li, M.; Zhang, Y.; Wu, X.; Zhang, C. Analysis of the Hydrogeochemical Characteristics and Origins of Groundwater in the Changbai Mountain Region via Inverse Hydrogeochemical Modeling and Unsupervised Machine Learning. *Water* **2024**, *16*, 1853. [CrossRef]
30. Pradhan, R.M.; Behera, A.K.; Kumar, S.; Kumar, P.; Biswal, T.K. Recharge and Geochemical Evolution of Groundwater in Fractured Basement Aquifers (NW India): Insights from Environmental Isotopes (Δ18O, Δ2H, and 3H) and Hydrogeochemical Studies. *Water* **2022**, *14*, 315. [CrossRef]
31. Ofterdinger, U.; MacDonald, A.M.; Comte, J.-C.; Young, M.E. Groundwater in Fractured Bedrock Environments: Managing Catchment and Subsurface Resources—An Introduction. *Geol. Soc. Lond. Spec. Publ.* **2019**, *479*, 1–9. [CrossRef]
32. Custodio, E. Groundwater in Volcanic Hard Rocks. In *Groundwater in Fractured Rocks*; CRC Press: Boca Raton, FL, USA, 2007; pp. 111–124.
33. Mohamed, E.A.; Worden, R.H. Groundwater Compartmentalisation: A Water Table Height and Geochemical Analysis of the Structural Controls on the Subdivision of a Major Aquifer, the Sherwood Sandstone, Merseyside, UK. *Hydrol. Earth Syst. Sci.* **2006**, *10*, 49–64. [CrossRef]
34. Samal, P.; Mohanty, A.K.; Khaoash, S.; Mishra, P. Hydrogeochemical Evaluation, Groundwater Quality Appraisal, and Potential Health Risk Assessment in a Coal Mining Region of Eastern India. *Water. Air Soil Pollut.* **2022**, *233*, 324. [CrossRef]
35. Piper, A.M. A Graphic Procedure in the Geochemical Interpretation of Water-analyses. *Eos Trans. Am. Geophys. Union* **1944**, *25*, 914–928. [CrossRef]
36. Davis, J.C. *Statistics and Data Analysis in Geology*, 2nd ed.; Sons, J.W., Ed.; Wiley New York: New York, NY, USA, 1986; Volume 646, ISBN 0471080799.
37. Schiavo, M.; Giambastiani, B.M.S.; Greggio, N.; Colombani, N.; Mastrocicco, M. Geostatistical Assessment of Groundwater Arsenic Contamination in the Padana Plain. *Sci. Total Environ.* **2024**, *931*, 172998. [CrossRef]
38. Khalidy, R.; Santos, R.M. Assessment of Geochemical Modeling Applications and Research Hot Spots—A Year in Review. *Environ. Geochem. Health* **2021**, *43*, 3351–3374. [CrossRef] [PubMed]
39. Parkhurst, D.L.; Appelo, C.A.J. Description of input and examples for PHREEQC version 3: A computer program for speciation, batch-reaction, one-dimensional transport, and inverse geochemical calculations. *US Geol. Surv. Tech. Methods* **2013**, *6*, 497.
40. Li, C.; Zhang, X.; Gao, X.; Li, C.; Jiang, C.; Liu, W.; Lin, G.; Zhang, X.; Fang, J.; Ma, L. Spatial and Temporal Evolution of Groundwater Chemistry of Baotu Karst Water System at Northern China. *Minerals* **2022**, *12*, 348. [CrossRef]
41. Walraevens, K.; Bakundukize, C.; Mtoni, Y.E.; Van Camp, M. Understanding the Hydrogeochemical Evolution of Groundwater in Precambrian Basement Aquifers: A Case Study of Bugesera Region in Burundi. *J. Geochem. Explor.* **2018**, *188*, 24–42. [CrossRef]
42. Maskooni, E.K.; Naseri-Rad, M.; Berndtsson, R.; Nakagawa, K. Use of Heavy Metal Content and Modified Water Quality Index to Assess Groundwater Quality in a Semiarid Area. *Water* **2020**, *12*, 1115. [CrossRef]
43. Jin, Q.; Feng, C.; Xia, P.; Bai, Y. Hardness-Dependent Water Quality Criteria for Protection of Freshwater Aquatic Organisms for Silver in China. *Int. J. Environ. Res. Public Health* **2022**, *19*, 6067. [CrossRef]
44. Ghourejili, S.; Yaghoubi, S.; Valizadeh Harzand, F.; Mousavi, Y.; Babapoor, A. Effects of Total Dissolved Solids on Pressure Drop and Net Driving Pressure in Different Designs of Brackish and Seawater Reverse Osmosis Systems. *Arab. J. Sci. Eng.* **2023**, *48*, 8785–8799. [CrossRef]
45. Basak, B.B.; Sarkar, B.; Biswas, D.R.; Sarkar, S.; Sanderson, P.; Naidu, R. *Bio-Intervention of Naturally Occurring Silicate Minerals for Alternative Source of Potassium: Challenges and Opportunities*, 1st ed.; Elsevier Inc.: Amsterdam, The Netherlands, 2017; Volume 141.
46. Zhang, X.; Zhao, R.; Wu, X.; Mu, W. Hydrogeochemistry, Identification of Hydrogeochemical Evolution Mechanisms, and Assessment of Groundwater Quality in the Southwestern Ordos Basin, China. *Environ. Sci. Pollut. Res.* **2022**, *29*, 901–921. [CrossRef]

47. Lu, C.; Cheng, W.; Yin, H.; Li, S.; Zhang, Y.; Dong, F.; Cheng, Y.; Zhang, X. Study on Inverse Geochemical Modeling of Hydrochemical Characteristics and Genesis of Groundwater System in Coal Mine Area—A Case Study of Longwanggou Coal Mine in Ordos Basin. *Environ. Sci. Pollut. Res.* **2024**, *31*, 16583–16600. [CrossRef]
48. Boyd, C.E. pH, Carbon Dioxide, and Alkalinity. In *Water Quality*; Springer: Cham, Switzerland, 2000; pp. 105–122.
49. Alex, R.; Kitalika, A.; Mogusu, E.; Njau, K. Sources of Nitrate in Ground Water Aquifers of the Semiarid Region of Tanzania. *Geofluids* **2021**, *2021*, 29–31. [CrossRef]
50. Cavalcante, R.P.; Malvestiti, J.A.; Júnior, J.P.D.; Dantas, R.F. Modeling Carbonate/Bicarbonate and Nitrate Disturbance during Secondary Effluent Disinfection by UV/H_2O_2 and UV/Ozone. *Water Sci. Technol.* **2022**, *86*, 2943–2962. [CrossRef] [PubMed]
51. Edet, A. Hydrogeology and Groundwater Evaluation of a Shallow Coastal Aquifer, Southern Akwa Ibom State (Nigeria). *Appl. Water Sci.* **2017**, *7*, 2397–2412. [CrossRef]
52. Gao, P.; Fan, K. Sulfur-Oxidizing Bacteria (SOB) and Sulfate-Reducing Bacteria (SRB) in Oil Reservoir and Biological Control of SRB: A Review. *Arch. Microbiol.* **2023**, *205*, 162. [CrossRef] [PubMed]
53. Xiao, S.; Fang, Y.; Chen, J.; Zou, Z.; Gao, Y.; Xu, P.; Jiao, X.; Ren, M. Assessing the Hydrochemistry, Groundwater Drinking Quality, and Possible Hazard to Human Health in Shizuishan Area, Northwest China. *Water* **2023**, *15*, 1082. [CrossRef]
54. Nigate, F.; Yenehun, A.; Belay, A.S.; Van Camp, M.; Walraevens, K. Hydrogeochemical Processes and Groundwater Evolution in Complex Volcanic Highlands and Alluvio-Lacustrine Deposits (Upper Blue Nile), Ethiopia. *Environ. Sci. Pollut. Res.* **2023**, *30*, 63953–63974. [CrossRef]
55. Hayashi, M. Temperature-Electrical Conductivity Relation of Water for Environmental Monitoring and Geophysical Data Inversion. *Environ. Monit. Assess.* **2004**, *96*, 119–128. [CrossRef]
56. Ngole-Jeme, V.M. Fire-Induced Changes in Soil and Implications on Soil Sorption Capacity and Remediation Methods. *Appl. Sci.* **2019**, *9*, 3447. [CrossRef]
57. Garrels, R.M.; Christ, C.L. *Solutions, Minerals, and Equilibria*; Harper & Row: Manhattan, NY, USA, 1965; ISBN 0877353336.
58. Drever, J.I. *The Geochemistry of Natural Waters: Surface and Groundwater Environments*; Prentice Hall: Eaglewood Cliffs, NJ, USA, 1997.
59. Alfarrah, N.; Walraevens, K. Groundwater Overexploitation and Seawater Intrusion in Coastal Areas of Arid and Semi-Arid Regions. *Water* **2018**, *10*, 143. [CrossRef]
60. Zhu, C.; Lu, P. Alkali Feldspar Dissolution and Secondary Mineral Precipitation in Batch Systems: 3. Saturation States of Product Minerals and Reaction Paths. *Geochim. Cosmochim. Acta* **2009**, *73*, 3171–3200. [CrossRef]
61. Bennett, G.; Van Reybrouck, J.; Shemsanga, C.; Kisaka, M.; Tomašek, I.; Fontijn, K.; Kervyn, M.; Walraevens, K. Identification of Low Fluoride Areas Using Conceptual Groundwater Flow Model and Hydrogeochemical System Analysis in the Aquifer System on the Flanks of an Active Volcano: Mount Meru, Northern Tanzania. *Sci. Total Environ.* **2022**, *814*, 152682. [CrossRef]
62. Fenta, M.C.; Anteneh, Z.L.; Szanyi, J.; Walker, D. Hydrogeological Framework of the Volcanic Aquifers and Groundwater Quality in Dangila Town and the Surrounding Area, Northwest Ethiopia. *Groundw. Sustain. Dev.* **2020**, *11*, 100408. [CrossRef]
63. WHO. *Guidelines for Drinking-Water Quality Addendum*; WHO: Geneva, Switzerland, 2017; Volume 109.
64. Bennett, G.; Van Reybrouck, J.; Shemsanga, C.; Kisaka, M.; Tomašek, I.; Fontijn, K.; Kervyn, M.; Walraevens, K. Hydrochemical Characterisation of High-Fluoride Groundwater and Development of a Conceptual Groundwater Flow Model Using a Combined Hydrogeological and Hydrochemical Approach on an Active Volcano: Mount Meru, Northern Tanzania. *Water* **2021**, *13*, 2159. [CrossRef]
65. Chang, W.; Wan, J.; Tan, J.; Wang, Z.; Jiang, C.; Huang, K. Responses of Spring Discharge to Different Rainfall Events for Single-Conduit Karst Aquifers in Western Hunan Province, China. *Int. J. Environ. Res. Public Health* **2021**, *18*, 5775. [CrossRef] [PubMed]

water

MDPI

Systematic Review

Advanced Analytical Approaches for Phenolic Compounds in Groundwater: A PRISMA Systematic Review

Giancarla Viviana Santa-Cruz-Mérida [1], Juliana Ikebe Otomo [1], Diego Ricardo Araoz-Prado [2], Elaine Aparecida Rodrigues [1,3,*], Delvonei Alves de Andrade [1] and Oscar Vega Bustillos [1]

1 Nuclear and Energy Research Institute IPEN-CNEN, Av. Prof. Lineu Prestes, 2242 Butantã, São Paulo 05508-000, Brazil; santacruz.viviana@usp.br (G.V.S.-C.-M.); julianaikebe@gmail.com (J.I.O.); delvonei@usp.br (D.A.d.A.); ovega@ipen.br (O.V.B.)
2 Independent Researcher, São Paulo 04153-040, Brazil; araozpradodr@gmail.com
3 Environmental Research Institute IPA, Av. Professor Frederico Hermann Junior, 345, São Paulo 05259-900, Brazil
* Correspondence: elainearodrigues@usp.br; Tel.: +55-11-97064-2857

Abstract

Groundwater is of vital significance to achieve the 2030 Agenda and its Sustainable Development Goals (SDGs); nevertheless, it remains vulnerable to contamination from phenolic compounds. This systematic review critically evaluates the chromatographic and spectrometric techniques applied in real-case groundwater contamination studies involving phenolic compounds. It highlights advances, limitations, and research gaps. A comprehensive research strategy retrieved a total of 345 publications from the Scopus and Web of Science databases, of which 52 (from the years 1984 to 2023) met the PRISMA criteria. The review identified 107 phenolic compounds across 25 countries. Temporal and spatial analyses, source identification, and pollutant profiles were systematically assessed per country. The study also mapped priority (n = 11) and hazardous (n = 27) phenolic compounds according to the USEPA, comparing their detectability across analytical methods. The most frequently applied techniques were LC-MS and LC-MS/MS, while GC-MS demonstrated competitive performance when combined with derivatization. The study found that contamination was predominantly reported in the Global North, with industrial and domestic sewage being the main sources. The highest concentrations reported were 40×10^6 ng/L for 4-methylphenol and 50×10^6 ng/L for phenol, both of which are hazardous substances typically linked to industrial/domestic effluents and the use of pesticides. These findings highlight the imperative for regulatory measures and international scientific collaboration to enhance groundwater monitoring, particularly in the Global South, where data scarcity persists.

Keywords: mass spectrometry; gas chromatography; water quality; 2030 Agenda; phenol

1. Introduction

The 2030 Agenda was adopted in 2015 by the United Nations (UN). It is an action plan consisting of 17 Sustainable Development Goals (SDGs), as a roadmap for people to have peace and prosperity, to end poverty, and to protect the planet [1]. Although universal access to sanitation and clean water is one of the goals of the 2030 Agenda (SDG 6) [1,2], the complexity of its implementation poses a significant barrier to achieving sustainable development [3–5]. Countries, regions, and cities experiencing rapid economic development are primarily responsible for the increase in water demand [6]. Currently, around half of the world's population experiences severe water shortages for at least part of the year [7].

Water shortages are affected by hydro-climatic conditions, socioeconomic factors, government policies, and an increase in human demand for the resource [3,8]. Most of this increase occurs in cities, countries, and regions experiencing rapid economic development, especially in emerging economies [6]. Groundwater represents approximately 99% of all liquid freshwater on the planet and plays an essential role in supplying human consumption, agricultural irrigation, and industrial processes [6,9]. Groundwater is responsible for supplying around 25% of all water used for irrigation and half of the freshwater used for domestic purposes [9]; it is also particularly important in arid and semi-arid regions where surface water is limited [10].

With 58.3% of the world's population living in urban areas (4.77 billion people) [11], it is estimated that half of them are supplied by groundwater [9]. Aquifer systems play a fundamental role in combating poverty and in socioeconomic development, as well as being essential for strengthening socio-environmental resilience in the face of climate change [9,12]. Although groundwater has some relative geological protection, its contamination can be difficult to reverse and results in significant socio-environmental impacts and risks to public health [9,10,12,13]. Among the main environmental pollutants and sources of water contamination, phenolic compounds stand out, with more than 60 different phenols already identified and quantified in aqueous matrices around the world [14].

These compounds are common substances in nature from the metabolism of plants; their chemical structure has a specific characteristic, which comprises an aromatic ring with one or more hydroxyl substituents [15]. The replacement of the hydrogen of the aromatic ring with a halogen functional group, alkyl, phenyl, and benzyl, among others, gives rise to a vast group of phenolic compounds [16]. For example, nitrophenols, which are present in the atmosphere due to the gases generated by burning fuel in vehicles, and chlorophenols, which are used in disinfectants, antiseptics, and various household products [17].

Phenolic compounds are also widely synthesized in the industry for coal conversion, phenolic resin, the coke oven, petrochemicals, textiles, leather, fiberglass, paint, the paper industry, adhesives, pharmaceuticals, soap, paper, and wood preservatives [17–19]. Sources of contamination include but are not limited to industrial effluents, landfill leaching, agricultural runoff, and wastewater discharges [19,20]. In addition, pesticides and herbicides derived from phenols pose additional risks to the contamination of aquatic ecosystems [19].

The toxicity of these compounds is widely documented, and they are classified as priority pollutants due to their association with endocrine disorders, damage to the central nervous system, circulatory system failures and carcinogenic potential, in addition to being biologically accumulated in organisms through the food chain [14,18,20–22]. In contact with water, these compounds are transformed into new fractions that can pose even more risks than their original molecules [19].

Detecting phenolic compounds in groundwater is challenging due to the large number of known and unknown compounds and their varied physicochemical characteristics, which include low volatility, high polarity, and high solubility in water [14,23]. Phenols have very low concentrations and are affected by factors such as dilution by flow with other sources, seasonal climatic variations, pH, chemical–physical changes in general, and sediment resuspension [14].

Conventional and emerging methods for their detection and quantification have been reviewed in the literature and include hyphenated techniques developed from the coupling of two or more analytical techniques, for example, gas chromatography–mass spectrometry (GC-MS), high-performance liquid chromatography–tandem mass spectrometry (HPLC-MS/MS), inductively coupled plasma mass spectrometry (ICP-MS), and time-of-flight mass spectrometry (TOF-MS) [14,19,22,24]. Generally, the first analytical technique separates or increases the resolution, and the second technique characterizes the sample [24].

Although the academic production on advanced analytical techniques is diverse, the contamination of groundwater matrices by phenolic compounds is complex and involves the hydrogeological conditions of the aquifer and the diversity of degradation sources.

This varied production also highlights the need for a synthesis of knowledge applied to real groundwater samples since the lack of a systematic review comparing the efficiency and suitability of the techniques can compromise the recognition of the best approach for real phenolic compound contamination scenarios.

Thus, the study aims to conduct a Systematic Literature Review, following the Preferred Reporting Items for Systematic Review and Meta-Analyses (PRISMA) protocol [25], to investigate chromatographic and spectrometric techniques applied in real cases of groundwater contamination by phenolic compounds. The adoption of the PRISMA protocol enabled a careful and comparative evaluation of the available methods to contribute to the selection of the most effective approaches for monitoring and remediating this problem.

2. Materials and Methods

The review has been carried out following the Preferred Reporting Items for Systematic Reviews and Meta-Analyses (PRISMA 2020) guidelines. The literature search was initially conducted in October and November 2024 and subsequently updated in March 2025. The objective of the study is to identify, categorize, and compare the analytical approaches used to detect phenolic compounds in groundwater. Considering it is a systematic review, this study relies entirely on secondary data extracted from peer-reviewed publications.

2.1. Eligibility Criteria

The PRISMA protocol was followed to reduce bias and ensure the quality of the work [25–27]. The Population, Interest and Context (PI-Co) mnemonic was used to formulate the research question; this mnemonic is normally used to develop Systematic Literature Review (SLR) questions based on qualitative review or synthesis [28]. Thus, the following were defined: population—groundwater contaminated by phenolic compounds; interest—use of chromatographic and spectrometric analytical techniques for identification and quantification; context—detection of phenolic contaminants in environmental samples, with a focus on real groundwater. Based on this, general research questions were structured for the study: (1) What chromatographic and spectrometric techniques have been used to analyze phenolic compounds in real groundwater? (2) In which location was the study carried out, and what is the environmental context of the site? (3) What are the main limitations and advances of the techniques used?

Studies using chromatography and spectrometry techniques to analyze phenolic contaminants in real groundwater were included. The articles selected for this review were published in any period if they were available in their full version in one of the languages of technical mastery of the research team (English, Portuguese, or Spanish). The exclusion criteria for the studies were as follows: (1) object of study other than groundwater and/or phenols and their compounds; (2) simulated scenarios or synthetic groundwater; (3) studies that did not use spectrometry and chromatography techniques to analyze phenols; (4) no specification of the study site; (5) identification of phenolic compounds without their proper quantification.

2.2. Information Sources

The Web of Science (WoS) and Scopus repositories were selected as the preferred databases for collecting articles for this research. Created in 1964, WoS is the oldest database, with coverage in 2020 of around 75 million academic data points, more than 1.5 billion references cited in 254 disciplines [29], and coverage of around 34,000 journals [30]. Although each database presents at least some set of journals covered exclusively, this number is small for WoS [29], which also presents a low proportion of absence of highly cited documents [31]. Given the relevance to science of local and regional journals that address social needs, in 2015, WoS deepened its reach, maintaining a balance between subjects and regions [30].

Scopus was launched in 2004 by Elsevier as a database of abstracts and citations with peer-reviewed scientific content that has more than 77.8 million records [29,32]. Scopus is among the largest curated bibliographic databases, whose indexed material is selected through a process of content screening followed by continuous re-evaluation [29,32]. With more than 23,000 active journal titles, Scopus indexes 66.07% more unique journals compared to WoS [29].

Although sources such as Google Scholar, Dimensions, and Microsoft Scholar can be considered more comprehensive, their search functionalities have several limitations that make complex Boolean searches in their web interface and metadata export difficult [33]. From this perspective, WoS and Scopus remain the main sources of publication metadata and impact indicators [34,35]. As extracting data from different databases can produce different results due to the varying coverage of journals in the two sources [29], to obtain a more representative sample, both WoS and Scopus were selected as the primary sources for the studies.

2.3. Search and Selection Strategies

A comprehensive search was carried out, with no restrictions on period or language, using a combination of descriptors and the Boolean operators "AND" and "OR", with the application of the operator (*) to capture variations of the words; the search string used is shown in Box 1. In the WoS database, the search was conducted on 24 October 2024, while for the Scopus database, the search was conducted on 5 November 2024. The searches in the databases were updated on 14 March 2025. In both databases, the search was carried out with institutional registration and access by the University of São Paulo.

Box 1. Search string used in the scientific databases Web of Science and Scopus on 24 October and 5 November 2024, respectively.

WoS	("phenol*") (Topic) and (" groundwater*" or " ground water*" or "subsurface water*" or "Underground water*") (Topic) and ("spectrometry" OR "chromatography") (Topic)
Scopus	(TITLE-ABS-KEY (("phenol*")) AND TITLE-ABS-KEY (("groundwater*" OR "ground water*" OR "subsurface water*" OR "Underground water*")) AND TITLE-ABS-KEY (("spectrometry" OR "chromatography")))

The search identified 151 records in WoS and 281 in Scopus published between 1968 and 2025. The results were saved in lists in the respective databases, exported as a file with a <Tab delimited file> "TXT" structure for WoS and "CVC" format for Scopus. The main data (index base, authors, year, article title, source title, language, document type and DOI) were converted and organized in a Microsoft Excel spreadsheet, with an identifier number for each record. Two variables were added for validation: "evaluation", with the values <included/excluded>, and "exclusion criteria", with the values <study objective other than groundwater or phenols/simulated scenarios or synthetic waters/study site not identified/full document not located/unavailable in English-Portuguese-Spanish/does not use spectrometry-chromatography/duplicate study> (Figure 1).

All 432 records retrieved were accessed to identify the complete studies in English, Portuguese, or Spanish. In this evaluation, 87 duplicate records were excluded, leaving a total of 345 studies for analysis of their convergence with the research topic and eligibility concerning the established criteria.

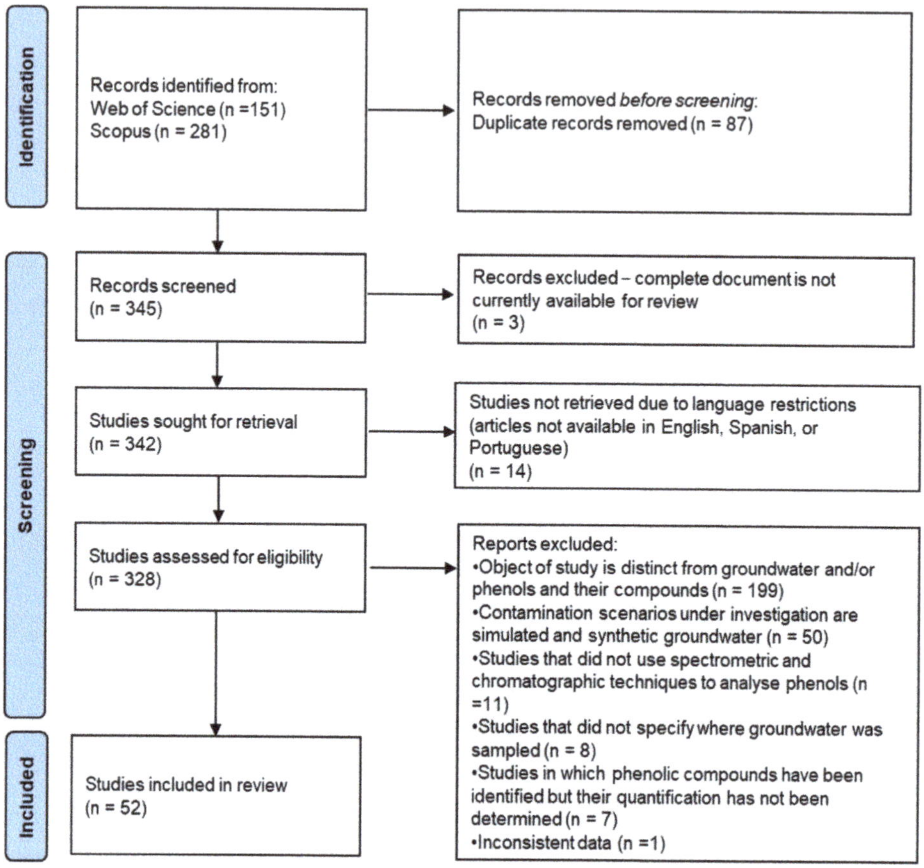

Figure 1. PRISMA 2020 flow diagram (adapted for this study).

The studies were analyzed in their entirety. This analysis was carried out independently by two researchers to reduce individual bias and increase the reliability of the review. All the analyses were validated and certified by a third researcher with more experience in the field. After evaluation, of the 345 articles analyzed, 52 remained for technical review: 23 from Scopus only, 8 from WoS only, and 21 studies indexed in both databases (Table S1—Supplementary Material).

2.4. Data Extraction, Evaluation and Synthesis

The WoS and Scopus databases were reassessed. The lists of studies initially retrieved from the two databases were edited to keep only the studies selected as eligible. The complete metadata from each database was exported and converted into "XLS" format files. These two files were merged into a single database, with normalization of the equivalent categories, and exported into "TXT" and "XLS" formats.

Data extraction was conducted systematically, using a standardized matrix for collection [36]. The matrix was structured into categories related to the following: (i) bibliographic information (title, authors, year of publication); (ii) study objectives; (iii) sample characteristics (type and specificity of groundwater; socioeconomic distribution (Global North and South); environmental context; type of contamination); (iv) analytical methods and techniques employed (experimental conditions, detection limits and instrumentation); (v) results and main findings (concentration and types of phenolic compounds, performance of analytical techniques and practical relevance); (vi) quality aspects (assessment of risk of bias and reported limitations).

The extracted data were organized in a Microsoft Excel spreadsheet and coded to facilitate analysis. Initially, open coding was carried out by assigning descriptive labels to the extracted data (variables). The codes were then grouped or ungrouped and normalized using axial coding to connect related themes [36,37]. This methodological approach was used to standardize the data and to guarantee the quality of its extraction.

To minimize errors and biases, the data were extracted by one researcher, and the consistency of the extraction was validated by two independent researchers. The variables corresponding to the chemical elements, analyte extraction techniques, and analytical techniques were standardized by the researcher with the most experience in trace analysis and environmental chemistry and certified by the researcher who carried out the extraction.

As the identification of several common names and synonyms can lead to duplications or omission of relevant data when analyzing information from different sources, to reduce bias and avoid ambiguities caused by different names, the Chemical Abstracts Service (CAS) number was used in the standardization of phenolic compounds. When compounds with various names were found in different articles, the most common one was selected and identified with the CAS number in the database for this study. The CAS number was searched on the PubChem website [38].

The risk of bias was assessed for the 52 studies included in the sample. Although all of them presented evidence that the analyses were carried out on real groundwater, it was not possible to identify the environmental context of the collection site or the precise location of the sites sampled for 40% (N = 21) of them. However, the transparency of the analytical methods and the availability of complementary data were considered when assessing the risk of sampling bias. As the compounds detected were identified with their respective CAS number, and the sources of pollution and the analytical, extraction, and derivatization techniques were duly standardized and certified by expert researchers in the field, the risk of bias was considered low, with a high level of robustness of the findings. Although the findings of this review may be subject to publication bias, this risk is also considered low given that the global research ecosystem increasingly relies on data from WoS and Scopus [34], which were the two repositories used to collect the articles.

2.5. Classification of Specific Phenolic Compounds

In 1970, the Environmental Protection Agency (USEPA) was created for the protection of human health and to safeguard the natural environment on US territory [39]. Environmental regulations are updated periodically, organized, and codified in the Code of Federal Regulations (CFR), which is structured into 50 titles, each covering a specific area of federal regulation. To implement and enforce environmental regulations, the USEPA has developed test procedures (analytical methods) for measuring regulated pollutants, which are contained in regulations codified at Title 40 CFR [40].

Title 40 covers regulations on environmental protection, which include topics such as water and air quality, solid and hazardous waste management, toxic substances, and chemical control [40]. Globally, the USEPA is recognized as one of the most influential environmental regulatory agencies [41]. With an adequate budget, a wealth of technical knowledge, and a variety of available technologies [41], the USEPA has an institutional leadership position in published research on environmental monitoring [42], contributing to policies in other nations. For this RSL, Chapter I (Environmental Protection Agency) of Title 40 was examined in its Subchapter D Water Programs (Part 116); Subchapter J Superfund, Emergency Planning, and Community Right-to-Know Programs (Part 302); and Subchapter N Effluent Guidelines and Standards (Part 423) [40].

The designation of hazardous substances for the Clean Water Act (CWA) is made in Part 116 of the 40 CRF, which establishes the framework for regulating the discharge of pollutants into U.S. waters to restore and maintain their chemical, physical, and biological integrity [43,44]. Substances classified as hazardous by the USEPA, when discharged in any quantity into jurisdictional waters, present an imminent and substantial endangerment to public health or welfare [45]. With the designation of a chemical (element and compound) as a hazardous substance by the CWA, its reportable quantity is established by regulation, and discharge in quantities that may be harmful to navigable waters and waters of the contiguous zone is prohibited; a total of 326 substances are currently classified as hazardous [45].

Part 302 of 40 CFR deals with "Designation, reportable quantities, and notifications" in compliance with the Comprehensive Environmental Response, Compensation, and Liability Act of 1980 (CERCLA). This act establishes prohibitions and requirements regarding closed and abandoned hazardous waste sites. It governs liability for hazardous releases and establishes a trust fund for cleanup when no responsible party can be identified [46]. CERCLA also specifies "reportable quantities" (RQs) for each substance, which are the minimum release amounts that require notification to the USEPA [45].

In addition to hazardous substances, priority pollutants are also regulated by the CWA. For these, Appendix A of 40 CFR Part 423, designated as the Priority Pollutants List, lists 126 priority pollutants that are regulated by the USEPA and for which analytical test methods have been developed [47]. The phenolic compounds found in the 52 studies in this review were classified based on these environmental regulations established by the USEPA.

2.6. Data Analysis and Synthesis

The data were normalized for co-occurrence analysis, which aims to explore the keywords that co-occur in two or more articles and that represent the main content of the studies by describing the topics covered by a knowledge domain. A keyword network shows the structure of the body of knowledge by presenting the inter-relationships and organization of research topics [48,49].

VOSviewer software version 1.6.20 was used to analyze the co-occurrence of keywords. After selecting the keywords described by the author, 156 terms were extracted for normalization (standardization of synonyms and different forms of the same word). Exported in a "TXT" file for processing, the normalization resulted in 116 terms for analysis (Table S2—Supplementary Material). The adjusted terms were entered into a synonym file with instructions for the analytical software to replace all instances of a keyword (embedded terms) with another (label) [50].

In the keyword co-occurrence analysis, keywords were identified that co-occur more frequently and have greater link strength. Thus, keywords with a minimum occurrence of two times and a minimum link strength of two were considered, resulting in 29 matches. VOSviewer's "overlay visualization" function was used to help detect the evolutionary trend of the research topics investigated.

Microsoft Excel, a tool widely used for compiling, organizing, and visualizing data, was used to process and analyze the data collected. The extracted data were initially systematized in electronic spreadsheets for consistent treatment of the information to guarantee its accuracy. Based on the structural organization of the database, to meet the objectives of each synthesis devised by the research team, Python v.3.1 was used to support parameterization and Generative Artificial Intelligence (GAI) to explore graphical representations of the information. Multiple interactions were carried out to improve the resulting graphical summaries, which served as the basis for the preparation of final infographics built by an illustrator using specialized graphic publishing tools to ensure a clear and informative visual representation of the study's findings.

In the temporal and socioeconomic distribution of the studies and the phenolic compounds identified in groundwater, the Global North and South structure [51,52] was considered, based on the list of countries in the Global South [53]. The heatmap technique was used for the graphical summaries of contamination sources and their relationship with groups of phenolic compounds and the distribution of regulated contaminants by country, with the data initially organized in spreadsheets to standardize and certify the information. Using IAG to identify the most favorable graphic structure, the data systematized in a Microsoft Excel spreadsheet was reworked by an illustrator using graphic editing tools. The intersection between the substances regulated by 40 CFR Parts 116, 302, and 423 [44,46,47] was established by parameterizing the data in a Microsoft Excel spreadsheet and applying it to a Venn diagram, which allows visualization of the overlap between the lists of regulated substances. The graph was developed by an illustrator and validated by the research team to ensure that all the components were correctly categorized and that they adequately represented the similarities and divergences between the regulations analyzed.

For the priority compounds that simultaneously presented LD and LQ data, the research team prepared a scatter plot generated with Python v.3.1 to allow visualization of these limits between the analytical techniques employed. A scatter plot was also drawn up for a comparative analysis of the GC-MS analytical technique with and without derivatization.

The efficiency of the recovery techniques (%) of priority phenolic compounds, which simultaneously presented LD and LQ data and which did not use derivatization, was analyzed using the Sunburst chart to visualize the hierarchy between the methodologies applied. This chart was made up of three levels of information: (i) the central ring, showing the extraction techniques used (LPME, SPE, and online SPE); (ii) the middle ring, showing the analytical techniques used for quantification (LC-MS/MS; GC-MS, HPLC-UV and LC-UV); and (iii) the outer ring, showing the recovery values (%) for the priority phenolic compounds analyzed. The graphical summary of the maximum concentrations of priority phenolic compounds per country was established in a bubble chart adjusted from the data organized and systematized in Microsoft Excel spreadsheets.

The meta-analyses and summaries of information were complemented by thematic descriptive narrative analyses.

3. Results and Discussion

3.1. Temporal Evolution, Sources of Contamination and Global Distribution of Groundwater Contamination by Phenolic Compounds

With 345 studies previously selected in the searches carried out in the two referential scientific databases (WoS and Scopus), the number of articles that fell within the scope of this review was relatively low. Only 15.07% (N = 52) were considered eligible; they covered the years 1984 to 2023 (Figure 2).

Between 1984 and 2000—a period of around 20 years—only eleven studies were identified in just seven countries. Canada [54], Croatia [55], Slovakia [56], Poland [57], and Germany [58] have one publication related to each country, and Denmark [59,60], has two publications; the USA stood out with four studies in this same time frame [61–64].

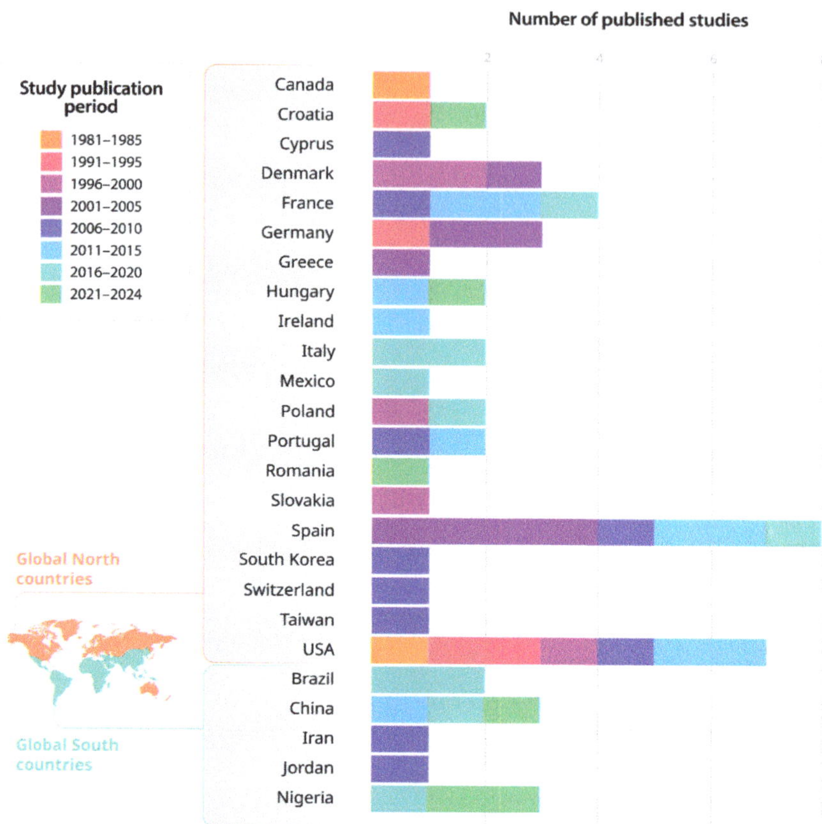

Figure 2. Temporal and socioeconomic distribution of studies on phenol contamination in groundwater from 1984 to 2023, with emphasis on the Global North and South.

The use of chromatographic methods to detect phenolic compounds in groundwater in different regions of the world before the 2000s suggests that chromatographic methods were already widely accepted. This result also shows that groundwater contamination was also a problem of primary concern, for example, in Denmark, where the supply of drinking water is almost entirely based on the use of groundwater susceptible to contamination by polar pesticides [59]. In the USA, wastewater represents a major source of pollution and an important source of groundwater contamination. For example, in Cape Cod, Massachusetts, drinking water is entirely supplied by groundwater and is especially vulnerable to wastewater discharge [64].

The advance in the sensitivity and specificity of analytical techniques over the decades is evident. In older studies evaluating the sites of Southington, Norwich, Watertown, Danbury, and Sheshire in the USA [61], the identification of phenolic compounds was based only on GC-MS without derivatization, while studies developed in the late 1990s already used LC-MS and chemical derivatization to improve the detection of phenols at trace levels [59,64]. In this same period, leachates from landfills and industries were mapped as critical sources of contamination by phenolic compounds [54,55,62–64], with studies also addressing agricultural waste and runoff as a source of contamination [57,59,61].

In the years between 2001 and 2020, significant evidence was found of the expansion of studies on phenol contamination in groundwater, with 37 studies carried out in 20 countries, the majority belonging to the Global North. Spain made the largest contribution, with eight publications [65–72]. There were three studies in the USA [73–75] and France [76–78] in the continental area and one study in the French overseas territories of Guadeloupe and Martinique (Atlantic Ocean), Reunion and Mayotte (Indian Ocean) and French Guiana (continental area in South America) [79].

With two publications in each country, studies were identified in Germany [80,81], Italy [82,83] and Portugal [84,85]. They were also found in Cyprus [86], Denmark [87], Greece [88], Hungary [89], Ireland [90], Switzerland [91], Taiwan [92], South Korea [93], Poland [94] and Mexico [95]. In the same period, countries in the Global South also assessed groundwater contamination by phenolic compounds: Brazil [96], Karak, Jordan [97], Henan province, China [98] and Iran [99].

Between 2021 and 2024, only five studies were identified that met the criteria of this review, with investigations in Dongying, in the Yellow River estuary, in China [100], in the Danube River basin [101], in the city of Ibadan [102] and Osun [103] both in Nigeria, where groundwater was found to be contaminated by multiple organic compounds [102], and in Monterrey Nuevo León, México [95].

It is interesting to note that no publications from the Global South were identified until 2005. In addition to this gap, evidence was also found of gaps in the details of the research site. Of the total number of studies selected (N = 52), 28.85% (N = 15) had an insufficient location (with superficial references to the site and its environmental context). This location was considered adequate for 40.4% of the studies (N = 21), which indicated the sites investigated but without specifying their hydrogeological and environmental characteristics. Only 19.2% of the studies (N = 10) provided detailed information on the sites sampled, which would make it possible to support comparative research between studies or additional complementary analyses on the same sites.

These studies with detailed information evaluated the locations of Zeitz, in Germany [80], Rosário do Catete in Sergipe, Brazil [104], Guiyang City in Guizhou Province, China [105], Zagreb in Croatia [55], Catalonia, Barcelona region, in Spain [71], Southington, Norwich, Watertown, Danbury and Sheshire and Cape Cod in Massachusetts, USA [61,64], Milan in Italy [83], and Ibadan and Osun, Oyo and Lagos States, in Nigeria [102,103].

In the analysis of keyword co-occurrence (Figure 3), where the color gradient (blue to yellow) indicates the temporal evolution of searches, topics such as "phenols", "groundwater", and "pesticides" showed connections throughout the period analyzed, suggesting consolidated research themes.

Concern about pesticides in groundwater, for example, was also identified throughout the period investigated, since the 1990s. These studies pointed to municipal and industrial sewage and agricultural runoff in Gdansk in Poland [57] and agricultural runoff in Denmark [59] as sources of contamination. In Denmark as well, water-soluble chlorinated phenols were reported to be associated, among other sources, with the degradation of pesticides [87], while in Spain, in the regions of Catalonia, Galicia, and Andalusia, minimal impact of herbicides, phthalates, alkylphenols, and BPA has been demonstrated in spring water intended for bottling, suggesting the effectiveness of measures to protect water from any anthropogenic impact [70].

More recent topics, such as "emerging contaminants", reflect the growing concern about emerging pollutants. Thus, state-of-the-art broad-scope target screening was carried out, with the detection of 33 emerging compounds that qualified as potential pollutants specific to the Danube River basin [101]. In Milan, Italy, more than 80 emerging contaminants were investigated [83]. Endocrine-disrupting compounds, which constitute a broad group of contaminants of emerging concern, have been studied in the northeastern region of Emilia-Romagna, also in Italy [82]; in agricultural areas of Spain, close to large cities [65];

and in the Barcelona region, in Saint Joan Despí [67]. In this same location, possible trends in the seasonal distribution of estrogens, pesticides, and BPA have been investigated, with evidence that groundwater pollution is quite persistent and that the rate of degradation in this environment is very slow [68].

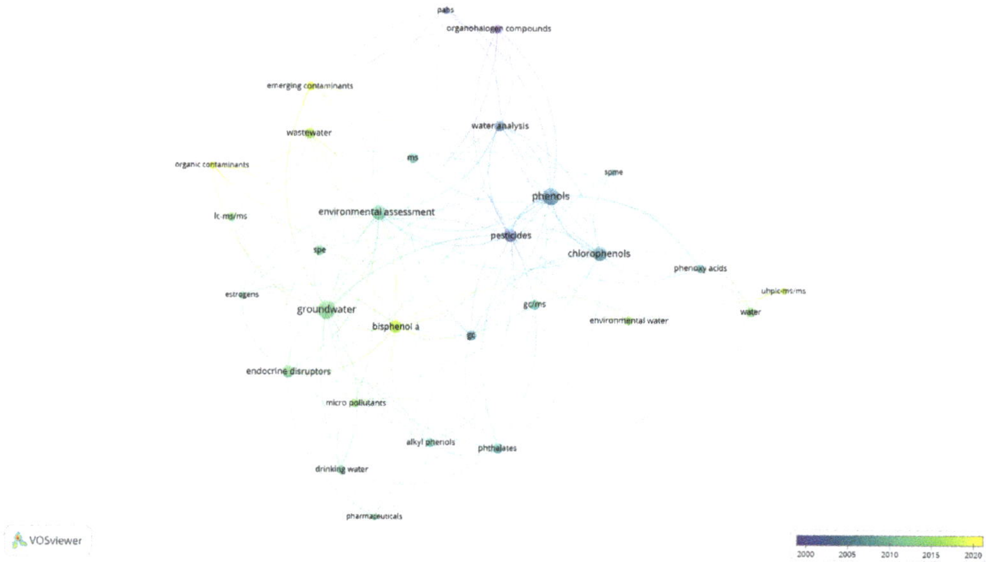

Figure 3. Network of keyword co-occurrence in groundwater contamination studies: phenolic compounds—emerging pollutants and analytical techniques.

The keyword co-occurrence network also highlighted recent themes, such as "mass spectrometry" and "LC-MS/MS", which represent advances in analytical techniques and the consequent increase in precision in the detection of phenolic compounds, allowing for more sensitive and selective analyses. Advanced analytical methods such as UPLC-MS/MS, for example, were used to assess groundwater contamination in Shengiu County, China, with detection limits of 0.1 ng/L for the identification of 2,4-dibromophenol (2,4-DBP), 2,4,6-tribromophenol (2,4,6-TBP), 4-bromophenol and 4-bromo-3-methylphenol [98], as well as the determination of 4-chloro-2-methylpenol (2,4-MCP) in Ireland [90].

Countries from the Global South made a small contribution, with only seven studies carried out between 2001 and 2020, developed in Brazil [96,104], China [98,105], Jordan [97], Iran [99] and Nigeria [103]. In the following interval (2021–2023), studies from the Global South assessed groundwater contamination by phenolic compounds at sites located in China [100] (Hai-Liang and Tong-Na, 2022) and Nigeria [102,103].

A total of 306 occurrences were counted in the Global North and 45 in the Global South, considering "occurrences" of all the phenolic compounds analyzed, even if repeated, in the different studies reviewed. A total of 115 occurrences of maximum detected concentrations were reported, with 94 in the Global North and 21 in the Global South (Figure 4).

Of the total occurrences in the Global North (N = 94), 48 different phenolic compounds were identified. Of these, 10 are considered priority compounds [47]—2,4,6-trichlorophenol, 2,4-dimethylphenol, 2,4-dichlorophenol, 2,4-dinitrophenol, 2-methyl-4,6-dinitrophenol, 2-chlorophenol, 4-chloro-3-methylphenol, 4-nitrophenol, pentachlorophenol and phenol—and they showed a range of concentrations from 4 ng/L (2,4-dichlorophenol) to 40,000,000 ng/L (phenol) in Ireland [90] and USA [63], respectively.

Figure 4. Global distribution of studies between 1984 and 2023 and phenolic compounds reported in the investigated groundwater.

Twenty-one occurrences that were observed in the Global South correspond to five different phenolic compounds (2,4,6-trichlorophenol, 2,4-dimethylphenol, 2,4-dinitrophenol, 2-methyl-4,6-dinitrophenol and phenol) that are considered priority compounds (Part 423 40 CFR) [47], with 2,4-dinitrophenol having the highest maximum concentration of 1,313,000 ng/L in Nigeria [103].

One of the factors affecting water scarcity is water contamination, which mainly impacts cities and countries in emerging economies experiencing rapid economic growth [3,6,8]. However, in these same regions, the use of advanced techniques for assessing groundwater contamination by phenolic compounds is still limited. This gap suggests the need to improve environmental monitoring in under-represented regions, where the absence of systematic data can hinder the implementation of effective regulatory policies.

As few phenols are addressed in global water quality guidelines, regulation is a major challenge to mitigating risks to human health and the environment [14]. For the compounds classified as priority compounds, 2,4,6-trichlorophenol and pentachlorophenol exceeded the limit of 1500 ng/L and 30 ng/L, respectively [106], established in section 304 (a) of the CWA [107]. With quantified concentrations above 30 ng/L, pentachlorophenol was detected in six investigations [56,79–81,87,89]. Concentrations below the LQ or LD were also reported [57,62,84,104], as were investigations that did not detect the substance [88,92, 101], and in these studies, the LDs or LQs were greater than 30 ng/L. Only three studies had quantification results or LDs and LQs below 30 ng/L [64,101,104].

The reported concentration of 2,4-dimethylphenol in Denmark was 2,090,000 ng/L [60]. In two studies, conducted in Denmark [87] and Germany [81] 2,4-dichlorophenol was reported to be above 300 ng/L, while 2-chlorophenol was detected above 100 ng/L in three of the seven studies that evaluated it [80,81,89].

The World Health Organization's Guidelines for Drinking Water Quality [108] and the Europe Commission's Phenol—Summary Risk Assessment Report [109] set a limit concentration for phenol of 1000 ng/L. Among the investigations that analyzed this substance [54,55,57,60–63,67,80,81,84,87–89,96,103], five studies reported concentrations lower than the LD or LQ, although they exceeded the 1000 ng/L concentration [54,67,84,88,96]. Regarding the concentrations detected, 56% of the studies (N = 9) reported values above 1000 ng/L [60–63,80,81,87,89,103].

Although the effects of environmental management in a river basin are related to how pollution control cooperates in the upstream and downstream regions [110], only one study has evaluated legacy and emerging contaminants in the Danube River Basin, with collections from sites in Romania, Croatia, and Hungary [101].

In France, overseas departments (DOM) are a major issue for the implementation of the EU Groundwater Directive (GDW) due to methodological limitations, knowledge gaps, and the absence of specific indicators for these environments, as well as the lack of local analytical laboratories and geographical isolation [79]. To address these difficulties, analyses were conducted in five French overseas territories, determining the occurrence of 36 organic micropollutants, including pesticides that exceeded the groundwater quality limit of 100 ng/L. The results indicated the stability of the substances and reinforced the importance of packaging and transportation conditions for samples collected outside the laboratory [79]. As similar challenges can affect countries and locations in the Global South, including a lack of laboratory infrastructure, geographical isolation, and the absence of specific indicators, the strategies developed have the potential to contribute to overcoming these limitations.

Groups of phenol compounds, alkylphenols and chlorophenols were identified with greater diversity and concentration correlated with contamination by industrial and municipal effluents, pesticides and fuels (Figure 5). Phenol is an aromatic organic compound used in chemical skin peels, anesthetics and disinfectants; in lower concentrations, it makes up formulations for antiseptics, lotions, ointments, paints, polishes, adhesives, lacquers, varnishes and solvents [111]. Alkylphenols are derivatives of phenols and comprise a wide range of compounds used as emulsifiers for polymerization, detergents, pesticides and the processing of wool and metals [112]. The investigations conducted by the 52 studies in this review suggest that the alkylphenols detected come from a variety of contamination sources and are strongly associated with industrial discharges and urban effluents, pesticides, fuels and endocrine disruptors, while the most significant contamination sources for phenols were industrial effluents and pesticides.

Chlorophenols are a class of chlorinated halogenated phenols, made up of 24 types of chlorophenols. A total of 12 substances in this group have been reported as hazardous by 40 CFR Part 302 [46]: 2,3,4,6-tetrachlorophenol, 2,3,4-trichlorophenol, 2,3,5-trichlorophenol, 2,3,6-trichlorophenol, 2,4,5-trichlorophenol, 2,4,6-trichlorophenol, 3,4,5-trichlorophenol, 2,4-dichlorophenol, 2,6-dichlorophenol, 2,4-dichlorophenoxyacetic acid, 2-chlorophenol and pentachlorophenol [113]. These compounds stand out for their variety of contaminants, with the greatest representation in industrial discharges, pesticides, herbicides, antiseptics, disinfectants, urban effluents and the degradation of natural substances [113].

The results have also shown a direct link with public health and environmental protection policies. In France, for example, monitoring of hormones, pharmaceuticals, pesticides, perfluoroalkyl substances (PFASs), phthalates, and alkylphenols (APs) at the ultra-trace level was carried out to analyze the potential presence of emerging contaminants in bottled mineral waters [78]. Among the 330 compounds investigated in 40 bottled waters, no phthalate, hormone, or pharmaceutical substance was quantified. On the other hand, 19 compounds were quantified in 11 samples at the low ng/L level, 11 of which were herbicides or their metabolites, 6 PFAS, and 2 APs. The answers obtained suggest that the natural geological protection of the aquifers exploited has not managed to prevent the presence of some PFASs, herbicides, and their metabolites in some samples, even at levels well below the applicable French regulations [78]. Thus, although benefiting from long-term protection policies in their watersheds and their natural geological protection, French bottled spring waters (SPWs) or natural mineral waters (NMWs) are still subject to deterioration by emerging contaminants [114].

Figure 5. Correlation matrix between sources of contamination and groups of phenolic compounds detected in groundwater studies using advanced analytical techniques.

3.2. Phenolic Compounds Evaluated and Detected by Groundwater Studies

The detection and quantification of phenols and their compounds in groundwater matrix have evolved with the development of new equipment; these developments have made it possible to improve analytical techniques and appropriate sample preparation methods. In Denmark, for example, an extensive study was carried out with 7671 samples collected from 1115 screens between 1993 and 2001 for the National Groundwater Monitoring Program (GRUMO). Phenol and chlorophenol data were obtained using GC with ECD, FID, or MS detection [87]. These different detection techniques coupled with GC are because phenolic compounds have different functional groups, with brominated and chlorinated phenolic compounds being more suitable for detection via ECD and alkylated compounds via FID. In addition to being suitable for most phenolic groups, MS detection is a complementary tool for identifying and quantifying lower concentrations of these pollutants (Table 1).

In a study from Denmark, the phenolic compounds identified included phenol, methyl-phenols, dimethyl-phenols (2,4-dichlorophenol; 2,6-dichlorophenol) and pentachlorophenol [87]. Phenol was reported most frequently, in 14.6% of the 159 screens examined, while methyl- and dimethyl-phenols were detected in less than 1% of the several hundred screens examined; dichlorophenols showed concentrations below the potability limit of 0.1 μg/L, and pentachlorophenol was found in 7 of the 1090 screens examined. Possible sources of contamination include the production of plastics and detergents by the chemical industry and impurities in pesticides (for chlorophenols). Other sources of pollution suggested were creosote, a substance widely used to preserve railway sleepers, which contains cresol and phenol, also used in old gas plant installations, pentachlorophenol (a component used to impregnate wood between 1956 and 1979), as well as the degradation of organic matter, which can give rise to some phenols [87].

The GC-FID technique has been used to detect and quantify phenolic compounds in different environmental contexts [54,60–62,97]. In the USA, in the towns of Southington, Cheshire, Norwich, Watertown, and Danbury in Connecticut, the contamination of groundwater by leachate from municipal landfills that received domestic and industrial waste was analyzed. Phenol and alkylphenols (3-ethylphenol, 4-ethylphenol, 3-methylphenol, 4-methylphenol) were reported, with phenol showing one of the highest concentrations (1,200,000 ng/L) [61]. In groundwater adjacent to three ferrous smelting landfills in Wisconsin, also in the USA, phenol was detected above the limit of quantification, with a significant concentration of 540,000 ng/L, while chlorophenols, nitrophenols, alkylphenols, nitrocresols, and chlorocresols were detected at very low levels (<LQ), which may indicate either residual presence or limitations of the analytical methodology [62].

Studies conducted in the Middle East, in the city of Karak, Jordan, evaluated the absorption of organic compounds in different plant species resulting from the use of wastewater for irrigation in agriculture. Consequently, the study evaluated irrigation as a potential source of groundwater contamination, with a total concentration of 1000 to 4000 ng/L being reported for the sum of the phenolic compounds analyzed, with o-cresol standing out as having the highest reported concentration (780 ng/L) [97]. In Canada, groundwater located near a tailings pond was also investigated, but the phenolic compounds evaluated were below the LD of 1000 ng/L [54].

GC-GC-TOFMS was used in California, USA, to examine 22 groundwater samples collected from five sites impacted by fuel spills [74]. As the technique was applied for the non-targeted analysis of polar petroleum metabolites, it was possible to detect an additional 760 compounds, including acids, esters, alcohols, ketones, aldehydes, and phenols, with the determination of 2,4-dimethylphenol, 2,3,5-trimethylphenol, 4-cumylphenol, 4-pentylphenol, 4-tert-amylphenol, 4-tert-butylphenol, 4-nonylphenol). Although phenols were reported, as they were not identified at multiple sites and were present in very low concentrations (<LOQ), these findings suggest that phenols are not the predominant metabolites in oil biodegradation environments [74].

Table 1. Analytical and extraction techniques were used to determine phenolic compounds in real water samples from different locations.

Analytical Technique	Extraction Technique	Phenolic Group	LD Range (ng/L)	LQ Range (ng/L)	Detected Concentration Range (ng/L)	Pollution Source	Country (Authors)
GC-ECD; GC-FID; GC-MS	-	Chlorophenols	-	-	20 to 340	Natural substance degradation, Fuel, Pesticides	Denmark [87]
		Alkylphenols			600 to 4200		
		Alkylphenols ethoxylates			-		
		Phenols			70 to 5100		
GC-FID	TCLP, Purge and trap, LLE, SPE and Derivatization	Alkylphenols	1000 to 10,000	2000 to 3000	180 to 2.09 × 10^7	Creosote	Canada [54], Jordan [97], USA [61,62], Denmark [60]
		Chlorocresols	-	40,000	<LQ	Ferrous foundry	
		Chlorophenols	-	40,000 to 180,000	710 to 1740	Landfill leaching, Municipal wastewater, Industrial wastewater, Pesticides	
		Nitrocresols	-	80,000	640		
		Nitrophenol	-	60,000 to 140,000	<LQ		
		Phenols	1000 to 10,000	30,000	11,000 to 1.2 × 10^6	Treated wastewater	
GC-GC-TOFMS	-	Alkylphenols	-	1 × 10^6 to 11 × 10^6	<LOQ	Fuel	USA [74]
GC-MS	CLSA, direct injection, extraction and derivatization, LLE, LPME Clean up and derivatization, MASE, LPME, OASIS-SPE (offline), SPE, SPE (online), SPME, SPME (online), TD-SBSE, SPE LLE, SPME	Alkylphenols	1 to 100,000	2 to 10,000	2.2 to 50 × 10^6	Alcohols, Carboxylic acids, Chlorination, Endocrine disrupters (EDCs), Fuel, Industrial wastewater, Jet fuel (JP8), Mining, Municipal wastewater, Natural substance degradation, Pesticides, Pharmaceuticals, Phenolic compounds, Phenoxy alkanoic acids	Brazil [104], China [100,105], France [78], Germany [80,81], Hungary [89], Nigeria [102], Portugal [84], South Korea [93], Spain [65,66,70,71], Taiwan [92], USA [63,64,73], Poland [94]
		Alkylphenols ethoxylates	3.3 to 86.7	15.6 to 260	-		
		Biphenols	1.1 to 2.9	4.2 to 10,000	14 to 33		
		Bisphenols	3 to 1370	1 to 16.2	6 to 6880		
		Bromophenols	190	570	<LQ		
		Chlorocresols	74 to 200	220	7000 to 113,910		
		Chlorophenols	1.4 to 4100	2 to 12,000	2.08 to 216,200		
		Halogenated Phenols	10	-	17,000		
		Nitrophenols	2.2	6.6	<LQ		
		Phenols	0.05 to 100,000	0.167 to 28,000	720 to 40 × 10^6		
GC-NPD/GC-ECD	SPE-LLE	Nitrophenols	10,000 to 30,000	-	15,000 to 88,000	Explosive and Heavy metals	Germany [58]
		Nitrocresols	10,000 to 30,000	-	15,000 to 31,000		
HPLC-SPE-NMR-TOF-MS	SPE by ^{1}H NMR	Bisphenols	-	-	12,000 to 13,000	Explosives, Heavy metals	Switzerland [91]
		Nitrophenols	-	-	-		
LC-MS	SPE, Profexs and Prospekt, SPE Oasis	Alkylphenols	52.9 to 264.7	-	374.2 to 20,000	Endocrine disrupters (EDCs), Industrial wastewater, Pesticides, Municipal wastewater	Denmark [59], Italy [82], Spain [67,68], USA [75], Mexico [95]
		Alkylphenols ethoxylates	-	-	2500		
		Bisphenols	0.61 to 6.3	2,02	5 to 25,000		
		Chlorocresols	34.9 to 38.5	-	-		
		Chlorophenols	3 to 87.7	-	-		
		Nitrocresols	3 to 7	-	-		
		Phenols	1216.2	-	-		

Table 1. *Cont.*

Analytical Technique	Extraction Technique	Phenolic Group	LD Range (ng/L)	LQ Range (ng/L)	Detected Concentration Range (ng/L)	Pollution Source	Country (Authors)
LC-MS/GC-MS	LLE	Alkylphenols	-	-	31 to 78	Municipal wastewater, Industrial wastewater,	Croatia [55]
		Alkylphenols ethoxylates	-	-	10		
		Phenols	-	-	120 to 280		
LC-MS-MS	SPE, SPE (negative and positive mode) SPE (on line) SPE (on line), SPME (on line) SPE Oasis (on line and off line) SPME	Alkylphenols	3.32 to 220	1 to 10	4.9 to 106	Bromophenols, Emerging contaminants (ECs), Endocrine disrupters (EDCs), Industrial wastewater, Laboratory contamination, Municipal wastewater, Organophosphate flame retardants (OPFRs), Pesticides, Pharmaceuticals, Phenoxyacetic acid herbicides	China [98], Cyprus [86], France [76,77,79], Ireland [90], Italy [83], Portugal [85], Romania—Croatia—Hungary [101]
		Alkylphenols ethoxylates	0.7 to 38.8	2 to 10	2.2 to 211.9		
		Bisphenols	0.93 to 6.29	1 to 10	6.29 to 7400		
		Bromocresols	0.1	0.25	-		
		Bromophenols	0.1 to 2.5	0.25 to 5	1.29 to 15.57		
		Chlorocresols	0.1	0.2	5 to 1364		
		Chlorophenols	0.08 to 1.2	0.2 to 4.9	1 to 1461		
		Nitrophenols	0.06	0.17	-		
		Phenols	1.17	3.5	-		
		Phthalene	1.33	4	-		
LC-UV	A-SPE; Extraction and derivatization; Molecularly imprinted polymers (MIP); LC-Sorption C18; LPME SPE (online)	Alkylphenols	10 to 4190	1200 to 11,600	1 to 1230	Municipal wastewater, Industrial wastewater, Acid herbicides, Pesticides, Chemical waste, Phenolic compounds	Brazil [96], Spain [69,72], USA [64], Greece [88], Poland [57], Iran [99], Nigeria [103]
		Alkylphenols ethoxylates	1950 to 7900	-	-		
		Bisphenols	-	-	22 to 9000		
		Chlorocresols	100	-	-		
		Chlorophenols	50 to 21,000	70,000	75,000 to 407,000		
		Nitrocresols	13,000	-	-		
		Nitrophenols	50 to 17,000	58,000	210,000 to 1,313,000		
		Phenols	400 to 19,000	1400 to 24,000	119,000 to 552,000		

Notes: Abbreviations: ^{1}H NMR: Proton Nuclear Magnetic Resonance, A-SPE: Admicellar Solid-Phase Extraction, C18: Octadecyl, CLSA: Closed-Loop Striping Analysis, DCM: Dichloromethane, ECs: Emerging Contaminants, EDCs: Endocrine-Disrupting Chemicals, GC-ECD: Gas Chromatography with Electron Capture Detector, GC-FID: Gas Chromatography with Flame Ionization Detector, GC-GC-TOFMS: two-dimensional Gas Chromatography with Time-of-Flight Mass Spectrometry, GC-NPD: Gas Chromatography—Nitrogen–phosphorus Detector. GC-MS: Gas Chromatography–Mass Spectrometry, HPLC-SPE-NMR-TOF-MS: High-Performance Liquid Chromatography coupled to Solid-Phase Extraction and Nuclear Magnetic Resonance and Time-of-Flight Mass Spectrometry, JP8: Jet Propulsion, OPFRs: Organophosphate Flame Retardants, LC: Liquid chromatography, LC-MS/GC-MS: Liquid Chromatography–Mass Spectrometry/Gas Chromatography–Mass Spectrometry, LC-MS-MS: Liquid Chromatography–tandem Mass Spectrometry, LC-UV: Liquid Chromatography-Ultraviolet, LLE: Liquid–Liquid Extraction, LPME: Liquid-Phase Microextraction, MASE: Microwave-Assisted Solvent, MIP: Molecularly imprinted polymer, NMR: Nuclear Magnetic Resonance, SPE: Solid-Phase Extraction, SPME: Solid-Phase Microextraction, TCLP: Toxicity Characteristic Leaching Procedure, TD-SBSE: Thermodesorption System—Stir Bar Sorptive Extraction, USA: United States of America.

Nineteen studies conducted in twelve different countries have used GC-MS as an analytical technique using various extraction techniques. Studies were carried out in Germany [80,81], Brazil [104], China [100,105], South Korea [93], Spain [65,66,70,71], USA [63,64,73], France [78], Hungary [89], Nigeria [102], Portugal [84], Taiwan [92] and Poland [94]. In more recent studies, for example, low or non-detectable levels of phenols have been reported in Nigeria, using acetylation to increase the volatility of phenols in GC-MS analysis [102], and in China, using silylation, which improves the thermal stability and spectrometric response of phenolic compounds [100]. On the other hand, the highest concentrations were determined in domestic and industrial landfills in the United States, with reported concentrations of 40 to 50 $\times$ 10^6 ng/L by direct injection of the sample into the GC-MS [63].

In another example, two residential areas in Catalonia (Spain) were investigated in an episode of drinking water contamination due to a strong sweet and paint/solvent odor in the water, with reports of nausea associated with its consumption. The organic compound 3-(trifluoromethyl)phenol (CAS number 98-17-9) has been detected at concentrations of up to 17,000 ng/L in groundwater [71]. Used as a raw material, this compound is an intermediate in the synthesis of pesticides, medicines, resins, dyes and other products. As well as causing serious burns to the skin and damage to the eyes, 3-(trifluoromethyl)phenol is harmful if it is inhaled, ingested, or comes into contact with the skin; it is also harmful to aquatic life, with long-lasting effects [115].

The HPLC-SPE-NMR-TOF-MS technique is an advanced analytical approach that combines various instrumental techniques for the separation, purification and structural identification of chemical compounds. Highly sensitive and specific, it was used in Switzerland for a comprehensive non-targeted analysis of contaminated groundwater at a former munition destruction site [91]. In this same study, a high concentration of PBA was detected, in the range of 12,000 to 13,000 ng/L, due to this compound being used in the ammunition to stabilize the propellant charge.

LC-MS is a powerful analytical tool used to separate, identify and quantify chemical compounds in a variety of complex matrices. In Denmark, its use determined seven acid herbicides in groundwater, four of which were considered hazardous (2,4-dichlorophenol, 2,4-dichlorophenoxyacetic acid, 2-methyl-4,6-dinitrophenol, and dinoseb); however, all the compounds evaluated remained below the detection limit, which ranged from 3 to 26 ng/L [59]. In Barcelona, Spain, a concentration of approximately 1600 ng/L of 4-nonylphenol, which is considered an endocrine disruptor, was reported in the groundwater that serves as the city's second source of water [67]. Selected endocrine-disrupting compounds were also analyzed in groundwater from Romagna, northern Italy, with BPA occurring below the MQL [82]. A study in the USA identified the residential septic system as a potential contaminant of local groundwater by detecting concentrations of nonylphenol at levels of 20,000 ng/L [75]. In a groundbreaking study conducted in Mexico, the presence of BPA was identified in groundwater at concentrations of 25,000 ng/L. The concern that motivated the study was the reuse of municipal water for irrigation [95].

In Croatia, even though the Sava River is an essential ecological, economic and social resource, both the watercourse and its entire basin upstream of Zagreb have been under severe pressure in recent decades. As early as the 1990s, the city was densely occupied and one of the most industrialized in the region [116]. This process has affected both the river and the aquifer, which is directly connected to the Sava and which supplies the city, as part of the domestic and industrial effluent is discharged into the watercourse [55]. In the groundwater of Zagreb, the compound 4-nonylphenol at a concentration of 78 ng/L, 10 ng/L of monoethoxylated 4-nonylphenol, and 280 ng/L of phenol were identified using LC-MS/GC-MS techniques [55].

The LC-UV technique is the most widespread but is less sensitive than LC-MS and LC-MS/MS. This same technique was used in studies carried out in Poland [57], Spain [69,72], the USA [64], Brazil [96], Greece [88], Iran [99] and Nigeria [103]. The study conducted in Nigeria stands out due to the concentration of 1,313,000 ng/L of 2,4-Dinitrophenol found in groundwater used for public water supply. The concern is exacerbated by the fact that the region lacks proper water treatment, posing a significant health risk to the population.

On the other hand, in studies using the LC-MS/MS technique [76–79,83,85,86,90,98,101], LDs were obtained in the range of 0.3 ng/L for 2,4-dichlorophenoxyacetic acid, 0.7 ng/L for 2,4-dichlorophenol, 1.2 ng/L for 2,4,5-trichlorophenol [90], 0.06 ng/L for 2,4-dinitrophenol and 0.13 ng/L for pentachlorophenol [101]. All of these are considered hazardous by 40 CFR Part 302 [46]. These same compounds have shown maximum concentrations of up to 418 ng/L for pentachlorophenol, for example [101]. Concentrations above 1000 ng/L have also been reported for compounds such as 4-chloro-2-methylphenol and mecocrop in Ireland [90] and BPA in France [79], reaching 7400 ng/L. In other words, these compounds are present in the environment at concentrations below the LDs that many techniques achieve.

Several environmental and methodological factors have been shown to exert a significant influence on the outcomes of analytical procedures. These factors encompass conditions related to the collection, storage and handling of samples. The selection of extraction methodology (e.g., LLE, SPE, or SPME), solvents and chromatographic parameters can also have a significant impact on outcomes. To minimize such variability, researchers must adhere to strict protocols, including the use of appropriate containers (e.g., amber glass bottles for organic compounds), maintaining refrigeration until extraction and employing basic laboratory quality control practices (e.g., using properly decontaminated materials).

Furthermore, adherence to method validation procedures is essential to ensure data reliability. Such procedures include calibration curves, precision and accuracy testing, determination of LD and LQ, recovery studies and the use of certified reference standards. Despite the fact that analytical methods are designed to standardize results, a certain degree of dependence on operator skill remains, particularly during sample preparation and extraction steps. Consequently, the implementation of comprehensive training programs and the execution of reproducibility checks are of paramount importance.

In addition to analytical performance, economic and operational factors constitute crucial considerations in the selection of techniques for the monitoring of phenolic compounds in groundwater. While methods such as LC-MS/MS and GC-MS/MS offer high sensitivity and selectivity, they require expensive instrumentation, specialized laboratory infrastructure and highly trained personnel. These constraints can significantly limit their application in contexts where resources are limited, particularly in countries of the Global South.

Conversely, less sophisticated configurations, such as standard GC or LC systems, though less sensitive, are more economically accessible and may be suitable for routine analyses or preliminary screening, especially in regions with limited analytical infrastructure. The extant literature illustrates this distribution, with advanced techniques more commonly associated with institutions and countries that possess greater financial and technical capacity. Consequently, the selection of analytical methods involves, among other aspects, a balance between technical rigor and financial or logistical feasibility, an aspect that is particularly relevant for environmental monitoring programs and evidence-based public policy planning in under-resourced areas.

3.3. Priority Pollutants and Hazardous Compounds Identified in Contaminated Groundwater

In the 52 studies reviewed, 107 specific phenolic compounds were identified. These substances were classified based on the regulations set out in Parts 116, 302, and 423 of the 40 CFR [44,46,47], to determine the substances of greatest concern, to allow for a more consistent risk assessment. A total of 27 of the 107 phenolic compounds are classified as hazardous substances (40 CFR Part 302) [46]. Among these 27 substances, 17 compounds are also regulated as hazardous under the CWA (40 CFR Part 116) [44], and 11 are additionally designated priority toxic pollutants (40 CFR Part 423) [47] (Figure 6).

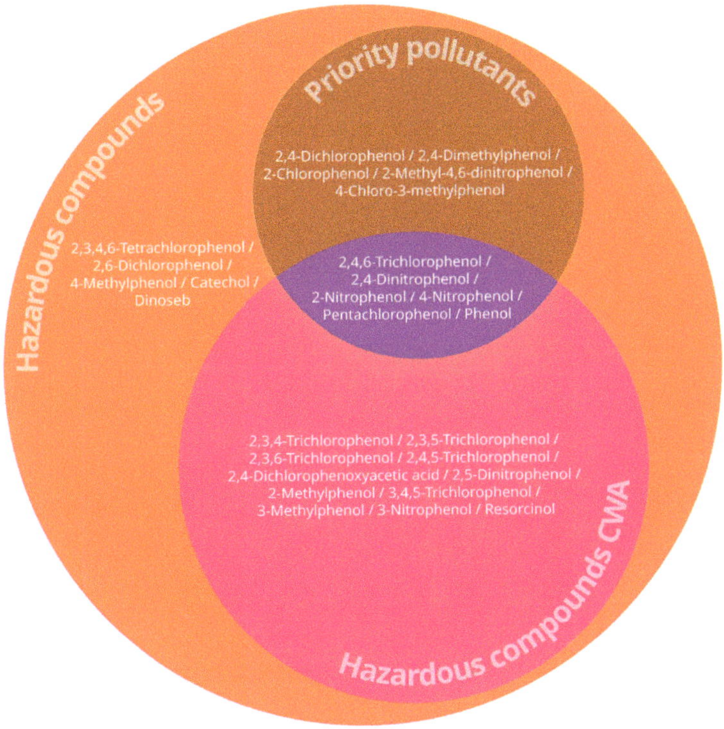

Figure 6. Phenolic compounds identified in groundwater and their classification under USEPA regulations.

The identification and classification of phenolic compounds contribute to understanding the danger they pose to groundwater quality. In addition, this classification enables a more precise assessment of the risks to the environment and the health of the population. Evidence of potential and actual risks was found, especially for compounds that are classified simultaneously in the three regulations analyzed from 40 CFR (Figure 7). The Supplementary Material details the risks for the 27 compounds identified in the studies and classified as hazardous substances (Table S4—Supplementary Material).

The compounds 2,4,6-trichlorophenol, 2,4-dinitrophenol, 2-nitrophenol, 4-nitrophenol, pentachlorophenol, and phenol are regulated as hazardous in three concurrent lists of the 40 CFR. Part 423 of the CFR [47] sets out the substances that should receive priority consideration in the development of effluent limitation guidelines. This regulation is in line with the Clean Water Act (CWA), provided for in Part 304(b) [107], which establishes criteria and guidelines for the control of pollutants to protect water quality in the USA. Part 116 designates hazardous substances that may pose an imminent risk to human health and aquatic life [44], while Part 302 [46] lists the substances that must be monitored in domestic and industrial effluents and that may affect the quality of the receiving bodies of water. 2,4,6-trichlorophenol and pentachlorophenol belong to the class of chlorinated phenols, are extremely toxic and can cause cancer; these substances are widely used in pesticide and herbicide formulations, with pentachlorophenol being restricted to certified consumers [38,117].

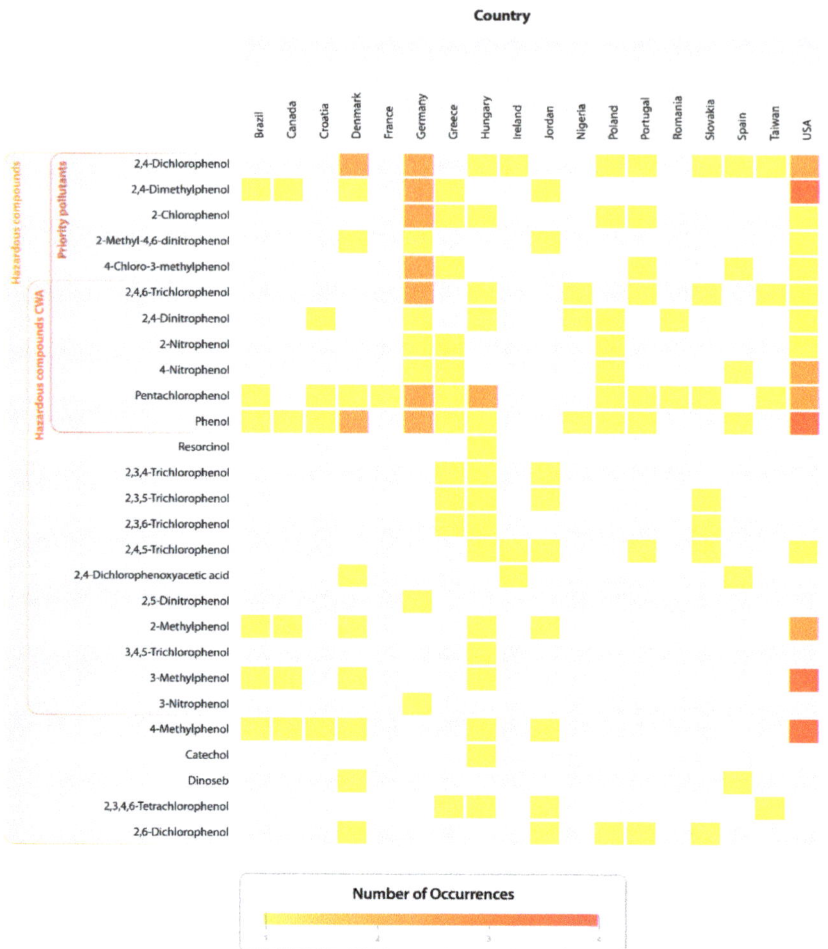

Figure 7. Distribution of the occurrence of hazardous and priority phenolic compounds identified in groundwater.

In the USA, 2,4-dinitrophenol was widely used in the 1930s to treat obesity; a few years later, in 1938, its use was discontinued due to its relationship with several fatalities and the occurrence of cataracts. It is currently used in herbicide, fungicide and insecticide formulations [118]. In addition, 2,4-dinitrophenol can be formed by the nitrification of 2-nitrophenol and 4-nitrophenol. Nitrophenols are pollutants with mutagenic potential and are widely present in the atmosphere. As they are the most abundant compounds in rainwater, they cause concern due to their phytotoxic activities and ability to penetrate plant tissues and enter aquatic environments [119]. In industry, nitrophenols are intermediates to produce dyes, pigments, pharmaceuticals, rubber components, photographic components and pesticides, including fungicides and wood preservatives [120]. Phenols are widely used in various personal care products, cleaning products, paints and waxes, among others.

Even in low concentrations, due to their easy availability in these products, care should be taken with their exposure [111].

3.4. Performance of Analytical Techniques for the Detection and Quantification of Priority Phenolic Compounds in Groundwater

In a specific investigation conducted for the priority compounds observed in this research, studies that simultaneously presented LQ and LD data were selected for analysis of the sensitivity of the techniques, comparison between methods and their relevance for monitoring contamination.

The visualization of the LD and LQ for the priority compounds shows significant differences depending on the analytical technique used (Figure 8). The main graph (right) covers a wide range of values and demonstrates the separation between the groups of points, suggesting that the GC-MS and HPLC-UV analytical techniques tend to result in LD and LQ values that are considerably higher than those observed with other techniques.

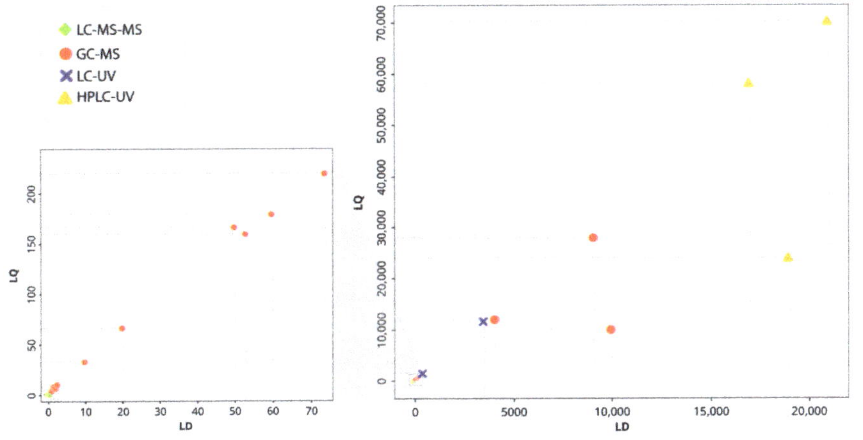

Figure 8. Comparison of the detection limits (LD) and quantification limit (LQ) for four analytical techniques used in studies on groundwater contamination by priority phenolic compounds: LC-MS/MS (green), GC-MS (orange), LC-UV (blue) and HPLC-UV (yellow).

The secondary graph (left) highlights the low concentration range, where LC-MS/MS and GC-MS have their lowest limits, highlighting the sensitivity of these methods compared to LC-UV and HPLC-UV, which operate in a much higher detection range. In particular, the LC-MS/MS technique generally shows greater sensitivity, with lower values for both LD and LQ in this subset of data. As an example, a study conducted with 42 groundwater samples collected from various locations in Ireland had detection limits between 0.08 and 4.7 ng/L for the 18 analytes evaluated, with all compounds detected below the limits of 0.1 ng/L allowed in EU drinking water legislation [90]. The secondary graph also shows the distribution of some GC-MS points, suggesting greater variability in detection and quantification limits. GC-MS reported values ranging from close to zero [64] to the higher values highlighted in the primary graph (right) [84]. This variability may be the result of different operating conditions and/or the use of different sample preparation techniques to be analyzed by GC-MS.

Derivatization is the process of transforming an analyte so that it is detectable in instrumental analytical methods and, specifically, in GC. The technique makes it possible to improve aspects of GC by adapting compounds so that they can be analyzed, producing good resolution, peak symmetry of the compound and detectability because of the interaction between the analyte and the GC detector. The derivatization reagent is the substance used to chemically modify a compound to produce a new compound with properties suitable for GC analysis [121].

Derivatization has been employed as an additional technique to extraction techniques in studies conducted in Cape Cod in Massachusetts, USA [64], Zeitz, Germany [80], Tiszaug and Budapest, Hungary [89], Rosário do Catete, Brazil [104], Dongying, China [100], Ibadan, Nigeria [102], western parts of Slovakia [56] and the northeast region of Poland [94] (Figure 9).

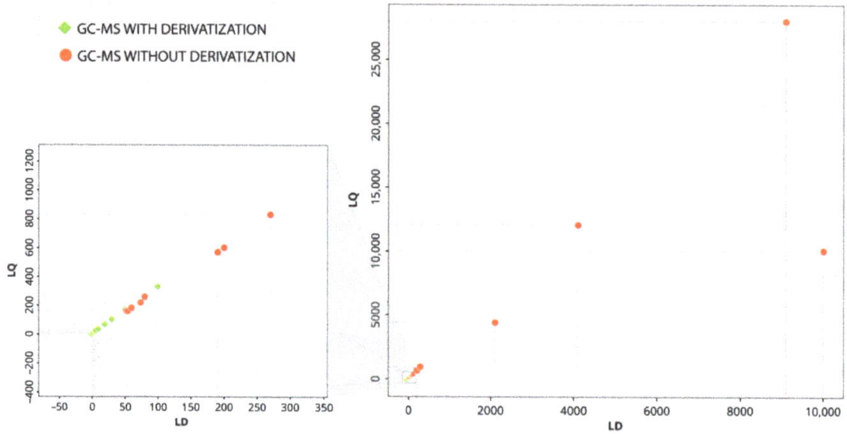

Figure 9. Comparison of the GC-MS analytical technique with and without derivatization in studies on groundwater contamination by phenolic compounds.

The use of GC-MS without derivatization resulted in higher LD and LQ values compared to the use of derivatization, which may suggest a lower capacity of the method to detect and quantify phenols at low concentrations [73,84,92]. Because of this, greater dispersion was observed in the values obtained without derivatization compared to studies with derivatization, which may indicate greater variability in the performance of the technique for different phenolic compounds. In contrast, reduced LQ and LD values have been reported with the use of derivatization [64,89], with evidence of an improvement in the analytical efficiency of the method. This behavior can be explained by the greater volatility and thermal stability of phenolic derivatives, with reduced interference and increased detector sensitivity.

The findings reinforce that derivatization in GC-MS is an efficient technique for reducing detection and quantification limits, making a significant contribution to the analysis of phenolic compounds in groundwater, especially for monitoring trace levels. Among the derivatization reagents used are BSTFA (bis(trimethylsilyl)trifluoroacetamide) with TMCs at different concentrations [64,100,104] and trimethylsilyl-N,N-dimethylcarbamate (TMSDMC) [89]. Both reagents revealed the technique's ability to optimize the detection and quantification of a wide variety of phenols, chlorophenols, alkylphenols, ethoxylates, and bisphenols.

When evaluating the influence of different derivatization reagents on the determination of LD and LQ in the analysis of phenolic compounds by GC-MS, it was observed that the choice of derivatizer directly affects analytical sensitivity (Figure 10).

The reagents examined, BSTFA + 1% TMCS, BSTFA + 10% TMCS and TMSDMC, showed a positive (and expected) correlation between LD and LQ, since LQ is, by definition, a higher value than LD.

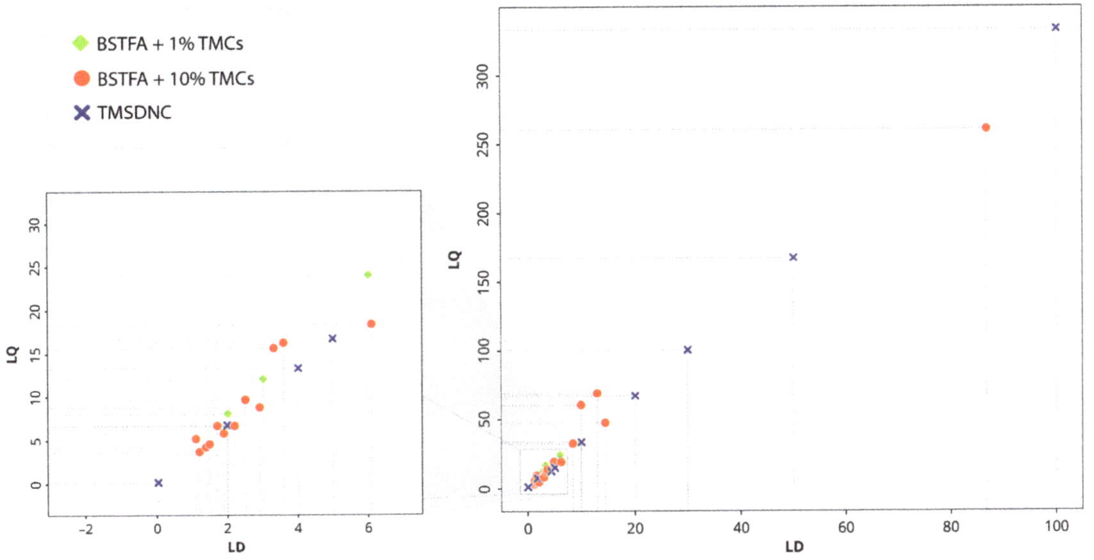

Figure 10. Effect of different derivatization reagents on the detection and quantification limits of phenolic compounds by GC-MS.

BSTFA + 1% TMCS [100] and BSTFA + 10% TMCS [64] showed very similar results, with close LD and LQ values, especially for lower concentrations of the phenolic groups: chlorophenols, biphenols, alkylphenols, nitrophenols and bisphenols. This suggests that varying the percentage of TMCS between 1% and 10% when using BSTFA does not have a significant impact on the sensitivity of the technique for most of the compounds analyzed. However, for one of the compounds in the alkylphenol ethoxylates group, the use of BSTFA + 10% TMCS resulted in a considerably higher LD and LQ [64].

On the other hand, TMSDMC showed values for some compounds from the alkylphenol and phenol groups that were very similar to BSTFA but also had much higher LD and LQ values, especially for chlorophenol concentrations [89], which indicates a lower analytical sensitivity when compared to BSTFA-based reagents. The difference in performance is more pronounced at higher concentrations.

Studies have demonstrated the effectiveness of GC-MS in determining phenolic compounds using, for example, different sample preparation methods. Liquid–liquid extraction (LLE) [55,58,60,93,100,102], solid-phase extraction (SPE) [58,59,65–70,75,76,78,79,82,83,85, 86,88–91,93,97–99,101,103–105], liquid-phase microextraction (LPME) [92,96] and stir bar sorptive extraction (SBSE) [80] have been reported. Membrane-assisted solvent extraction (MASE) [81], solid-phase microextraction (SPME) [77,78,84], closed-loop strip analysis (CLSA) [71], derivatization [54,62,64,80,89,100,102,104] and other approaches have also been used.

The GC-MS technique with derivatization has been optimized for the determination of alkylphenols, alkylphenol ethoxylates and bisphenols [64,100,102,104]. Derivatization is also convenient for the simultaneous determination of a wide variety of phenols and chlorophenols [80,89].

These checks not only help to evaluate the performance of different analytical techniques but also help to select the most suitable technique according to the specific needs of each investigation. In addition, the analysis of data dispersion suggests that the choice of analytical technique can have a significant impact on the ability to detect and quantify priority compounds at low concentrations.

Determining the efficiency of extraction techniques is also a critical factor when examining phenols in groundwater environmental matrices, as it directly influences the accuracy and reliability of the results. Thus, the recovery (%) of an analyte reflects the ability of the extraction method to isolate the compound of interest in the matrix without significant losses or interferences that could compromise its quantification.

For the studies that reported priority phenolic compounds, which simultaneously presented LD and LQ data and did not employ derivatization [90,92,96,101,103], different extraction approaches combined with different analytical techniques were compared (Figure 11).

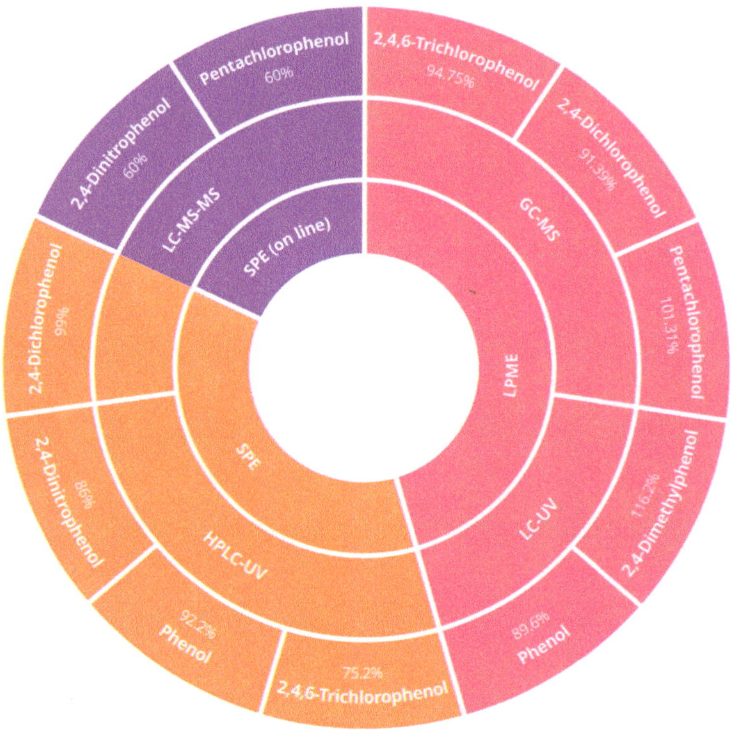

Figure 11. Efficiency of extraction techniques for priority phenolic compounds: recovery rates across different analytical methods.

The three concentric rings represent the efficiency of the extraction techniques in the recovery (%) of priority phenolic compounds. The extraction techniques (inner ring) were used to isolate the phenolic compounds from the environmental matrix before analysis and included LPME, SPE and online SPE. Different coupled analytical techniques (middle ring) were used for the identification and quantification of the extracted compounds (LC-MS/MS; GC-MS; LC-UV and HPLC-UV), while the recovery efficiency per compound (outer ring) displays the percentage recovery for each priority phenolic compound, associated with a given combination of extraction and analysis.

The SPE-LC-MS-MS technique demonstrated high efficiency in the extraction of 2,4-dichlorophenol (99% recovery) [85], while in the SPE (online)-LC-MS/MS technique, pentachlorophenol and 2,4-dinitrophenol showed lower efficiencies (60%) [101].

The results of the LPME-GC-MS technique showed good efficiency, with recoveries of 94.75% for 2,4,6-trichlorophenol, 91.39% for 2,4-dichlorophenol and 101.31% pentachlorophenol [92]. Although pentachlorophenol showed a recovery of slightly more than 100% [92], as this response can be attributed to small experimental variations or matrix interferences, further investigation is recommended.

With recoveries that varied significantly between the compounds analyzed, 89.60% for phenol and 116.20% for 2,4-dimethylphenol [96], the LPME-LC-UV technique also showed recovery above 100% for 2,4-dimethylphenol [96], which suggests further investigation due to the possibility of interferents or errors in quantification.

The SPE-HPLC-UV technique showed high efficiency in the recovery of phenol (92.2%) and good efficiency in the recovery of 2,4-dinitrophenol (86%) and 2,4,6-trichlorophenol (75.2%).

When examining the maximum concentrations of priority compounds identified by country, phenol was quantified at 40,000,000 ng/L using GC-MS in landfill monitoring in the USA [63] (Figure 12).

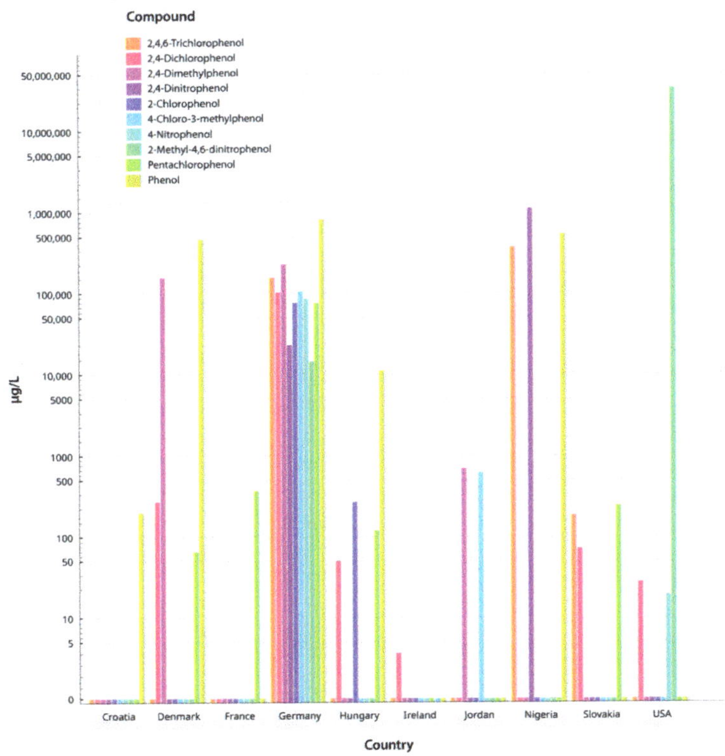

Figure 12. Spatial distribution and maximum concentrations of priority phenolic contaminants in groundwater.

Maximum concentrations of 35 ng/L were obtained for 2,4-dichlorophenol and 26 ng/L for pentachlorophenol in Cape Cod, Massachusetts, associated with contamination by urban effluents [64].

In Germany, for example, in addition to phenol in high concentration (866,140 ng/L), high concentrations of 2,4-dimethylphenol (304,010 ng/L), 2,4,6-trichlorophenol (216,200 ng/L), 4-chloro-3-methylphenol (113,910 ng/L), 2-chlorophenol (79,740 ng/L) and pentachlorophenol (79,500 ng/L) have also been reported [81]. For these results, GC-MS and membrane-assisted solvent extraction (MASE) were used, without the need for derivatization or complex sample preparation for extraction [81] (Figure 12).

As water is key to achieving sustainable development, peace, and prosperity [6], its integrated management is central to both the governance of the resources worldwide and the implementation of the 2030 Agenda [122]. Progress towards achieving the SDG may be severely constrained by the continued decline of nature [123–125], especially considering that more than half of the 169 SDG targets are dependent on nature for their achievement [125]. Thus, the depletion of natural resources jeopardizes human well-being and adds to the list of challenges to be overcome, especially in terms of water scarcity and its costs both for people and for maintaining the ecosystems that sustain life [4].

In this context, the establishment of regulations, robust indicators and continuous monitoring of water resources play a fundamental role, especially for groundwater. Despite their importance for human supply and ecosystems, aquifer systems remain invisible [9,126] and subject to overexploitation and contamination, particularly by phenolic compounds.

Indications have been found that the adoption of advanced analytical techniques makes it possible to detect and quantify these contaminants sensitively, contributing to the formulation of evidence-based public policies for the protection and monitoring of aquifer systems. However, to advance its implementation, especially in the Global South, international scientific and technical cooperation can play a strategic role in the sharing of knowledge. This perspective is fundamental for tackling the cumulative legacy of Euro-American colonialism, imperialism, and capitalism and the resulting socioeconomic, political and geographical inequalities [5].

In Nigeria, for example, although groundwater is essential for supplying rural areas and is vulnerable to several vectors, there have been few studies examining the level of EDs [102]. In Brazil, the vulnerability of water sources and supply systems to EDs has highlighted the importance of including these compounds in national controls to guarantee quality water supply for the population [104].

The expansion of advanced analytical technologies, especially in remote locations, can also be limited by the short-term stability of substances during transportation, temperature control from the sampling point to the analytical laboratory and the stability of target substances under laboratory conditions [79]. These difficulties become even more relevant in countries, cities and regions where laboratory infrastructure and research investments are more restricted, especially in locations in the Global South.

4. Limitations of the Study

Twenty-three percent of the studies were not analyzed because they did not specify the study site, because they dealt with simulated scenarios or synthetic waters or because they identified phenolic compounds without quantifying them, even using spectrometric and chromatographic techniques, which may represent a bias in the research results. In addition, the databases selected for the research, as well as the descriptors used, can carry some bias, possibly reducing the representativeness of the sample.

Keyword co-occurrence analysis may also be limited by terms that were not attributed to the authors or were absent from the documents evaluated.

Another possible restriction refers to the heterogeneity in data reporting between the included studies, which made direct comparison and meta-analysis of the results difficult, specifically concerning parameters such as LD, LQ, recovery percentage (%) and derivatization, among others.

As environmental, operational and methodological factors have been demonstrated to influence analytical results, it can be assumed that the authors of the studies reviewed took the necessary precautions to ensure the robustness and credibility of their analytical procedures. Nevertheless, not all of the articles reviewed explicitly mentioned whether full validation was carried out or whether standardized protocols, especially for manual steps, were adopted.

Although priority was given to articles in English, Portuguese and Spanish, which are languages in which the research team has technical mastery, only 4% (N = 14) of the total studies retrieved (N = 342) were in other languages. Thus, the possible exclusion of works from this set is unlikely to introduce significant bias in the sample. Even so, this limitation could restrict the inclusion of any studies published in other languages, including those from regions under-represented in the global scientific literature. Similarly, the other eligibility criteria adopted may also have favored the inclusion of studies predominantly conducted in the Global North, which would result in a possible socioeconomic bias.

The inconsistent terminology of phenolic compounds (use of synonyms) may have affected the sensitivity of the search. Thus, it is possible that studies were not identified due to variability in the nomenclature used. Although this review was conducted according to the PRISMA 2020 guidelines, it was not prospectively registered. However, all methodological steps were documented to ensure the transparency of the research and its reproducibility.

5. Conclusions

This review covered a broad period of approximately 40 years of research (1984 to 2023), despite the low number of eligible studies. The temporal evolution of the publications analyzed suggests an increase in concern about groundwater contamination by phenolic compounds over the years. In addition, the socioeconomic scope of the research has widened, with countries in both the Global North and South, although scientific production has been more expressive in the former.

The studies highlighted the diversity of analytical techniques used in the detection and quantification of phenolic compounds, with differences in their sensitivity, resolution and suitability for a given class of chemical compounds. Among the methods identified, LC-MS and LC-MS/MS stood out as the main analytical approaches, with extremely low detection limits (LD) and quantification limits (LQ) in the ng/L range. GC-MS, in turn, proved to be competitive when associated with derivatization with BSTFA + TMCs before analysis. In addition, in terms of sample preparation methods, liquid-phase extraction (LPE) was the most widely used technique, with satisfactory recovery rates.

The lack of specific regulations for phenolic compounds in aqueous matrices at a global level, especially in groundwater, reinforces the urgent need for national regulatory agencies to define maximum permissible limits. In addition, these regulations should cover a comprehensive number of phenolic compounds, particularly those classified as hazardous, since concentrations over 1000 ng/L have been reported in several studies, reaching a maximum concentration of 50×10^6 ng/L.

The main contributors to the presence of these contaminants in the aquifer systems investigated include domestic and industrial wastewater, as well as pesticides with phenolic compounds in their formulation. This scenario represents a significant challenge for achieving SDG 6, which aims to guarantee the availability and sustainable management of water and sanitation for all. Furthermore, considering the interdependence of the goals set out in the 2030 Agenda, the implementation of coordinated efforts to achieve them in full is both desirable and necessary.

From a practical point of view, the implementation of advanced analytical techniques, especially in countries of the Global South, must be accompanied by strategies that ensure operational viability and equitable access to these technologies, given the structural, logistical and financial challenges that can hinder their adoption. Evidence from this research also highlights the importance of carefully selecting analytical methods for determining phenolic compounds in groundwater. Knowledge about the efficiency and limitations of different techniques can guide the choice of the best approaches for early detection of contamination.

In addition to the technical evaluation, the identification and quantification of the risk associated with phenolic compounds, based on the regulations established by reference agencies, are fundamental for prioritizing areas of intervention, establishing safety limits and subsidizing strategic decisions in cases of contamination.

In terms of public policy, it is advisable to establish stricter and more harmonious regulatory standards between different countries, as well as encourage technology transfer and partnerships, including international ones, to strengthen local capacities for monitoring and mitigating contamination. The implementation of funding mechanisms aimed at laboratory modernization and the development of national monitoring programs is essential. Likewise, promoting research and innovation in new analytical techniques and specialized technical training is indispensable for training highly qualified human resources and for disseminating good practices in sustainable groundwater management.

Finally, the search for more sensitive analytical methods for detecting phenolic compounds in low concentrations represents a promising area for future research. Technological and scientific advances in this area could contribute to more accurate monitoring and the development of innovative remediation techniques for the preservation and recovery of contaminated aquifers.

Supplementary Materials: The following supporting information can be downloaded at: https://www.mdpi.com/article/10.3390/w17081173/s1, Table S1: 52 unified records: 23 from Scopus only, 8 from WoS only and 21 studies indexed in both databases; Table S2: Authors' keywords incorporated into the same label for co-occurrence analysis—Synonym Table for VOSviewer; Table S3: Bibliography on the presence of phenolic compounds in groundwater identified and quantified by chromatographic and spectrometric analytical techniques; Table S4: Characteristic of the 27 Hazardous compounds classified with 40 CFR-Part 302; Table S5: Toxicity and Environmental Regulations classification of the 27 Hazardous compounds; Table S6: Classification of studies that used Derivatization with GC-MS analytical technique; Table S7: Comparison of the efficiency of extraction techniques that did not use derivatization to analyze priority compounds.

Author Contributions: Conceptualization, G.V.S.-C.-M., E.A.R. and O.V.B.; Methodology, G.V.S.-C.-M., D.R.A.-P., E.A.R., D.A.d.A. and O.V.B.; Software, D.R.A.-P.; Validation, G.V.S.-C.-M., J.I.O. and E.A.R.; Formal analysis, G.V.S.-C.-M., D.R.A.-P., J.I.O. and E.A.R.; Investigation, G.V.S.-C.-M., J.I.O., D.R.A.-P. and E.A.R.; Resources, D.R.A.-P., E.A.R., D.A.d.A. and O.V.B.; Data curation, G.V.S.-C.-M., J.I.O., D.R.A.-P. and E.A.R.; Writing—original draft preparation, G.V.S.-C.-M., J.I.O. and E.A.R.; Writing—reviewer and editing, G.V.S.-C.-M., J.I.O. and E.A.R.; Visualization, G.V.S.-C.-M., J.I.O., D.R.A.-P. and E.A.R.; Supervision, E.A.R.; Project administration, G.V.S.-C.-M. and E.A.R.; Funding acquisition, D.A.d.A. and O.V.B. All authors have read and agreed to the published version of the manuscript.

Funding: This research was funded by "Coordination for the Improvement of Higher Education Personnel" (CAPES, 01). The APC was funded by TVGD Holding.

Data Availability Statement: All the data are available in the manuscript and its Supplementary Materials.

Acknowledgments: We thank Luisa Sadeck for her support in the illustrations. During the preparation of this manuscript, the authors used GPT (OpenAI, GPT-4, 2025 version) to support graphical structure definitions, and assist in the generation of initial drafts for figures and graphics summaries. The authors have reviewed and edited the output and take full responsibility for the content of this publication.

Conflicts of Interest: The authors declare no conflicts of interest. The funders had no role in the design of the study; in the collection, analysis, or interpretation of data; in the writing of the manuscript; or in the decision to publish the results.

Abbreviations

The following abbreviations are used in this manuscript:

μg/L	Microgram per Liter;
^{1}H NMR	Proton Nuclear Magnetic Resonance;
2,4,6-TBP	2,4,6-Tribromophenol;
2,4-MCP	4-chloro-2-methylpenol;
APs	Alkylphenols;
A-SPE	Admicellar Solid-Phase Extraction;
BPA	Bisphenol A;
BSTFA	Bis(trimethylsilyl)trifluoroacetamide;
C18	Octadecyl;
CAS	Chemical Abstracts Service;
CERCLA	Comprehensive Environmental Response, Compensation and Liability Act of 1980;
CFR	Code of Federal Regulations (United States);
CG	Gas Chromatography;
GC-MS	Gas Chromatography;
CLSA	Closed-Loop Striping Analysis;
CWA	Clean Water Act;
DBP	Dibromophenol;
DCM	Dichlorometano;
DOM	Overseas Departments;
DSDG	Division for Sustainable Development Goals;
EC	Europe Commission;
ECs	Emerging Contaminants;
ECD	Electron Capture Detector;
EDs	Endocrine Disruptors;
EDCs	Endocrine-Disrupting Chemicals;
FID	Flame Ionization Detector;
GC	Gas Chromatography;
GC-ECD	Gas Chromatography with Electron Capture Detector;
GC-FID	Gas Chromatography with a Flame Ionization Detector;
GC-GC-TOFMS	Two-Dimensional Gas Chromatography with Time-of-Flight Mass Spectrometry;
GC-NPD	Gas Chromatography–Nitrogen–Phosphorus Detector;
GC-MS	Gas Chromatography–Mass Spectrometry;
GC-MS/MS	Gas Chromatography–Tandem Mass Spectrometry;
GRUMO	The National Groundwater-Monitoring Program;
GWD	The EU Groundwater Directive;
HPLC	High-Performance Liquid Chromatography;
HPLC-MS/MS	High-Performance Liquid Chromatography–Tandem Mass Spectrometry;
HPLC-SPE-NMR-TOF-MS	High-Performance Liquid Chromatography coupled to Solid-Phase Extraction and Nuclear Magnetic Resonance and Time-of-Flight Mass Spectrometry;
IAG	Generative Artificial Intelligence;
ICP-MS	Inductively Coupled Plasma Mass Spectrometry;
IPCC	Intergovernmental Panel on Climate Change;
JP8	Jet Propulsion;
LC	Liquid Chromatography;
LC-MS	Liquid chromatography–Mass Spectrometry;

LC-MS/GC-MS	Liquid chromatography–Mass Spectrometry-Gas Chromatography–Mass Spectrometry;
LC-MS/MS	Liquid Chromatography–Tandem Mass Spectrometry;
LC-UV	Liquid Chromatography-Ultraviolet;
LD	Detection Limit;
LOD	Limit of Detection;
LOQ	Limit of Quantification;
LQ	Quantification Limit;
LLE	Liquid–Liquid Extraction;
LME	Liquid-Phase Micro Extraction;
LPME	Liquid-Phase Micro Extraction;
LPME-GC-MS	Liquid-Phase Micro Extraction–Tandem Mass Spectrometry;
LPME-LC-UV	Liquid-Phase Micro Extraction-Liquid Chromatography-Ultraviolet;
MASE	Microwave-Assisted Solvent;
MIP	Molecularly Imprinted Polymer;
MQL	Method Quantification Limit;
MS	Mass Spectrometry;
MTBE	Methyl Tertiary Butyl Ether;
ng/L	Nanogram per Liter;
NIH	National Institutes of Health;
NLM	National Library of Medicine;
NMR	Nuclear Magnetic Resonance;
NMW	Natural Mineral Water;
ONU	Organização das Nações Unidas;
OPFRs	Organophosphate Flame Retardants;
PFASs	Perfluoroalkyl Substances;
PICo	Population, Interest and Context;
PRISMA	Preferred Reporting Items for Systematic Review and Meta-Analyses;
R	Recovery (%);
RQ	Reportable Quantity;
SBSE	Stir Bar Sorptive Extraction;
SDG	Sustainable Development Goal;
SERS	Surface Enhanced Raman Spectroscopy;
SLR	Systematic Literature Review;
SPE	Solid-Phase Extraction;
SPE-LC-MS-MS	Solid-Phase Extraction–Liquid Chromatography–Tandem Mass Spectrometry;
SPE-LC-APCI-MS	Solid-Phase Extraction–Liquid Chromatography Atmospheric Pressure Chemical Ionization–Mass Spectrometry;
SPE (on line)-LC-MS-MS	Solid-Phase Extraction (on line)–Liquid Chromatography–Tandem Mass Spectrometry;
SPME	Solid-Phase Microextraction;
SPME-GC-MS	Solid-Phase Microextraction with Gas Chromatography–Mass Spectrometry;
SPW	Spring Water
TBP	Tribromophenol;
TCLP	Toxicity Characteristic Leaching Procedure;
TD-SBSE	Thermodesorption System–Stir Bar Sorptive Extraction;
TMCS	Trimethylchlorosilane;
TMSDMC	Trimethylsilyl-N,N-dimethylcarbamate;
TOF-MS	Time-of-Flight Mass Spectrometry;
TRI	Toxics Release Inventory;

UN	United Nations;
UPLC	Ultra-Performance Liquid Chromatography;
UPLC-MS/MS	Ultra-Performance Liquid Chromatography–Tandem Mass Spectrometry;
UHPLC	Ultra-High-Performance Liquid Chromatography;
USA	United States of America;
US CFR	United States Code of Federal Regulations;
USEPA	United States Environmental Protection Agency;
WHO	World Health Organization;
WoS	Web of Science

References

1. UN. General Assembly. Transforming Our World: The 2030 Agenda for Sustainable Development. 2015. Available online: https://www.un.org/en/development/desa/population/migration/generalassembly/docs/globalcompact/A_RES_70_1_E.pdf (accessed on 23 February 2025).
2. UN-DSDG. Division for Sustainable Development Goals. 2025. Available online: https://sdgs.un.org/about (accessed on 23 February 2025).
3. Huang, Z.; Yuan, X.; Liu, X. The key drivers for the changes in global water scarcity: Water withdrawal versus water availability. *J. Hydrol.* **2021**, *601*, 126658. [CrossRef]
4. Rodrigues, E.A.; Ferreira, M.L.; de Carvalho, A.R.; Bustillos, J.O.W.V.; Victor, R.A.B.M.; Sodré, M.G.; de Andrade, D.A. Land, water, and climate issues in large and megacities under the lens of nuclear science: An approach for achieving sustainable development goal (SDG11). *Sustainability* **2022**, *14*, 13646. [CrossRef]
5. Rodrigues, E.A.; Carvalho, A.R.D.; Ferreira, M.L.; Victor, R.A.B.M.; Luca, E.F.D.; Rocha, G.C.D.; Andrade, D.A.D. Beyond the Rising Tide: Towards Effective Climate Policy in Coastal Urban Centers. *Land* **2024**, *13*, 2071. [CrossRef]
6. UN. *The United Nations World Water Development Report 2024: Water for Prosperity and Peace*; UNESCO: Paris, France, 2024; 153p. Available online: https://unesdoc.unesco.org/ark:/48223/pf0000388948 (accessed on 23 February 2025).
7. IPCC. Summary for policymakers. In *Climate Change 2023: Synthesis Report*; Lee, H., Romero, J., Eds.; Contribution of Working Groups I II and III to the Sixth Assessment Report of the Intergovernmental Panel on Climate Change; IPCC: Geneva, Switzerland, 2023; pp. 1–34. Available online: https://www.ipcc.ch/report/ar6/syr/downloads/report/IPCC_AR6_SYR_SPM.pdf (accessed on 23 February 2025).
8. Gleick, P.H.; Cooley, H. Freshwater scarcity. *Annu. Rev. Environ. Resour.* **2021**, *46*, 319–348. [CrossRef]
9. UN. *The United Nations World Water Development Report 2022: Groundwater: Making the Invisible Visible*; UNESCO: Paris, France, 2022; 225p, Available online: https://unesdoc.unesco.org/ark:/48223/pf0000380721 (accessed on 23 February 2025).
10. Li, P.; Karunanidhi, D.; Subramani, T.; Srinivasamoorthy, K. Sources and consequences of groundwater contamination. *Arch. Environ. Contam. Toxicol.* **2021**, *80*, 1–10. [CrossRef]
11. UN-Habitat. *World Cities Report 2022: Envisaging the Future of Cities*; UN-Habitat: Nairobi, Kenya, 2022; 422p, Available online: https://unhabitat.org/sites/default/files/2022/06/wcr_2022.pdf (accessed on 23 February 2025).
12. Ravindiran, G.; Rajamanickam, S.; Sivarethinamohan, S.; Karupaiya Sathaiah, B.; Ravindran, G.; Muniasamy, S.K.; Hayder, G. A Review of the Status, Effects, Prevention, and Remediation of Groundwater Contamination for Sustainable Environment. *Water* **2023**, *15*, 3662. [CrossRef]
13. Kumar, R.; Qureshi, M.; Vishwakarma, D.K.; Al-Ansari, N.; Kuriqi, A.; Elbeltagi, A.; Saraswat, A. A review on emerging water contaminants and the application of sustainable removal technologies. *Case Stud. Chem. Environ. Eng.* **2022**, *6*, 100219. [CrossRef]
14. Ramos, R.L.; Moreira, V.R.; Amaral, M.C.S. Phenolic compounds in water: Review of occurrence, risk, and retention by membrane technology. *J. Environ. Manag.* **2024**, *351*, 119772. [CrossRef] [PubMed]
15. Ayad, R.; Akkal, S. Phytochemistry and biological activities of algerian Centaura an related genera. *Stud. Nat. Prod. Chem.* **2019**, *63*, 357–414. [CrossRef]
16. Rutala, W.A.; Weber, D.J. Disinfection, Sterilization, and Control of Hospital Waste. In *Mandell, Douglas, and Bennett's Principles and Practice of Infectious Diseases*, 8th ed.; Elsevier Saunders: Amsterdan, The Netherlands, 2015; Volume 2, pp. 3294–3309. [CrossRef]
17. Silah, H.; Erkmen, C.; Unal, D.N.; Uslu, B. Sensing of phenol and chlorophenols using carbon nanotubes modified glassy carbon electrode. In *Sensing of Deadly Toxic Chemical Warfare Agents, Nerve Agent Simuants, and Their Toxicologial Aspects*; Elsevier: Amsterdam, The Netherlands, 2023; Chapter 13; pp. 297–329. [CrossRef]
18. Abdollahi, M.; Hassani, S.; Derakhshani, M. Phenol. In *Philip Wexler, Encyclopedia of Toxicology*, 3rd ed.; Academic Press: Cambridge, MA, USA, 2014; pp. 871–873. [CrossRef]

19. Ahmaruzzaman, M.; Mishra, S.R.; Gadore, V.; Yadav, G.; Roy, S.; Bhattacharjee, B.; Kumari, K. Phenolic Compounds in Water: From Toxicity and Source to Sustainable Solutions-An Integrated Review of Removal Methods, Advanced Technologies, Cost Analysis, and Future Prospects. *J. Environ. Chem. Eng.* **2024**, *12*, 112964. [CrossRef]
20. Sarkar, S.; Gill, S.S.; Das Gupta, G.; Kumar Verma, S. Water toxicants: A comprehension on their health concerns, detection, and remediation. *Environ. Sci. Pollut. Res.* **2022**, *29*, 53934–53953. [CrossRef] [PubMed]
21. Xie, X.; Shi, J.; Pi, K.; Deng, Y.; Yan, B.; Tong, L.; Jiang, G. Groundwater quality and public health. *Annu. Rev. Environ. Resour.* **2023**, *48*, 395–418. [CrossRef]
22. Deng, Z.H.; Li, N.; Jiang, H.L.; Lin, J.M.; Zhao, R.S. Pretreatment techniques and analytical methods for phenolic endocrine disrupting chemicals in food and environmental samples. *TrAC Trends Anal. Chem.* **2019**, *119*, 115592. [CrossRef]
23. Vaye, O.; Ngumbu, R.S.; Xia, D. A review of the application of comprehensive two-dimensional gas chromatography MS-based techniques for the analysis of persistent organic pollutants and ultra-trace level of organic pollutants in environmental samples. *Rev. Anal. Chem.* **2022**, *41*, 63–73. [CrossRef]
24. Vega-Bustillos, O.; Redígolo, M. *Princípios e Aplicações da Espectrometria de Massas*, 1st ed.; Scortecci: São Paulo, Brazil, 2024.
25. PRISMA. Welcome to the Preferred Reporting Items for Systematic Reviews and Meta-Analyses (PRISMA) Website. 2025. Available online: https://www.prisma-statement.org/ (accessed on 23 February 2025).
26. Moher, D.; Shamseer, L.; Clarke, M.; Ghersi, D.; Liberati, A.; Petticrew, M.; Prisma-P Group. Preferred reporting items for systematic review and meta-analysis protocols (PRISMA-P) 2015 statement. *Syst. Rev.* **2015**, *4*, 1. [CrossRef] [PubMed]
27. Page, M.J.; McKenzie, J.E.; Bossuyt, P.M.; Boutron, I.; Hoffmann, T.C.; Mulrow, C.D.; Moher, D. The PRISMA 2020 statement: An updated guideline for reporting systematic reviews. *BMJ* **2021**, *372*, 71. [CrossRef] [PubMed]
28. Lockwood, C.; Munn, Z.; Porritt, K. Qualitative research synthesis: Methodological guidance for systematic reviewers utilizing meta-aggregation. *JBI Evid. Implement.* **2015**, *13*, 179–187. [CrossRef]
29. Singh, V.K.; Singh, P.; Karmakar, M.; Leta, J.; Mayr, P. The journal coverage of Web of Science, Scopus and Dimensions: A comparative analysis. *Scientometrics* **2021**, *126*, 5113–5142. [CrossRef]
30. Birkle, C.; Pendlebury, D.A.; Schnell, J.; Adams, J. Web of Science as a data source for research on scientific and scholarly activity. *Quant. Sci. Stud.* **2020**, *1*, 363–376. [CrossRef]
31. Martín-Martín, A.; Orduna-Malea, E.; Thelwall, M.; López-Cózar, E.D. Google Scholar, Web of Science, and Scopus: A systematic comparison of citations in 252 subject categories. *J. Informetr.* **2018**, *12*, 1160–1177. [CrossRef]
32. Baas, J.; Schotten, M.; Plume, A.; Côté, G.; Karimi, R. Scopus as a curated, high-quality bibliometric data source for academic research in quantitative science studies. *Quant. Sci. Stud.* **2020**, *1*, 377–386. [CrossRef]
33. Martín-Martín, A.; Thelwall, M.; Orduna-Malea, E.; Delgado López-Cózar, E. Google Scholar, Microsoft Academic, Scopus, Dimensions, Web of Science, and OpenCitations' COCI: A multidisciplinary comparison of coverage via citations. *Scientometrics* **2021**, *126*, 871–906. [CrossRef] [PubMed]
34. Asubiaro, T.; Onaolapo, S.; Mills, D. Regional disparities in Web of Science and Scopus journal coverage. *Scientometrics* **2024**, *129*, 1469–1491. [CrossRef]
35. Pranckutė, R. Web of Science (WoS) and Scopus: The titans of bibliographic information in today's academic world. *Publications* **2021**, *9*, 12. [CrossRef]
36. Gioia, D. A systematic methodology for doing qualitative research. *J. Appl. Behav. Sci.* **2021**, *57*, 20–29. [CrossRef]
37. Corbin, J.; Strauss, A. *Basics of Qualitative Research: Techniques and Procedures for Developing Grounded Theory*, 4th ed.; Sage Publications: Thousand Oaks, CA, USA, 2014; 434p.
38. Kim, S.; Chen, J.; Cheng, T.; Gindulyte, A.; He, J.; He, S.; Li, Q.; Shoemaker, B.A.; Thiessen, P.A.; Yu, B.; et al. Atualização do PubChem 2025. *Nucleic Acids Res.* **2025**, *53*, D1516–D1525. [CrossRef]
39. USEPA. U.S. Environmental Protection Agency. 2025. Available online: https://www.epa.gov/ (accessed on 23 February 2025).
40. US-eCFR. Title 40. Title 40 Was Last Amended 2/07/2025. 2025. Available online: https://www.ecfr.gov/current/title-40 (accessed on 23 February 2025).
41. Guo, L.C.; Zeng, E.Y. Broadening the global reach of the United States environmental protection agency (USEPA) is vital to combating globalized environmental problems. *Environ. Sci. Technol.* **2010**, *44*, 6911–6913. [CrossRef] [PubMed]
42. Zhang, D.; Fu, H.Z.; Ho, Y.S. Characteristics and trends on global environmental monitoring research: A bibliometric analysis based on Science Citation Index Expanded. *Environ. Sci. Pollut. Res.* **2017**, *24*, 26079–26091. [CrossRef]
43. USEPA. Summary of the Clean Water Act. 2025. Available online: https://www.epa.gov/laws-regulations/summary-clean-water-act (accessed on 23 February 2025).
44. US-eCFR. Title 40, Part 116.4: Designation of Hazardous Substances. 2025. Available online: https://www.ecfr.gov/current/title-40/chapter-I/subchapter-D/part-116#116.4 (accessed on 27 February 2025).
45. USEPA. Hazardous Substance Designations and Release Notifications. 2025. Available online: https://www.epa.gov/epcra/hazardous-substance-designations-and-release-notifications (accessed on 23 February 2025).

46. US-eCFR. Title 40, Part 302.—Designation, Reportable Quantities, and Notification. 2025. Available online: https://www.ecfr.gov/current/title-40/chapter-I/subchapter-J/part-302 (accessed on 23 February 2025).
47. US-eCFR. Title 40 Part 423—Steam electric power generating point source category: Appendix A to Part 423–126 Priority Pollutants. 2025. Available online: https://www.ecfr.gov/current/title-40/chapter-I/subchapter-N/part-423 (accessed on 23 February 2025).
48. Bhuyan, A.; Sanguri, K.; Sharma, H. Improving the keyword co-occurrence analysis: An integrated semantic similarity approach. In Proceedings of the 2021 IEEE International Conference on Industrial Engineering and Engineering Management (IEEM), Singapore, 13–16 December 2021; IEEE: Piscataway, NJ, USA, 2021; pp. 482–487. [CrossRef]
49. Van Eck, N.J.; Waltman, L. Visualizing bibliometric networks. In *Measuring Scholarly Impact: Methods and Practice*; Springer International Publishing: Cham, The Netherland, 2014; pp. 285–320. [CrossRef]
50. Van Eck, N.J.; Waltman, L. VOSviewer Manual: Manual for VOSviewer Version 1.6.18. Universiteit Leiden/CWTS Meaningful metrics. 2022. Available online: https://www.vosviewer.com/documentation/Manual_VOSviewer_1.6.18.pdf (accessed on 17 February 2025).
51. Lees, N. The Brandt Line after forty years: The more North–South relations change, the more they stay the same? *Rev. Int. Stud.* **2021**, *47*, 85–106. [CrossRef]
52. Solarz, M.W. North–South, Commemorating the First Brandt Report: Searching for the contemporary spatial picture of the global rift. *Third World Q.* **2012**, *33*, 559–569. [CrossRef]
53. FCSSC—Finance Center for South-South Cooperation. Organization in Special Consultative Status with ECOSOC of the United Nations. Global South Countries (Group of 77 and China). 2025. Available online: http://www.fc-ssc.org/en/partnership_program/south_south_countries (accessed on 23 February 2025).
54. Hargesheimer, E.E.; Coutts, R.T.; MacKinnon, M.D. Characterization of simple phenols in oil sands extraction-process water. *Environ. Technol.* **1984**, *5*, 433–440. [CrossRef]
55. Ahel, M. Infiltration of organic pollutants into groundwater: Field studies in the alluvial aquifer of the Sava River. In *Organic Micropollutants in the Aquatic Environment: Proceedings of the Sixth European Symposium, Lisbon, Portugal, 22–24 May 1990*; Springer: Dordrecht, The Netherlands, 1990; pp. 423–427. [CrossRef]
56. Veningerová, M.; Prachar, V.; Kovacicová, J.; Uhnák, J. Analytical methods for the determination of organochlorine compounds: Application to environmental samples in the Slovak Republic. *J. Chromatogr. A* **1997**, *774*, 333–347. [CrossRef]
57. Biziuk, M.; Namieśnik, J.; Czerwiński, J.; Gorlo, D.; Makuch, B.; Janicki, W.; Wolska, L. Occurrence and determination of organic pollutants in tap and surface waters of the Gdańsk district. *J. Chromatogr. A* **1996**, *733*, 171–183. [CrossRef] [PubMed]
58. Wennrich, L.; Efer, J.; Engewald, W. Gas Chromatographic trace analysis of underivatized nitrophenols. *Chromatographia* **1995**, *41*, 361–366. [CrossRef]
59. Køppen, B.; Spliid, N.H. Determination of acidic herbicides using liquid chromatography with pneumatically assisted electrospray ionization mass spectrometric and tandem mass spectrometric detection. *J. Chromatogr. A* **1998**, *803*, 157–168. [CrossRef] [PubMed]
60. Johansen, S.S.; Hansen, A.B.; Mosboek, H.; Arvin, E. Identification of heteroaromatic and other organic compounds in ground water at Creosote-contaminated sites in Denmark. *Groundw. Monit. Remediat.* **2009**, *17*, 106–115. [CrossRef]
61. Sawhney, B.L.; Kozloski, R.P. Organic Pollutants in Leachates from Landfill Sites. *J. Environ. Qual.* **1984**, *13*, 349–352. [CrossRef]
62. Ham, R.K.; Boyle, W.C.; Engroff, E.C.; Fero, R.L. Organic compounds in ferrous foundry process waste leachates. *J. Environ. Eng.* **1993**, *119*, 34–55. [CrossRef]
63. Beihoffer, J.; Ferguson, C. Determination of selected carboxylic acids and alcohols in groundwater by GC—MS. *J. Chromatogr. Sci.* **1994**, *32*, 102–106. [CrossRef]
64. Rudel, R.A.; Melly, S.J.; Geno, P.W.; Sun, G.; Brody, J.G. Identification of alkylphenols and other estrogenic phenolic compounds in wastewater, septage, and groundwater on Cape Cod, Massachusetts. *Environ. Sci. Technol.* **1998**, *32*, 861–869. [CrossRef]
65. Lacorte, S.; Latorre, A.; Guillamon, M.; Barceló, D. Nonylphenol, octyphenol, and bisphenol a in groundwaters as a result of agronomic practices. *Sci. World J.* **2002**, *2*, 1095–1100. [CrossRef] [PubMed]
66. Latorre, A.; Lacorte, S.; Barceló, D. Presence of nonylphenol, octyphenol and bisphenol a in two aquifers close to agricultural, industrial and urban areas. *Chromatographia* **2003**, *57*, 111–116. [CrossRef]
67. López-Roldán, P.; López de Alda, M.J.; Barceló, D. Simultaneous determination of selected endocrine disrupters (pesticides, phenols and phthalates) in water by in-field solid-phase extraction (SPE) using the prototype PROFEXS followed by on-line SPE (PROSPEKT) and analysis by liquid chromatography-atmospheric pressure chemical ionisation-mass spectrometry. *Anal. Bioanal. Chem.* **2004**, *378*, 599–609. [CrossRef]
68. Rodriguez-Mozaz, S.; de Alda, M.J.L.; Barceló, D. Monitoring of estrogens, pesticides and bisphenol A in natural waters and drinking water treatment plants by solid-phase extraction–liquid chromatography–mass spectrometry. *J. Chromatogr. A* **2004**, *1045*, 85–92. [CrossRef]
69. Moral, A.; Sicilia, M.D.; Rubio, S.; Pérez-Bendito, D. Multifunctional sorbents for the extraction of pesticide multiresidues from natural waters. *Anal. Chim. Acta* **2008**, *608*, 61–72. [CrossRef]

70. Bono-Blay, F.; Guart, A.; de la Fuente, B.; Pedemonte, M.; Pastor, M.C.; Borrell, A.; Lacorte, S. Survey of phthalates, alkylphenols, bisphenol A and herbicides in Spanish source waters intended for bottling. *Environ. Sci. Pollut. Res.* **2012**, *19*, 3339–3349. [CrossRef]
71. Quintana, J.; Hernández, A.; Ventura, F.; Devesa, R.; Boleda, M.R. Identification of 3-(trifluoromethyl) phenol as the malodorous compound in a pollution incident in the water supply in Catalonia (NE Spain). *Environ. Sci. Pollut. Res.* **2019**, *26*, 16076–16084. [CrossRef]
72. Herrero-Hernández, E.; Rodrígues-Gonzala, E.; Andrade, M.S.; Sánchez-González, S.; Carabias-Martinéz, R. Occurrence of phenols and phenoxyacid herbicides in environmental waters using an imprinted polymer as a selective sorbent. *Sci. Total Environ.* **2013**, *454–455*, 299–306. [CrossRef] [PubMed]
73. Zemo, D.A.; O'Reilly, K.T.; Mohler, R.E.; Tiwary, A.K.; Magaw, R.I.; Synowiec, K.A. Nature and estimated human toxicity of polar metabolite mixtures in groundwater quantified as TPHd/DRO at biodegrading fuel release sites. *Groundw. Monit. Remediat.* **2013**, *33*, 44–56. [CrossRef]
74. Mohler, R.E.; O'Reilly, K.T.; Zemo, D.A.; Tiwary, A.K.; Magaw, R.I.; Synowiec, K.A. Non-targeted analysis of petroleum metabolites in groundwater using GC× GC–TOFMS. *Environ. Sci. Technol.* **2013**, *47*, 10471–10476. [CrossRef]
75. Swartz, C.H.; Reddy, S.; Benotti, M.J.; Yin, H.; Barber, L.B.; Brownawell, B.J.; Rudel, R.A. Steroid estrogens, nonylphenol ethoxylate metabolites, and other wastewater contaminants in groundwater affected by a residential septic system on Cape Cod, MA. *Environ. Sci. Technol.* **2006**, *40*, 4894–4902. [CrossRef]
76. Baugros, J.B.; Giroud, B.; Dessalces, G.; Grenier-Loustalot, M.F.; Cren-Olivé, C. Multiresidue analytical methods for the ultra-trace quantification of 33 priority substances present in the list of REACH in real water samples. *Anal. Chim. Acta* **2008**, *607*, 191–203. [CrossRef] [PubMed]
77. Dévier, M.H.; Le Menach, K.; Viglino, L.; Di Gioia, L.; Lachassagne, P.; Budzinski, H. Ultra-trace analysis of hormones, pharmaceutical substances, alkylphenols and phthalates in two French natural mineral waters. *Sci. Total Environ.* **2013**, *443*, 621–632. [CrossRef] [PubMed]
78. Le Coadou, L.; Le Ménach, K.; Labadie, P.; Dévier, M.H.; Pardon, P.; Augagneur, S.; Budzinski, H. Quality survey of natural mineral water and spring water sold in France: Monitoring of hormones, pharmaceuticals, pesticides, perfluoroalkyl substances, phthalates, and alkylphenols at the ultra-trace level. *Sci. Total Environ.* **2017**, *603*, 651–662. [CrossRef] [PubMed]
79. Vulliet, E.; Tournier, M.; Vauchez, A.; Wiest, L.; Baudot, R.; Lafay, F.; Cren-Olivé, C. Survey regarding the occurrence of selected organic micropollutants in the groundwaters of overseas departments. *Environ. Sci. Pollut. Res.* **2014**, *21*, 7512–7521. [CrossRef]
80. Montero, L.; Conradi, S.; Weiss, H.; Popp, P. Determination of phenols in lake and ground water samples by stir bar sorptive extraction–thermal desorption–gas chromatography–mass spectrometry. *J. Chromatogr. A* **2005**, *1071*, 163–169. [CrossRef]
81. Schellin, M.; Popp, P. Membrane-assisted solvent extraction of seven phenols combined with large volume injection–gas chromatography–mass spectrometric detection. *J. Chromatogr. A* **2005**, *1072*, 37–43. [CrossRef]
82. Pignotti, E.; Farré, M.; Barceló, D.; Dinelli, E. Occurrence and distribution of six selected endocrine disrupting compounds in surface and groundwaters of the Romagna area (North Italy). *Environ. Sci. Pollut. Res.* **2017**, *24*, 21153–21167. [CrossRef]
83. Castiglioni, S.; Davoli, E.; Riva, F.; Palmiotto, M.; Camporini, P.; Manenti, A.; Zuccato, E. Mass balance of emerging contaminants in the water cycle of a highly urbanized and industrialized area of Italy. *Water Res.* **2018**, *131*, 287–298. [CrossRef]
84. Simões, N.G.; Cardoso, V.V.; Ferreira, E.; Benoliel, M.J.; Almeida, C.M. Experimental and statistical validation of SPME-GC–MS analysis of phenol and chlorophenols in raw and treated water. *Chemosphere* **2007**, *68*, 501–510. [CrossRef]
85. Carvalho, A.R.M.; Cardoso, V.V.; Rodrigues, A.; Ferreira, E.; Benoliel, M.J.; Duarte, E.A. Occurrence and analysis of endocrine-disrupting compounds in a water supply system. *Environ. Monit. Assess.* **2015**, *187*, 1–8. [CrossRef]
86. Makris, K.C.; Snyder, S.A. Screening of pharmaceuticals and endocrine-disrupting compounds in water supplies of Cyprus. *Water Sci. Technol.* **2010**, *62*, 2720–2728. [CrossRef] [PubMed]
87. Juhler, R.K.; Felding, G. Monitoring methyl tertiary butyl ether (MTBE) and other organic micropollutants in groundwater: Results from the Danish National Monitoring Program. *Water Air Soil Pollut.* **2003**, *149*, 145–161. [CrossRef]
88. Papadopoulou-Mourkidou, E.; Patsias, J.; Papadakis, E.; Koukourikou, A. Use of an automated on-line SPE–HPLC method to monitor caffeine and selected aniline and phenol compounds in aquatic systems of Macedonia–Thrace, Greece. *Fresenius' J. Anal. Chem.* **2001**, *371*, 491–496. [CrossRef] [PubMed]
89. Kovács, Á.; Mörtl, M.; Kende, A. Development and optimization of a method for the analysis of phenols and chlorophenols from aqueous samples by gas chromatography–mass spectrometry, after solid-phase extraction and trimethylsilylation. *Microchem. J.* **2011**, *99*, 125–131. [CrossRef]
90. McManus, S.L.; Moloney, M.; Richards, K.G.; Coxon, C.E.; Danaher, M. Determination and occurrence of phenoxyacetic acid herbicides and their transformation products in groundwater using ultra high performance liquid chromatography coupled to tandem mass spectrometry. *Molecules* **2014**, *19*, 20627–20649. [CrossRef]

91. Godejohann, M.; Heintz, L.; Daolio, C.; Berset, J.D.; Muff, D. Comprehensive non-targeted analysis of contaminated groundwater of a former ammunition destruction site using 1H-NMR and HPLC-SPE-NMR/TOF-MS. *Environ. Sci. Technol.* **2009**, *43*, 7055–7061. [CrossRef]
92. Chung, L.W.; Lee, M.R. Evaluation of liquid-phase microextraction conditions for determination of chlorophenols in environmental samples using gas chromatography–mass spectrometry without derivatization. *Talanta* **2008**, *76*, 154–160. [CrossRef]
93. Shin, H.S. GC–MS determination of antioxidants in ground water contaminated with JP-8. *Chromatographia* **2007**, *66*, 893–897. [CrossRef]
94. Kapelewska, J.; Kotowska, U.; Karpinska, J.; Kowalczuk, D.; Arciszewska ASwirydo, A. Occurrence, removal, mass loading and environmental risk assessment of emerging organic contaminants in leachates, groundwaters and wastewaters. *Microchem. J.* **2018**, *137*, 292–301. [CrossRef]
95. Cruz-López, A.; Dávila-Pórcel, R.A.; de León-Gómez, H.; Rodríguez-Martínez, J.M.; Suárez-Vázquez, S.I.; Cardona-Benavides, A.; Castro-Larragoitia, G.J.; Boreselli, L.; Villalba, M.L.; Pinales-Munguía, A.; et al. Exploratory study on the presence of bisphenol A and bis(2-ethylhexyl)phthalate in the Santa Catarina river in Monterrey, N.L., Mexico. *Environ. Monit. Assess.* **2020**, *192*, 488. [CrossRef]
96. Dolatto, R.G.; Messerschmidt, I.; Pereira, B.F.; Martinazzo, R.; Abate, G. Preconcentration of polar phenolic compounds from water samples and soil extract by liquid-phase microextraction and determination via liquid chromatography with ultraviolet detection. *Talanta* **2016**, *148*, 292–300. [CrossRef]
97. Al Nasir, F.; Batarseh, M.I. Agricultural reuse of reclaimed water and uptake of organic compounds: Pilot study at Mutah University wastewater treatment plant, Jordan. *Chemosphere* **2008**, *72*, 1203–1214. [CrossRef] [PubMed]
98. Han, F.; Gao, Y.; Hu, F.; Yu, X.; Xie, H.; Li, H.; Zheng, W. Solid-phase extraction of seventeen alternative flame retardants in water as determined by ultra-high-performance liquid chromatography-tandem mass spectrometry. *J. Chromatogr. A* **2019**, *1602*, 64–73. [CrossRef] [PubMed]
99. Jafari, A.J.; Abasabad, P.; Salehzadeh, A. Endocrine disrupting contaminants ins water resources and sewage in Hamadan city of Iran. *Iran. J. Environ. Health. Sci. Eng.* **2009**, *6*, 89–96.
100. Hai-Liang, Y.; Tong-Na, Z. Simultaneous determination of nine C4–C9 alkylphenols and bisphenol A in environmental water using gas chromatography-mass spectrometry. *Chin. J. Anal. Chem.* **2022**, *50*, 100112. [CrossRef]
101. Ng, K.; Alygizakis, N.; Nika, M.C.; Galani, A.; Oswald, P.; Oswaldova, M.; Slobodnik, J. Wide-scope target screening characterization of legacy and emerging contaminants in the Danube River Basin by liquid and gas chromatography coupled with high-resolution mass spectrometry. *Water Res.* **2023**, *230*, 119539. [CrossRef]
102. Adebesin, T.N.; Lateef, S.A.; Oloruntoba, E.O.; Adejumo, M. Occurrence of bisphenol A, nonylphenol, octylphenol and heavy metals in groundwater from selected communities in Ibadan, Nigeria. *J. Water Health* **2023**, *21*, 740–750. [CrossRef]
103. Otitoju, O.B.; Alfred, M.O.; Ogunlaja, O.O.; Olorunnisola, C.G.; Olukanni, O.D.; Ogunlaja, A.; Omorogie, M.O.; Unuabonah, E.I. Pollution and risk assessment of phenolic compounds in drinking water sources from southwestern Nigeria. *Environ. Sci. Pollut. Res.* **2023**, *30*, 76798–76817. [CrossRef]
104. Maynard, I.F.N.; Cavalcanti, E.B.; da Silva, L.L.; Martins, E.A.J.; Pires, M.A.F.; de Barros, M.L.; Marques, M.N. Assessing the presence of endocrine disruptors and markers of anthropogenic activity in a water supply system in northeastern Brazil. *J. Environ. Sci. Health Part A* **2019**, *54*, 891–898. [CrossRef]
105. Tao, X.; Tang, C.; Wu, P.; Han, Z.; Zhang, C.; Zhang, Y. Occurrence and behaviour of nonylphenol and octylphenol in Nanming River, Guiyang City, China. *J. Environ. Monit.* **2011**, *13*, 3269–3276. [CrossRef]
106. USEPA. National Recommended Water Quality Criteria—Human Health Criteria Table. Environmental Protection Agency of United States. 2019. Available online: https://www.epa.gov/wqc/national-recommended-water-quality-criteria-human-health-criteria-table (accessed on 17 February 2025).
107. US-eCFR. Title 40 Part 304—Arbitration Procedures for Small Superfund Cost Recovery. 2025. Available online: https://www.ecfr.gov/current/title-40/chapter-I/subchapter-J/part-304 (accessed on 23 February 2025).
108. WHO—World Healty Organization. *Guidelines for Drinking-Water Quality*, 4th ed.; WHO: Geneva, Switzerland, 2017. Available online: https://www.ncbi.nlm.nih.gov/books/NBK579461/pdf/Bookshelf_NBK579461.pdf (accessed on 17 February 2025).
109. EC—European Commission. Phenol—Summary Risk Assessment Report. European Chemicals Bureau, Italy. 2006. Available online: https://echa.europa.eu/documents/10162/3e04f30d-9953-4824-ba04-defa32a130fa (accessed on 17 February 2025).
110. Huang, X. Transboundary watershed pollution control analysis for pollution abatement and ecological compensation. *Environ. Sci. Pollut. Res.* **2023**, *30*, 44025–44042. [CrossRef]
111. GOV.UK. Guidance Phenol: General information (Updated 1 October 2024): UK Health Security Agency. 2025. Available online: https://www.gov.uk/government/publications/phenol-properties-incident-management-and-toxicology/phenol-general-information (accessed on 17 February 2025).
112. Ye, L.; Guo, J.; Ge, R. Chapter Thirteen—Environmental Pollutants and Hydroxysteroid Dehydrogenases. *Vitam. Horm.* **2014**, *94*, 349–390. [CrossRef]

113. Bdanthadka, M.; Mehendale, H.M. Chlorophenols. In *Encyclopedia of Toxicology*, 3rd ed.; Elsiever: Amsterdam, The Netherlands, 2014; pp. 896–899. [CrossRef]
114. Lapworth, D.J.; Baran, N.; Stuart, M.E.; Manamsa, K.; Talbot, J. Persistent and emerging micro-organic contaminants in Chalk groundwater of England and France. *Environ. Pollut.* **2015**, *203*, 214–225. [CrossRef] [PubMed]
115. European Chemicals Agency. Information on Chemicals. 2025. Available online: https://echa.europa.eu/substance-information/-/substanceinfo/100.002.405 (accessed on 17 February 2025).
116. Bonacci, O.; Žaknić-Ćatović, A.; Roje-Bonacci, T. Significant Rise in Sava River Water Temperature in the City of Zagreb Identified across Various Time Scales. *Water* **2024**, *16*, 2337. [CrossRef]
117. Kim, J.; Choi, K.; Chung, D.S. Sample Preparation for capillary electrophoretic Applications. *Compr. Sampl. Sample Prep.* **2012**, *3*, 701–721. [CrossRef]
118. Janz, D.M. Dinitrophenols. In *Encyclopedia of Toxicology*, 3rd ed.; Elsevier: Amsterdam, The Netherlands, 2014; pp. 177–178. [CrossRef]
119. Vione, D.; Maurino, V.; Minero, C.; Duncianu, M.; Olariu, R.; Arsene, C.; Sarakha, M.; Mailhot, G. Assessing the transformation kinetics of 2- and 4-nitrphenol in the atmospheric aqueous phase. Implications for the distribution of both nitroisomers in the atmosphere. *Atmos. Environ.* **2009**, *43*, 2321–2327. [CrossRef]
120. NLM National Library of Medicine; National Institutes of Health; U.S. Department of Health and Human Services. Toxicological Profile for Nitrophenols: Chapter 1 Relevance to Public Health. 2025. Available online: https://www.ncbi.nlm.nih.gov/books/NBK601147/ (accessed on 17 February 2025).
121. Orata, F. Derivatization reactions and reagents for gas chromatography analysis. In *Advanced Gas Chromatography—Progress in Agricultural, Biomedical and Industrial Applications*; InTech: Rijeka, Croatia, 2012; pp. 83–156. ISBN 978-953-51-0298-4. [CrossRef]
122. Bilalova, S.; Newig, J.; Tremblay-Lévesque, L.C.; Roux, J.; Herron, C.; Crane, S. Pathways to water sustainability? A global study assessing the benefits of integrated water resources management. *J. Environ. Manag.* **2023**, *343*, 118179. [CrossRef]
123. IPBES. *Global Assessment Report on Biodiversity and Ecosystem Services of the Intergovernmental Science-Policy Platform on Biodiversity and Ecosystem Services*; Brondizio, E.S., Settele, J., Díaz, S., Ngo, H.T., Eds.; IPBES Secretariat: Bonn, Germany, 2019; 1148p. [CrossRef]
124. Brauman, K.A.; Garibaldi, L.A.; Polasky, S.; Aumeeruddy-Thomas, Y.; Brancalion, P.H.; DeClerck, F.; Verma, M. Global trends in nature's contributions to people. *Proc. Natl. Acad. Sci. USA* **2020**, *117*, 32799–32805. [CrossRef]
125. Hole, D.G.; Collins, P.; Tesfaw, A.; Barrera, L.; Mascia, M.B.; Turner, W.R. Make nature's role visible to achieve the SDGs. *Glob. Sustain.* **2022**, *5*, e8. [CrossRef]
126. Khiyat, Z. Groundwater in the Arab region: Making the invisible visible. *Desalination Water Treat.* **2022**, *263*, 204–206. [CrossRef]

MDPI AG
Grosspeteranlage 5
4052 Basel
Switzerland
Tel.: +41 61 683 77 34

Water Editorial Office
E-mail: water@mdpi.com
www.mdpi.com/journal/water

www.ingramcontent.com/pod-product-compliance
Lightning Source LLC
Chambersburg PA
CBHW061515210926
51850CB00008B/1922
9783725866762